Carbohydrate Chemistry for Food Scientists

Carbohydrate Chemistry for Food Scientists

Third Edition

James N. BeMiller
Whistler Center for Carbohydrate Research,
Department of Food Science,
Purdue University,
West Lafayette, Indiana

ELSEVIER
WP WOODHEAD PUBLISHING
An imprint of Elsevier
AACC INTERNATIONAL

Woodhead Publishing is an imprint of Elsevier
The Officers' Mess Business Centre, Royston Road, Duxford, CB22 4QH, United Kingdom
50 Hampshire Street, 5th Floor, Cambridge, MA 02139, United States
The Boulevard, Langford Lane, Kidlington, OX5 1GB, United Kingdom

Notices

Knowledge and best practice in this field are constantly changing. As new research and experience broaden our understanding, changes in research methods, professional practices, or medical treatment may become necessary.

Practitioners and researchers must always rely on their own experience and knowledge in evaluating and using any information, methods, compounds, or experiments described herein. In using such information or methods they should be mindful of their own safety and the safety of others, including parties for whom they have a professional responsibility.

To the fullest extent of the law, neither AACCI nor the Publisher, nor the authors, contributors, or editors, assume any liability for any injury and/or damage to persons or property as a matter of products liability, negligence or otherwise, or from any use or operation of any methods, products, instructions, or ideas contained in the material herein.

Library of Congress Cataloging-in-Publication Data
A catalog record for this book is available from the Library of Congress

British Library Cataloguing-in-Publication Data
A catalogue record for this book is available from the British Library

ISBN: 978-0-12-812069-9

Publisher: Andre Gerhard Wolff
Acquisition Editor: Megan R. Ball
Editorial Project Manager: Barbara Makinster
Production Project Manager: Prem Kumar Kaliamoorthi
Cover Designer: Mark Rogers

Typeset by TNQ Technologies

Carbohydrate Chemistry Preface

The third edition of this textbook—a textbook that grew out of my own experiences of teaching in the classroom—has been extensively rewritten. My goals were to update it with new material, to explain concepts more clearly, to make it more straightforward, readable, and easier to understand, to correct the mistakes that had crept into the second edition, and to remove some materials unrelated to food science. The basic structure of the material remains the same as described in the prefaces to the first two editions, but the material has been brought up to date. The book remains a beginning textbook for learning about carbohydrates as food ingredients, but it is intended to be thorough at that level.

It covers the portion of the vast field of carbohydrate chemistry that is important to food science and helpful to food scientists. The chemistry of food carbohydrates is presented in the context of ingredient manufacture and properties. Ingredient properties and structure–function relationships are presented in the context of uses, which include (1) the organic chemistry involved in the preparation of carbohydrate substances for use as ingredients, (2) the organic chemistry that could occur during preparation of a product at a food processing facility or final preparation of it prior to consumption, (3) the organic and physical chemistry that could be involved during product storage, (4) the physical chemistry of such attributes as flow properties that may be important during processing or use or to texture, and (5) fundamentals of the relationships of carbohydrates and human health.

The book begins with the simplest carbohydrates (the monosaccharides) and builds to oligosaccharides and then to polysaccharides (the most abundant and widely used carbohydrates) over the first five chapters. Following this basic foundation of food carbohydrate chemistry are chapters on individual polysaccharides or groups of closely related polysaccharides, beginning with starch (the most abundantly used carbohydrate in foods), followed by chapters on cellulose and hydrocolloids with similar backbone structures (the galactomannans and xanthan), the ionic and other hydrocolloids, dietary fiber and other health-related topics, and sweeteners. Each chapter on a hydrocolloid, or in some cases, sections of chapters, begins with the sources and the basic structure. It then progresses to the chemistry used to modify structures of native polysaccharides to expand their usefulness by altering their physicochemical properties and concludes with uses, tying applications to structures and properties. The goal is to develop an understanding of how structures determine physical properties, how physical properties determine functionalities, and how functionalities determine uses, so that knowing the relationships, one can select a carbohydrate that will best provide the required functionality (by knowing what physical properties are needed to provide the functionality needed and the structure needed to produce those properties) without memorizing the properties and uses of each individual carbohydrate.

Food carbohydrate chemistry is the focus, but discussion of the health aspects of carbohydrates (from a chemical standpoint) has been enlarged. Presentation of the Maillard browning reaction and related topics, viz., acrylamide formation and caramel color and flavor production, were made into a separate chapter. The very large chapter on starch has been divided into two chapters.

Once again, I included what I thought were the most important things in the chapter, that is, what I thought students should know and be able to do after study of each chapter. Lists of "Additional Resources" that will provide students with options for exploring topics more deeply have been updated and expanded. Overall, those who have studied this book should have a good understanding of the number and variety of carbohydrate ingredients available to the product developer and their attributes.

This edition is dedicated to Professor Roy Whistler, who originated the project and coauthored the first edition (1996) with me.

James N. BeMiller

Monosaccharides

1

Chapter Outline

Key information and skills that can be obtained from study of this chapter will enable you to:

1. Define and/or identify

chiral carbon atom	furanose ring
monosaccharide	anomeric carbon atom
saccharose group	anomers
dextrose	axial positions
aldose	equatorial positions
acyclic	glycoside
D sugar	glycosidic bond
L sugar	aglycon
ketose	deoxy sugar
pyranose ring	hygroscopic
Haworth projection	humectant

2. Identify and/or describe monosaccharides as to the kind of carbonyl group and number of carbon atoms (for example, aldotetrose, octulose, etc.).
3. Give Fischer and specific anomeric furanose and pyranose ring structures (Haworth projections) for the following: L-arabinose, D-xylose, D-galactose, D-glucose, D-mannose, D-fructose.
4. Make interconversions between names, acyclic structures, Haworth ring structures, and/or conformational ring structures of the sugars listed in objective 3.
5. Describe the chemical properties of hemiacetals and acetals.

Carbohydrate Chemistry for Food Scientists. https://doi.org/10.1016/B978-0-12-812069-9.00001-7

6. Show the interrelationships of D-glucose, D-mannose, and D-fructose via isomerization.
7. Describe the conditions under which isomerization occurs.
8. When given the name of a glycoside of a sugar listed in objective 3 (for example, ethyl β-D-galactopyranoside), give its chemical structure and vice versa.
9. Describe the two general conditions under which hydrolysis of glycosides occurs.
10. Describe the relationship of carbohydrates to humectancy and water activity.

Introduction

Carbohydrates are present in all living cells. Energy reaching Earth in the form of sunlight is transformed by land and marine plants into sugars, which are used partly near the point of synthesis (that is, in a green growing shoot) for construction of various plant components and structures, and in part they are transported to other locations of the plant where they are used to make other components. These plant components supply food and energy to all other forms of life. Some of the initial photosynthetic carbohydrate material is converted into other organic compounds, such as proteins, fats, and lignin. Most of the remaining carbohydrate is converted into polymers of sugars (called polysaccharides, which constitute more than 90% of the dry matter of plants and at least three-fourths of the dry weight of all living organisms). It is estimated that as much as 10^{12} tons of biomass are produced each year by photosynthesis.

Other than water, carbohydrates are the most common components of foods and the human diet (both as natural components and as added ingredients). Because milk contains carbohydrates (lactose and other oligosaccharides), carbohydrates are present in the very first food consumed by all mammals, including humans, and because they are constituents of all plant tissues, they are also present in the diets of almost all adult humans. They provide at least three-fourths of the caloric intake of humans on a worldwide basis. Starch, lactose, and sucrose are digested by normal humans, and they, along with D-glucose and D-fructose, are human energy sources. (Digestible carbohydrates, D-glucose, and D-fructose [along with proteins] contain about four calories per gram [dry weight], whereas fats and oils provide about nine calories per gram.)

Carbohydrates in foods are important, not only as energy sources but also as ingredients that impart texture and as dietary fiber that contributes to human health. Their use as food ingredients is large in terms of both quantities consumed and the variety of applications and products. Carbohydrate ingredients are abundant. They are inexpensive. They can be obtained from a variety of replenishable sources. They occur in diverse structures and degrees of polymerization. They are available in a large number of molecular sizes and shapes and in a variety of chemical and physical properties. They are amenable to both chemical and biochemical modifications (and in some cases physical and/or genetic modifications), and both modifications are employed to improve their properties and extend their use as food ingredients. They are nontoxic and, therefore, safe.

Naturally occurring and modified carbohydrates are used as ingredients in a wide variety of food products because of the wide range of functionalities they can impart. Therefore, much of the chemistry of carbohydrates is physical chemistry related to such properties as crispness, crystallization, emulsion and suspension stabilization, flow behavior, gel characteristics, gel formation, glass formation, glass transitions, humectancy, mouthfeel, water-holding capacity, and viscosity.[1] And carbohydrates are often used as ingredients because of the effects they have on other components of a formulation (the primary one being water, but including components such as proteins and other carbohydrates). Some (those that are categorized as dietary fiber or prebiotics) are added as ingredients for their health benefits. A few (and only a few) impart a sweet taste.

Enjoy your journey through the world of food carbohydrates. During your study, you will learn that only a few carbohydrates are sweet and used to impart sweetness; rather, most are not sweet and are used as ingredients to provide appearance, body, convenience, flavor, stability, texture, and other attributes (for example, color and flavor) to food products.[1] For example, you will learn that the primary reason for using sugar in a cake recipe is not to make the cake sweet but to provide the desired texture. You will also learn the science behind the provision of such functionalities. You will learn how chemical, enzyme-catalyzed, genetic, physical, and structural[1] modifications of naturally occurring carbohydrates are employed to improve the properties of the natural substances and to make new ingredients. And you will learn that carbohydrates not only occur in different structural types but also have different physiological effects (for example, digestible and nondigestible, metabolizable and nonmetabolizable, caloric, reduced caloric and noncaloric, and prebiotic and nonprebiotic), although that is not the main focus of this text on the chemistry of carbohydrates (Box 1.1).

Box 1.1 Some Uses of Carbohydrates

In addition to making up a high percentage of the human diet, carbohydrates provide us with clothing (cotton, rayons) and shelter material (wood). They are important in the manufacture of diverse products such as (in alphabetical order) adhesives, concrete, cosmetics, detergents, paints, paper, pharmaceutical products, printing inks and pastes, processed food products, and toothpastes, in oil and gas production, in ore recovery, and in many more applications. Carbohydrates are involved in many biological structures, including plant and bacterial cell wall structures, and in many processes, including fertilization, immunological responses, recognitions related to infections, and energy storage.

[1] Throughout this book, various lists of attributes, properties, applications, etc. are given in alphabetical order, rather than in order of importance.

Carbohydrates can be, and are, classified in several different ways. In the minds of food scientists, they are sometimes classified according to functionality (for example, as bulking agents, emulsion stabilizers, gel formers, suspending agents, viscosifiers, etc.). In this book, the traditional approach is taken, classifying them according to molecular size, beginning in this chapter with the smallest molecules.

Structures and nomenclature

Carbohydrate (a term derived from the German *kohlenhydrate* and the similar French *hydrate de carbone*) expresses the fact that most simple carbohydrates have the general elemental composition $C_x(H_2O)_y$ (that is, they are molecules that contain carbon atoms plus hydrogen and oxygen atoms in the same ratio as they occur in water). Their composition is related to the fact that they are produced by photosynthesis from carbon dioxide and water as indicated by the following unbalanced equation:

$$CO_2 + H_2O \rightarrow \text{a sugar} + O_2$$

However, the great majority of natural carbohydrate compounds found in living organisms do not have the simple empirical formula $C_x(H_2O)_y$. Rather most naturally occurring carbohydrates are oligomers (oligosaccharides [Chapter 3]) or polymers (polysaccharides [Chapter 4]) made by joining sugars with the simple empirical formula or sugars with modified structures related to the simple empirical formula. While lower–molecular weight carbohydrates for food use are often obtained by depolymerization of natural polymers, this book begins with a presentation of the simple sugars and builds from there to larger and more complex structures.

A characteristic of carbohydrates (which are also called saccharides) is that they contain chiral carbon atoms. A **chiral carbon atom** is a carbon atom that can exist in two different spatial arrangements (configurations). Chiral carbon atoms can be recognized easily because they are carbon atoms that have each of their four tetrahedral bonds connected to different atoms or groups of atoms. The two different arrangements of the four groups in space (configurations) are what are called nonsuperimposable mirror images (Fig. 1.1). In other words, one is the reflection of the other that one would see in a mirror, with everything on the right in one configuration on the left in the other and vice versa (Box 1.2).

Monosaccharides are carbohydrate molecules that cannot be broken down by hydrolysis[2] into simpler (smaller) carbohydrate molecules. Hence, monosaccharides are at times referred to as "simple sugars" or just :sugars," which infers that they are the simplest (smallest) of the carbohydrates. (The term saccharide is derived from saccharose, which is an old term for cane sugar. Now, it refers to any carbohydrate, especially a monosaccharide; but as subsequent chapters on oligo- and polysaccharides indicate, it can be applied to any size carbohydrate. *Mono* is derived from the Greek word for

[2] Hydrolysis refers to cleavage of a chemical bond with concurrent addition of the elements of water (H- and –OH) to the ends of the two new molecules formed by the cleavage.

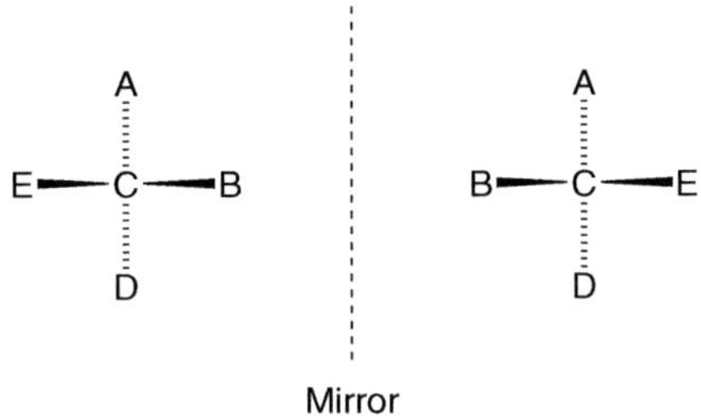

Figure 1.1 Chiral carbon atom. (A), (B), (D), and (E) represent different atoms or functional groups[2] attached to carbon atom (C). Wedges indicate chemical bonds projecting outward from the plane of the page; dashes indicate chemical bonds projecting into or below the plane of the page.

> **Box 1.2 Mirror Images**
>
> Nonsuperimposable mirror images can be illustrated by pressing the palms of your two hands together in front of you (fingers up). One is the reflection of the other that you would see in a mirror with everything that is on the right in one on the left in the other and vice versa, that is, your thumbs and each of your four fingers oppose each other. Now if you orient your hands in the same direction, for example, with the palms facing away from you, you will see that the thumbs are on opposite sides and that your two hands are different, that is, they are chiral and nonsuperimposable mirror images. You would soon find this out if you tried to put a glove for the left hand on the right hand, for example.

one. In chemistry, it often means containing only one, so the term monosaccharide means one saccharide or one sugar, indicating that it is a molecule composed of only one sugar unit and not of two or more sugar units joined together). Monosaccharides are the monomeric units of oligosaccharides (Chapter 3) and polysaccharides (Chapter 4), both of which contain more than one saccharide (sugar) unit and can be hydrolyzed to release their constituent monosaccharides. The common monosaccharides used as building blocks for oligo- and polysaccharides found in foods contain a group termed as the **saccharose group**.

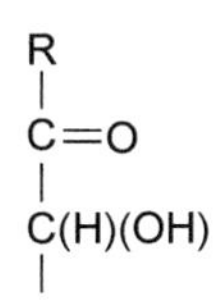

R = —H or —CH_2OH

The saccharose group, where R is a hydrogen atom (-H) (Aldoses) or a −CH2OH group (Ketoses)

Discussion of specific carbohydrate structures begins with D-glucose, the most common, most widely distributed, and most abundant carbohydrate (if all its combined forms are considered). D-Glucose (known commercially as **dextrose**) is a monosaccharide. D-Glucose is both a polyalcohol (polyhydroxy compound) and an aldehyde. It and all sugars containing an aldehydic group are classified as **aldoses** (Table 1.1). The prefix *ald-* indicates that they are aldehydes; the suffix *-ose* usually (but not always as you will find out) signifies a nonpolymeric carbohydrate (that is, a monosaccharide). D-Glucose contains six carbon atoms, making it a hexose (Table 1.1); more specifically, it is an aldohexose. When the structure of D-glucose is written in a vertical straight-chain fashion (termed an **acyclic** or open-chain structure) with the aldehydic group (position 1 [C1]) at the top and the carbon atom with the primary hydroxyl group attached to it at the bottom (at position 6 [that is, on C6]), it can be seen that all secondary hydroxyl groups are on carbon atoms C2, C3, C4, and C5. To make the determination/assignment of which side of the carbon chain the hydroxyl groups are on, a convention for orientation of the carbon chain is used. In this convention, the carbon chain is oriented so that each vertical (carbon-to-carbon) bond projects into the plane of the page and each horizontal bond projects outward from the plane of the page as in Fig. 1.1 (although in solution there is rotation about the vertical bonds that allows a hydroxyl group to be in any position with respect to the one above [or below] it, so that the molecules can actually assume a large number of different conformations [shapes]). Each of the four carbon atoms that have a secondary hydroxyl group attached to it (C2, C3, C4, C5) are chiral carbon atoms because each has four different substituents attached to it. Each chiral center has a mirror image (Fig. 1.1), and mirror image chiral carbon atoms are not superimposable on each other, just as a person's two hands are mirror images and are not superimposable.

Table 1.1 Classification of monosaccharides

Number of carbon atoms	Kind of carbonyl group: Aldehyde	Kind of carbonyl group: Ketone
3	triose	triulose
4	tetrose	tetrulose
5	pentose	pentulose
6	hexose	hexulose
7	heptose	heptulose
8	octose	octulose
9	nonose	nonulose

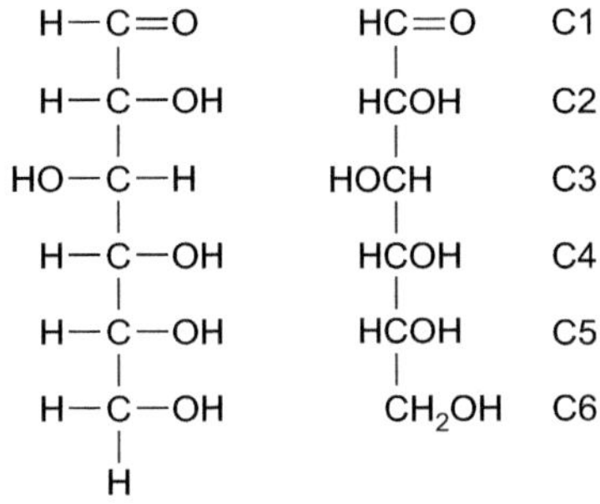

D-Glucose

Because each chiral carbon atom has a mirror image, there are 2^n (where n is the number of chiral carbon atoms) possible arrangements of these atoms. Thus, in a six-carbon aldose (of which D-glucose is one), in which there are four chiral carbon atoms, there are 2^4 or 16 different arrangements of the carbon atoms containing four secondary hydroxyl groups (theoretically allowing formation of 16 different six-carbon sugars with an aldehydic group). Eight of these sugars belong to what is known as the D series; eight are their mirror images and belong to the L series. (All 16 of these aldohexoses have the empirical formula $C_6H_{12}O_6$.) All sugars that have the hydroxyl group on the highest-numbered chiral carbon atom (C5 in the case of D-glucose) positioned on the right-hand side (using this projection and convention for the chain conformation) are termed **D sugars**, and all with the highest-numbered chiral atom having its hydroxyl group on the left are classified as **L sugars**. (Note that L-glucose is the complete molecular mirror image of D-glucose and not the substance with the opposite configuration of just C5. If only the configuration of C5 is changed, the product is L-idose [see below].) The structure of D-glucose is shown in its open-chain (acyclic) form (called the Fischer projection) with the carbon atoms numbered in the conventional manner. As is customary, the horizontal lines indicating the covalent chemical bonds to the hydrogen atoms and hydroxyl groups are omitted in the structure on the right. Because the lowermost carbon atom (C6 in the case of D-glucose) is not chiral, the relative positions of the atoms and groups attached to it need not be designated, and it is written as $-CH_2OH$. The great majority of carbohydrates found in foods are composed mostly of aldohexoses. Shown below is the aldopentose arabinose in both the D and L forms, both of which occur in nature. Glucose is found only in the form of D-glucose.

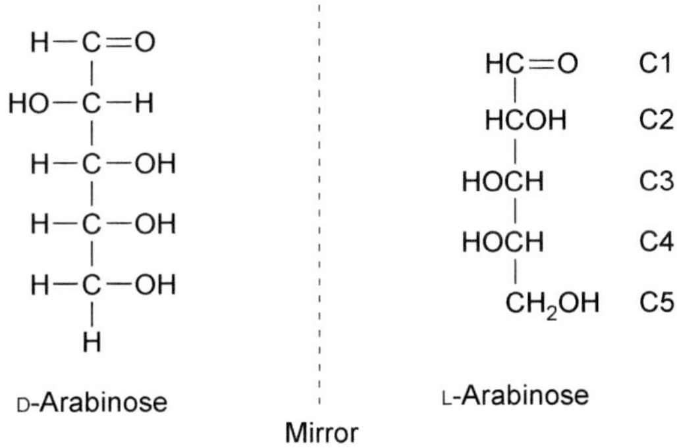

An organic chemist say that D-glucose and all other carbohydrate molecules as highly functionalized because there is a functional group[3] on each carbon atom. The complete functionalization of the backbone carbon atom structure with hydroxyl groups, a carbonyl oxygen atom (in the case of the simple sugars), and a high percentage of chiral carbon atoms are distinguishing features of carbohydrates.

D-Glucose, as its O6 phosphate ester[4] (D-glucose 6-phosphate), is the first sugar of photosynthesis. D-Glucose 6-phosphate (Chapter 2) is present in only small amounts because it is rapidly converted into sucrose (Chapter 3) for transport to various parts of the plant where it is used for synthesis of other substances. Because D-glucose is the monomeric building unit of cellulose (Chapter 8), it can be considered to be the most abundantly available organic compound on Earth (if its combined forms are considered). D-Glucose 6-phosphate is also used as an energy source in the plant's metabolism.

There are two aldoses containing three carbon atoms: D-glyceraldehyde and L-glyceraldehyde (D- and L-glycerose according to formal carbohydrate nomenclature, although these names are seldom used). Each possesses only one chiral carbon atom. Aldoses with four carbon atoms (the tetroses) have two chiral carbon atoms. Aldoses with five carbon atoms (the pentoses) have three chiral carbon atoms and comprise the second most common group of aldoses. Extending the series above six carbon atoms gives heptoses, octoses, and nonoses (seven, eight, and nine carbon atoms, respectively). Nine carbon atoms is the size limit of naturally occurring sugars. Only pentoses and hexoses are found in the common carbohydrates of food products. Development of the eight D-hexoses from D-glyceraldehyde is shown below using the Rosanoff shorthand projection (Fig. 1.2). In the Rosanoff projection, a circle represents the aldehydic group; horizontal lines indicate the location of each hydroxyl group on its chiral carbon atom, and the bottom of the vertical lines indicates the terminal, non-chiral hydroxymethyl (-CH_2OH) group (a primary hydroxyl group). Sugars whose names are in italics in Fig. 1.2 are commonly found in plants, almost exclusively in

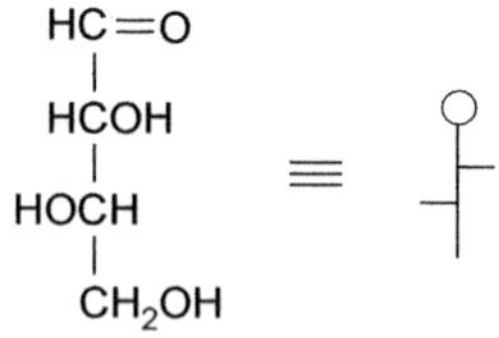

Figure 1.2 Relation of the Fischer projection to the Romanoff shorthand projection for L-threose.

[3] A functional group is an atom (other than a hydrogen atom) or a collection of chemically bonded atoms with a characteristic set of properties.

[4] O6 indicates that the phosphate group of the ester is attached to the oxygen atom on carbon atom number 6 (C6).

combined forms. Thus, they are present in our diets in combined forms. Only a small amount of D-glucose in the free monosaccharide form (section on Glycosides in this chapter) is present in natural foods (except for honey), and it is generally the only free aldose present.

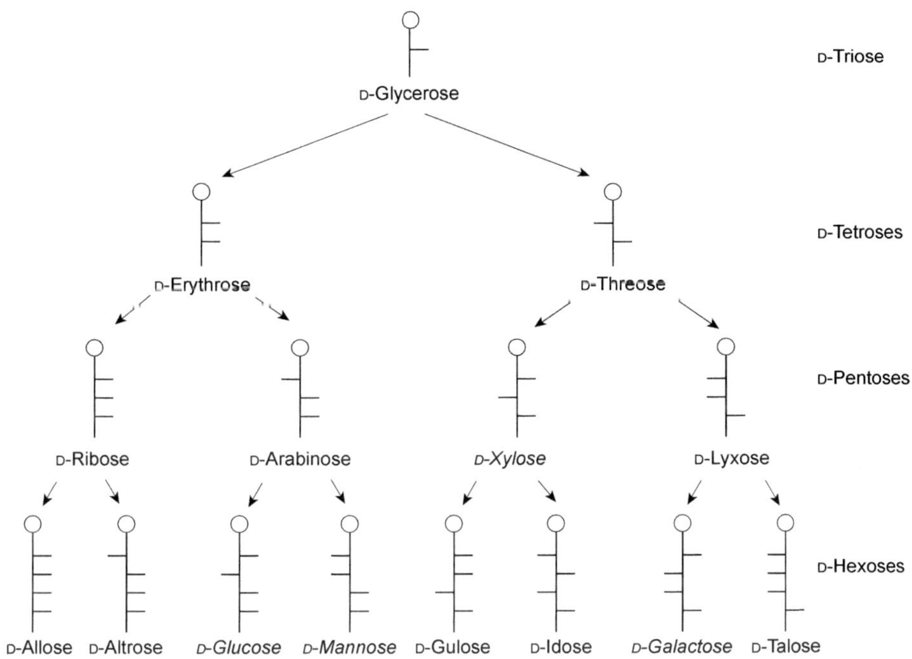

D-Glyceraldehyde occurs naturally as its O3 phosphate ester. Most other natural sugars, including the ubiquitous D-glucose, have the same configuration (D) of their highest-numbered chiral carbon atom (C5 in the case of a hexose). L-Sugars are much less numerous and abundant in nature than the D forms. The principal L-sugar found in foods is L-arabinose, which occurs in a combined form in some polysaccharides.

To make a molecular model of a sugar, two simple rules need to be followed. The first is to consider only one carbon atom at a time in copying a projected structure such as a Fischer projection or a Rosanoff structure. The second is to keep in mind that all horizontal bonds in the projected structure are envisioned as protruding toward you from the carbon atom, whereas vertical bonds are envisioned as protruding away from you.

So far aldoses (in which the carbonyl unit of the saccharose group is that of an aldehyde) have been discussed. In another type of monosaccharide, the carbonyl unit in the saccharose group is a ketone. These sugars are known as **ketoses**, with the prefix *ket-* identifying them as having a ketone group. D-Fructose is the prime example of this sugar group. D-Fructose is one of the two monosaccharide units of the disaccharide sucrose (Chapter 3) (the other being D-glucose) and makes up ca. 55% of a common commercial high-fructose syrup (Chapter 7) used in soft drinks. About 40% of the carbohydrates of honey are D-fructose.

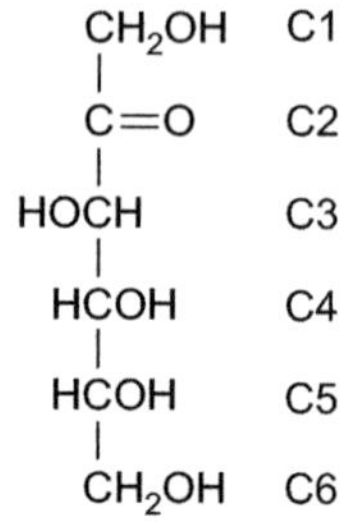

D-Fructose is the principal commercial ketose and the only one of importance in foods. (In the past, D-fructose was called both levulose and fruit sugar, but these designations are rarely used today.) D-Fructose has only three chiral carbon atoms (C3, C4, and C5). Thus, there are only 2^3 or 8 ketohexoses. The various ketotetroses, -pentoses, and -hexoses are related to nonchiral dihydroxyacetone. The suffix designating a ketose in systematic carbohydrate nomenclature is *—ulose* (Table 1.1). In systematic nomenclature, D-fructose is D-*arabino*-hexulose because its three chiral carbon atoms have the same configuration as those in D-arabinose. The Rosanoff projection of a ketopentose (pentulose) with the D-*threo* configuration (that is the configuration of the two chiral carbon atoms in D-threose) is given in Fig. 1.3 as another demonstration of the nomenclature principle.

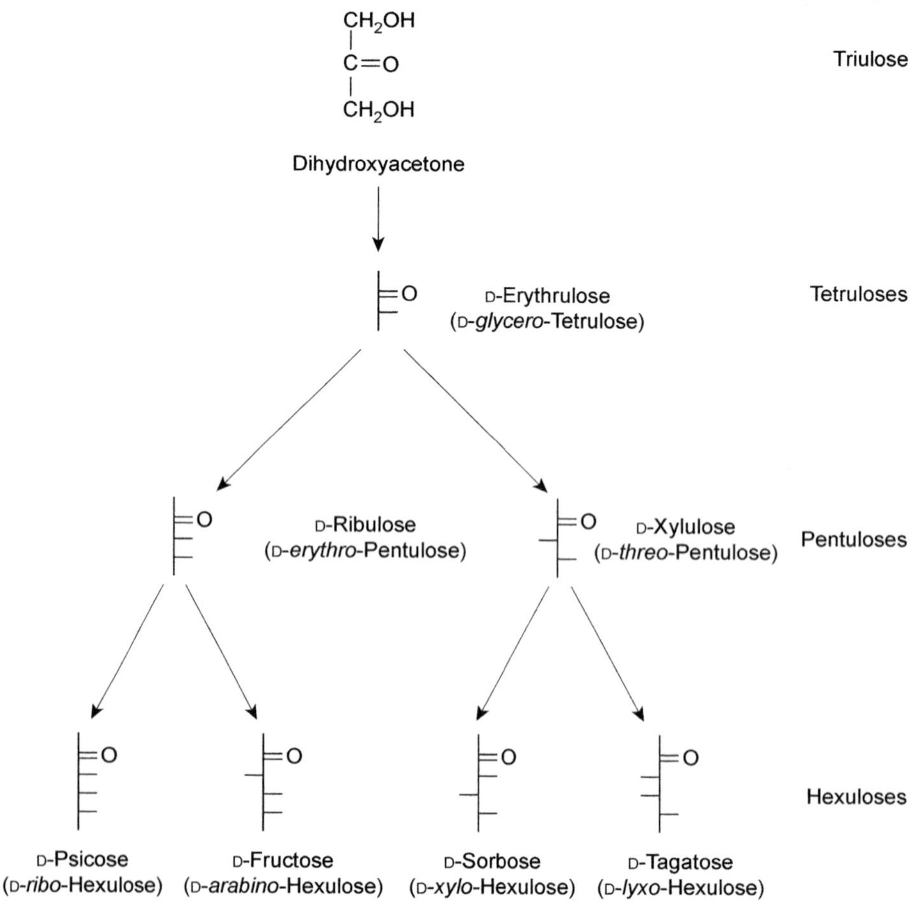

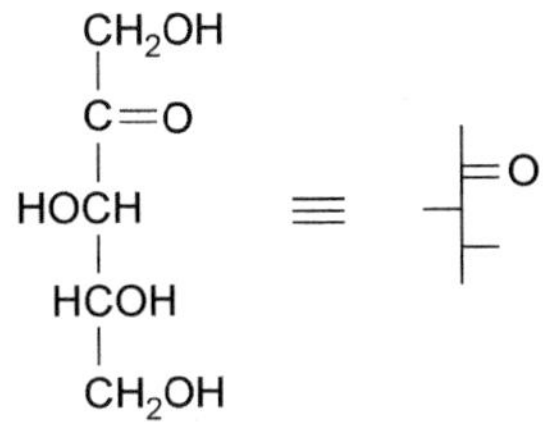

Figure 1.3 Rosanoff projection of a ketopentose (D-*threo*-pentulose, "D-xylulose") showing the configurations of the two chiral carbon atoms.

Isomerization

Simple aldoses and ketoses containing the same number of carbon atoms are isomers of each other (that is, a hexose and a hexulose both have the empirical formula $C_6H_{12}O_6$). Isomerization[5] of monosaccharides involves both the carbonyl group and the adjacent hydroxyl group. By isomerization, an aldose is converted into another aldose (with the opposite configuration of C2) and the corresponding ketose, and a ketose is converted into the corresponding two aldoses (Fig. 1.4). Therefore, by isomerization, D-glucose, D-mannose, and D-fructose can be interconverted (Fig. 1.5). Isomerization can be catalyzed by either a base or an enzyme (Chapter 7).

D-Tagatose, which is the ketose formed by isomerization of D-galactose in the same way as D-fructose is formed by isomerization of D-glucose (or D-mannose), is a commercially available, reduced-calorie sweetener (Chapter 19). Another commercially available, reduced-calorie ketose sweetener is called allulose by the company that makes it, although it is better known as D-psicose (see the scheme showing the four D hexuloses). Brown sugar, caramel, figs, maple syrup, molasses, and raisins contain small amounts of D-psicose.

Ring forms

Carbonyl groups of aldehydes are reactive and readily undergo nucleophilic attack by the unshared electrons of the oxygen atom of a hydroxyl group to produce a hemiacetal. By similar reactions, a carbonyl group of a ketone also produces a hemiacetal (sometimes specifically designated a hemiketal).

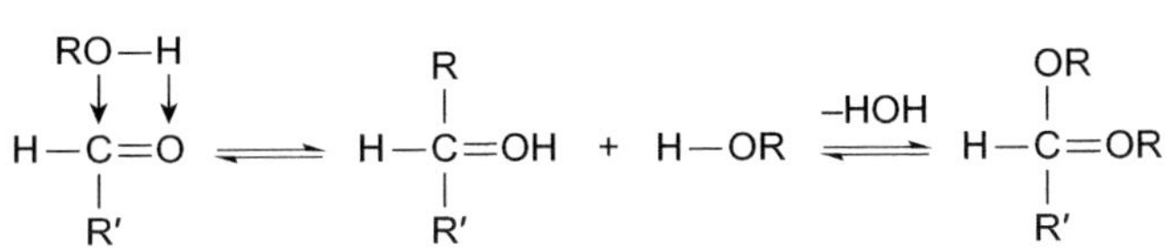

[5] Isomerization is the process of converting one isomeric molecule into another.

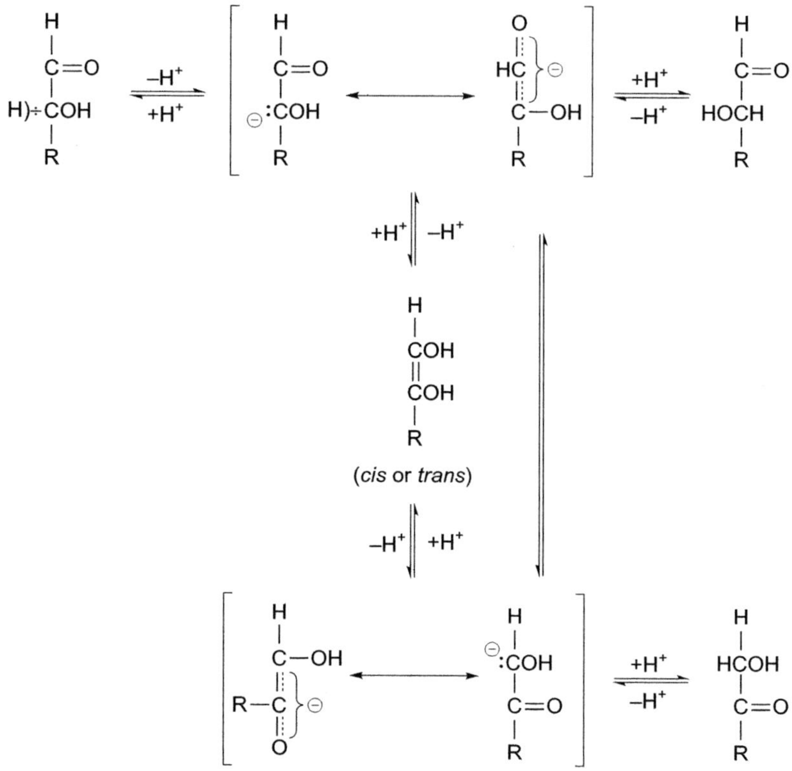

Figure 1.4 Monosaccharide isomerization.

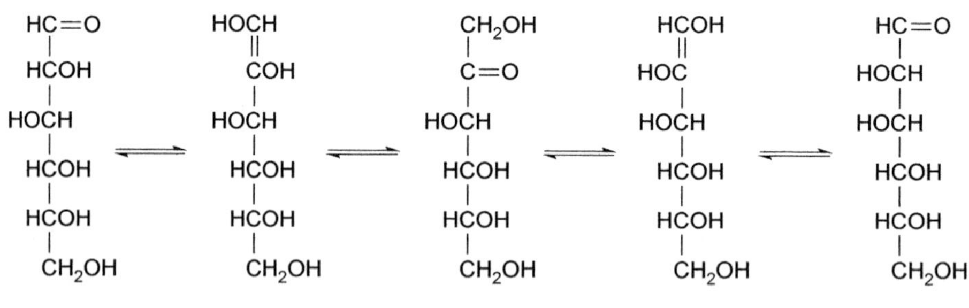

Figure 1.5 Interrelationship of D-glucose, D-mannose, and D-fructose via isomerization.

Hemiacetal formation can occur intramolecularly (that is, within the same aldose or ketose molecule) by reaction of the carbonyl group of a monosaccharide molecule with one of its own hydroxyl groups, as illustrated in Fig. 1.6 with D-glucose laid coiled on its side. The six-membered ring that results from reaction of an aldehydic group with the hydroxyl group at C5 is called a **pyranose** ring. Notice that, for the oxygen atom of the hydroxyl group at C5 to react to form the ring, C5 must rotate to bring its oxygen

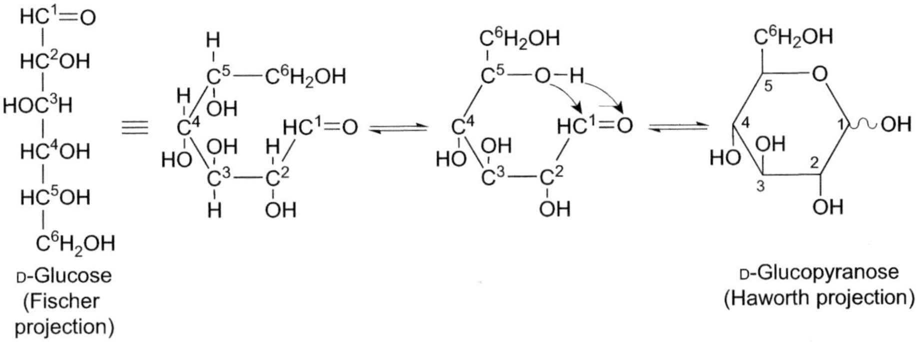

Figure 1.6 Relationship of the acyclic and pyranose ring (Haworth projection) structures of D-glucose.

atom upward. This rotation brings the hydroxymethyl group (C6) to a position above the plane of the ring. The representation of the D-glucopyranose ring used in Fig. 1.6 is called a **Haworth projection**.

To avoid clutter in writing ring structures, common conventions are adopted wherein ring carbon atoms are indicated by angles in the ring, and hydrogen atoms attached to carbon atoms are eliminated altogether (Fig. 1.7). A mixture of chiral forms (anomeric forms [see below]) is indicated by a wavy line.

Monosaccharides can also occur in a five-membered ring called the **furanose ring** (Fig. 1.8), but most free pentoses, hexoses, and heptoses occur in pyranose rings because the six-membered pyranose ring is more stable than the furanose ring (see below). Although the less stable furanose ring is less prevalent than the pyranose ring, it is found in several oligo- and polysaccharides (and in nucleic acids). The furanose ring is formed in the same way as the pyranose ring (Fig. 1.8).

When the carbon atom of the carbonyl group is involved in cyclic hemiacetal (pyranose or furanose ring) formation, it becomes chiral because four different groups are now attached to it. With D sugars, the configuration that has the hydroxyl group located below the ring is called the alpha form. For example, α-D-glucopyranose is D-glucose in the pyranose (six-membered) ring form with the hydroxyl group of the new chiral carbon atom (C1) in the alpha position (that is, below the plane of the ring). When the newly formed hydroxyl group at C1 is above the ring, it is in the beta position, and the structure is termed β-D-glucopyranose. This designation holds for all D sugars. The new chiral carbon atom is called the **anomeric carbon atom,** and the two anomeric forms (alpha [α] and beta [β]) are known as **anomers** and form an anomeric pair. For sugars in the L series, the opposite is true (that is, the anomeric hydroxyl group is up in the alpha anomer and down in the beta anomer), so that α-D-glucopyranose and α-L-glucopyranose are mirror images of each other.

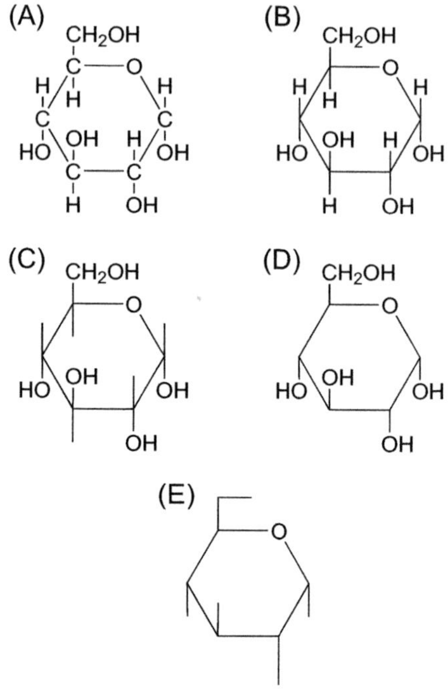

Figure 1.7 Five representations of the structure of α-D-glucopyranose. These two-dimensional representations of the pyranose ring are Haworth projections. Actually, the ring is not flat but puckered. (A) All C, H, and O atoms are shown. (B) C2, C3, C4, C5, and O5 are represented by a hexagon. (C) Hydrogen atoms attached to C1, C2, C3, C4, and C5 are indicated but not shown. (D) Hydrogen atoms attached to C1, C2, C3, C4, and C5 are omitted. (E) A skeletal structure in which the hydroxyl groups are also omitted, although their positions are indicated. Form D is the preferred form for ease of recognition and is used in this book.

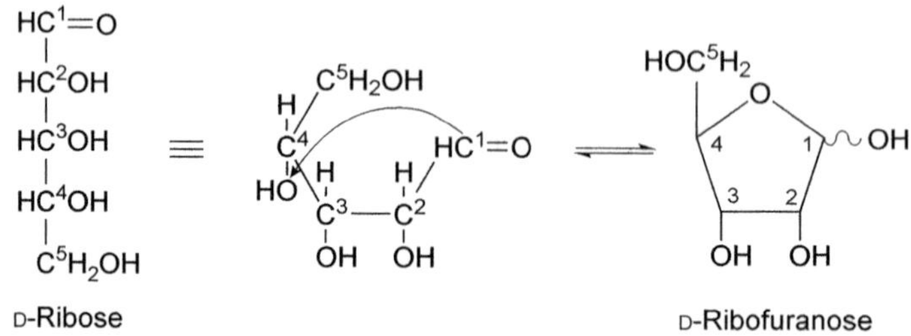

Figure 1.8 Formation of the furanose ring form of D-ribose.

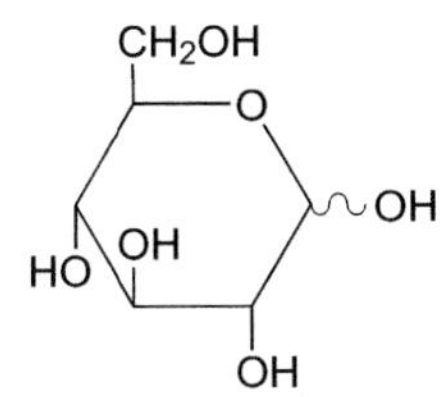

D-Glucopyranose

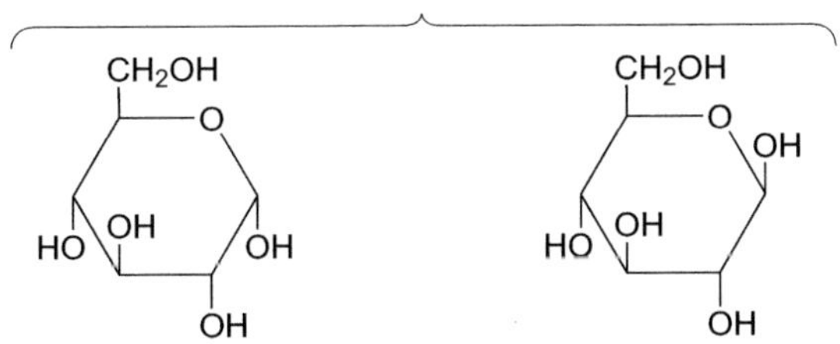

α-D-Glucopyranose β-D-Glucopyranose

However, pyranose rings are not flat with the attached groups sticking straight up and straight down as the Haworth representation suggests. Rather, they occur in a variety of shapes (conformations)—most commonly in one of two chair conformations (so-called because they are shaped somewhat like a chair). In a chair conformation, one bond on each carbon atom projects either up or down from the ring; these are called axial bonds and the attached atom or group is said to be in an **axial position**. The other bonds not involved in ring formation project out around the perimeter of the ring in what are called **equatorial positions** (Fig. 1.9), so they are called equatorial bonds. Equatorial bonds are either up or down with respect to the axial bonds, but not straight up or straight down with respect to the ring as are axial bonds.

The pyranose ring in a chair conformation can be pictured as a disk with about one-third of the disk bent slightly up and the one-third directly opposite the bent-up part bent slightly down. The equatorial positions are those that project outward around the perimeter (equator) of the structure, and the axial positions are those that project above and below the disk. The six-membered pyranose ring distorts the normal carbon and oxygen atom bond angles less than rings of other sizes do, so it is the predominate ring form. Strain is further lessened when the larger hydroxyl groups (that is, larger

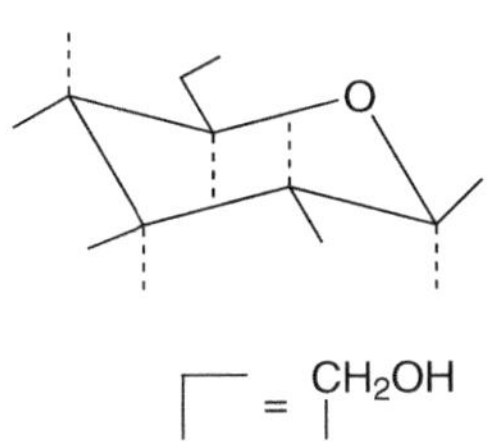

Figure 1.9 A pyranose ring showing the equatorial (*solid line*) and axial (*dashed line*) bond positions.

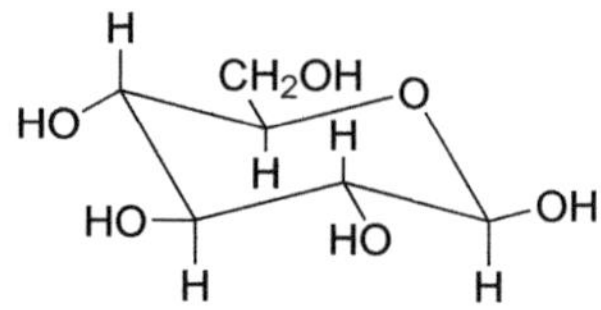

Figure 1.10 β-D-Glucopyranose in the $^4C_1$ conformation as drawn in structural formulas. All bulky groups are in equatorial positions, all hydrogen atoms in axial positions.

than hydrogen atoms) are separated maximally from each other by a ring conformation (shape) that arranges the greatest number of them in equatorial rather than axial positions, so rotation of carbon atoms takes place on their connecting bonds to swivel the hydroxyl groups to equatorial positions in so far as is possible.

Using β-D-glucopyranose as an example, C2, C3, C5, and the ring oxygen atom remain in a plane, but C4 is raised slightly above the plane and C1 is positioned slightly below the plane as in Figs. 1.9 and 1.10, giving the ring somewhat the shape of a chair. This conformation is designated the $^4C_1$ conformation. The notation C indicates that the ring is chair shaped; the superscript and subscript numbers indicate that C4 is above the plane of the ring formed by C2, C3, C5, and the ring oxygen atom and C1 is below the plane. There are two chair forms. The second ($^1C_4$) has all the axial and equatorial groups reversed, and therefore, has the hydroxyl groups of D-glucopyranose axial, causing nonbonding steric hindrance (interference). Because the $^1C_4$ conformation is a higher energy form, little, if any, D-glucopyranose ever exists in that conformation.

As noted, β-D-glucopyranose has all of its hydroxyl groups in the equatorial arrangement. Each is located either slightly above or slightly below the true equatorial position. In β-D-glucopyranose, the hydroxyl groups (all of which are in equatorial positions) alternate in an up-and-down arrangement, with that on C1 positioned slightly up and that on C2 positioned slightly down, and continuing around the ring. The bulky hydroxymethyl group ($-CH_2OH$; C6 in hexoses) is essentially always in a sterically free equatorial position.

In addition to the two chair forms, other conformations can exist, depending on the configuration of the sugar and any groups that may be attached to it. Thus, there are six boat conformations and six skew (twisted boat) shapes possible for the pyranose ring. Few molecules assume these higher energy forms, though in solution, they may occur momentarily.

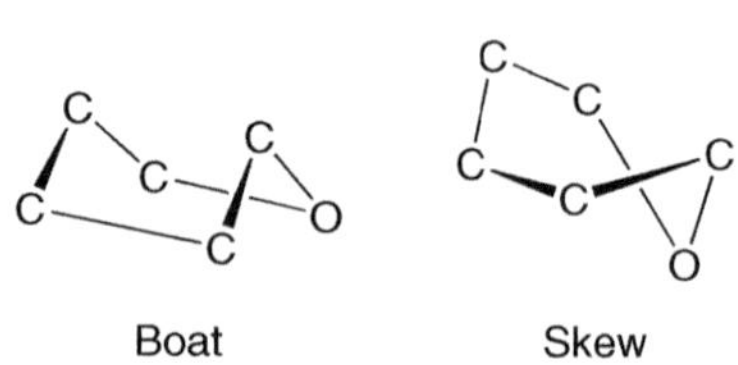

Six-membered sugar rings are rather stable if bulky groups such as hydroxyl groups and the hydroxymethyl group are in equatorial positions. Thus, α-D-glucopyranose (dextrose) dissolves in water to give a mixture containing the open-chain form and its five- and six-membered ring forms that rapidly equilibrate. At room temperature, the thermodynamically more stable six-membered (pyranose) ring forms predominate, followed by the five-membered (furanose) ring forms. Only traces of the seven-membered ring forms exist. The configuration of the anomeric carbon atom (C1 of aldoses) of each ring may be alpha or beta. The equilibrium ratio of the ring forms varies with the specific sugar and the temperature. For example, the ratio of α-D pyranose:β-D pyranose:α-D furanose:β-D furanose ring forms at room temperature is approximately 36:64:0:0 for D-glucose, 69:31:0:0 for D-mannose, 29:64:3:4 for D-galactose, and 60:35.5:2.5:0.5 for D-arabinose. The open-chain, aldehydic form constitutes only about 0.02% or less of the total forms (depending on pH and temperature); but because of rapid interconversion of the ring forms to the aldehydic form and vice versa (rapid opening and closing of the ring), a sugar can readily and rapidly react as if it were entirely in the aldehydic form.

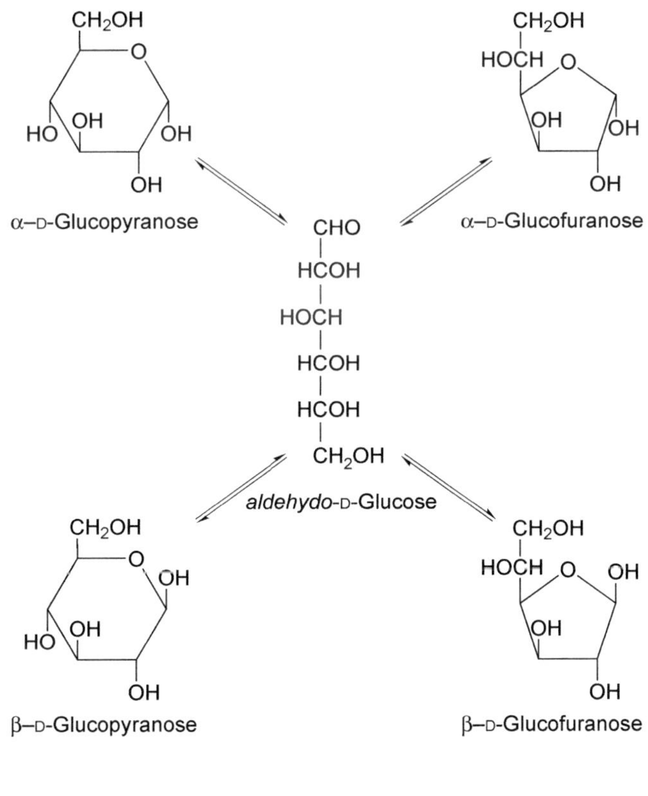

Pyranose ring forms are also the most thermodynamically stable forms of aldopentoses and ketohexoses. Although as a constituent of sucrose (Chapter 3), fructooligosaccharides (Chapter 3), and inulin (Chapter 10), D-fructose is in a five-membered

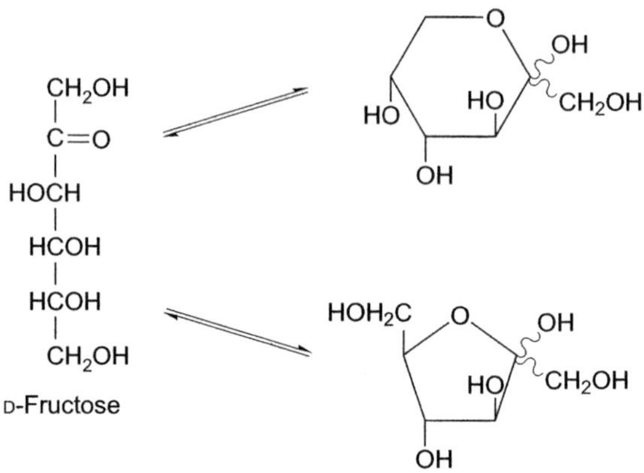

Figure 1.11 Interconversion of the acyclic (open-chain) and pyranose and furanose ring forms of D-fructose in an aqueous environment.

(furanose) ring form, crystalline D-fructose is β-D-fructopyranose. A room-temperature aqueous solution of D-fructose contains approximately 73% β-D-fructopyranose, 21% β-D-fructofuranose, and 6% α-D-fructofuranose (Fig. 1.11). β-D-Fructofuranose is the sweetest of these forms and is primarily responsible for D-fructose being 1.2–1.8 times sweeter[6] than sucrose on an equal weight basis (Chapter 19). (D-Fructose has 65%–95% of the sweetness[6] of sucrose on an equimolar basis.)

All sugars containing a carbonyl group (even though it might be in a hemiacetal ring form) are called reducing sugars (Chapter 2).

Glycosides

The hydroxyl group formed by formation of a pyranose or furanose ring can react further (by condensation) with the hydroxyl group of an alcohol to produce a full acetal. When a cyclic hemiacetal form of a monosaccharide reacts with an alcohol (under anhydrous conditions and in the presence of an acid catalyst) to produce a full acetal, the product is called a **glycoside**. The acetal linkage at the anomeric carbon atom is indicated by the *-ide* ending. In nature, the amount of free monosaccharides is very small, with almost all sugars attached to another molecule containing a hydroxyl group by means of what is called a **glycosidic bond**[7] or glycosidic linkage. Formation of both the hemiacetal rings and full acetals (glycosides) is shown in Fig. 1.12. (In the laboratory, the reaction occurs at elevated temperatures under anhydrous conditions in the presence of an acid, which catalyzes the reaction. However, the reaction depicted in Fig. 1.13 is not the way that living organisms make glycosides. In nature, glycosidic bonds are formed in aqueous environments in pathways involving several

[6] Depending on the pH and temperature of the system.

[7] The C—O—C bond connecting the sugar unit and the aglycon.

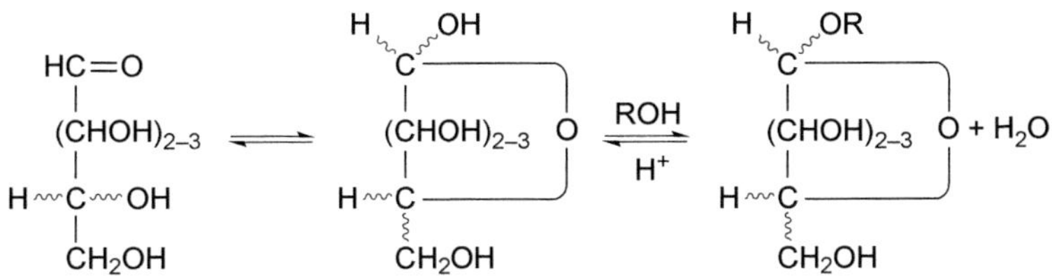

Figure 1.12 Formation of a pyranose or a furanose ring from a hexose (n = 3) or a pentose (n = 2), followed by formation of a glycoside (pyranoside or furanoside) by reaction with an alcohol in the presence of a source of hydrogen ions (protons).

intermediates with each reaction step catalyzed by an enzyme. Ultimately, the reaction is between a substrate containing a hydroxyl group and a carbohydrate activated at the anomeric carbon atom.) The squiggly lines in Fig. 1.12 indicate that (1) the configurations (arrangements of the hydroxyl groups) of the molecules is immaterial (that is, both reactions occur with all simple sugars with at least five carbon atoms regardless of the arrangement of their hydroxyl groups), and (2) both anomeric configurations (alpha and beta) are formed in these reactions.

Note that a second product of glycoside formation is water and that the reaction is reversible (that is, glycosides will undergo hydrolysis in the presence of water and an acid as the catalyst [elevated temperatures are usually also required]), which is why (in the laboratory) glycosides must be made under anhydrous conditions. Under hot, aqueous, acidic conditions (as may be encountered during preparation of processed foods), glycosidic bonds may be cleaved (that is, undergo hydrolysis), releasing aldoses and/or ketoses. Hydrolysis of glycosides may also be catalyzed by specific enzymes. These enzymes are generally much more specific for the nature of the specific sugar type, its ring size, and the configuration of the anomeric carbon atom than for the nature of the group bonded to the sugar to make a glycoside (which is called the **aglycon**). As examples, a β-glucosidase will catalyze the hydrolysis of any β-D-glucopyranoside (regardless of the aglycon), and a β-fructofuranosidase will catalyze the hydrolysis of any β-D-fructofuranoside, although specific parameters of enzyme-catalyzed hydrolysis may vary with the nature of the aglycon.

In the case of D-glucose reacting with methanol, the product is mainly methyl α-D-glucopyranoside, with less methyl β-D-glucopyranoside. (The methyl group in this case is the aglycon.) The two anomeric forms of furanosides are also formed, but being higher energy structures, their stability is lower than that of the six-membered pyranoside rings; so although they are formed initially in substantial amounts, they reorganize under the hot acidic conditions into the more stable forms and are present at equilibrium in comparatively low quantities.

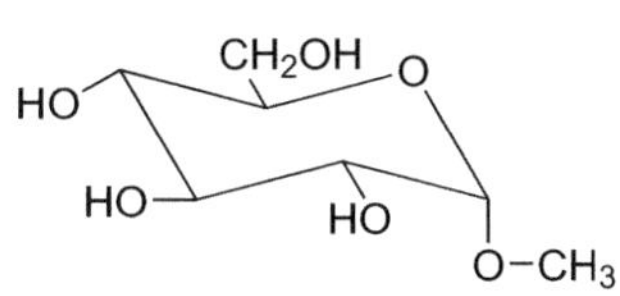

Methyl α-D-glucopyranoside

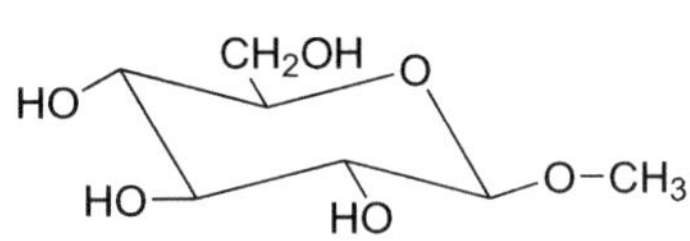

Methyl β-D-glucopyranoside

α-D-Glucopyranoside is the lowest energy form of common D-glucopyranosides and, hence, the most stable structure in aqueous solution, even though this places the aglycon group in an axial position. This phenomenon is called the anomeric effect. The reason for the anomeric effect is twofold: (1) In the α-D form, the dipoles of the nonbonding electrons of the ring oxygen atom and the oxygen atom at C1 are oriented so that one bisects the other, and this arrangement is more stable than the parallel orientation that occurs in the equatorial β-D-form. (2) The second, and perhaps stronger, reason is that one pair of the nonbonding electron orbitals of the ring oxygen atom can overlap favorably with the lone pair of electron orbitals on the back lobe of the C1 oxygen atom in the α-D form to give greater stability, a larger bond angle to the C—O bond, and greater bond length due to the weakening effect of the aglycon.

As already mentioned, glycosides undergo hydrolysis in acidic environments to yield a reducing sugar (Chapter 2) and a hydroxylated compound. Hydrolysis becomes more and more rapid as the temperature is raised.

Various enzyme preparations with glycosidase activities are used in fruit juice processing and wine making. (Some are present as contaminants in the enzyme preparations used for liquefaction to increase juice yields and for clarification of juices.) Glycosidases such as galactosidases and glucosidases effect hydrolysis of the naturally occurring glycosides, such as the anthocyanin pigment glycosides. Hydrolysis of these glycosides can have both positive and negative effects. Treatment with glycosidases often liberates volatile aromas and flavors that are normally present as nonvolatile glycosides. The resulting increase in flavor and aroma intensity is usually beneficial to fruit juice and wine producers. However, hydrolysis of some glycosides can have an adverse effect. For example, hydrolysis of feruloyl glucoside releases ferulic acid (Chapter 17), which is a precursor to a compound with an objectionable odor.

Other types of monosaccharides

Not all naturally occurring sugars have a hydroxyl group on every carbon atom (other than the carbonyl carbon atom). Sugars missing a hydroxyl group on one or more carbon atoms are known as **deoxy sugars**. 6-Deoxy-L-mannose has the alternative name L-rhamnose, which is the name almost always used. L-Rhamnose is a constituent of some plant polysaccharides, especially those related to pectin (Chapter 15). The sugar component of DNA is commonly called 2-deoxy-D-ribose or just deoxyribose, but it could also be called 2-deoxy-D-arabinose because C2 is no longer chiral. Because its only two chiral carbon atoms have the same configuration as those in D-erythrose, its proper name is 2-deoxy-D-*erythro*-tetrose.

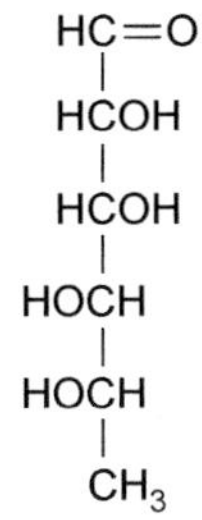

L-Rhamnose
(6-Deoxy-L-mannose)

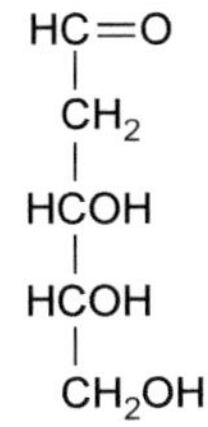

"2-Deoxy-D-ribose"
(2-Deoxy-D-*erythro*-tetrose)

A common sugar in animal polysaccharides, some bacterial polysaccharides, and the polysaccharide chitin (which is the structural component of insect and crustacean exoskeletons and some fungal cell walls) (Chapter 17) is D-glucosamine. D-Glucosamine belongs to a class of sugars known as aminodeoxy sugars. In them, the hydroxyl group is missing (deoxy) and has been replaced with an amino group, so the carbon atom is still chiral. The systematic name of D-glucosamine (2-amino-2-deoxy-D-glucose) indicates its exact structure, including the configuration of C2.

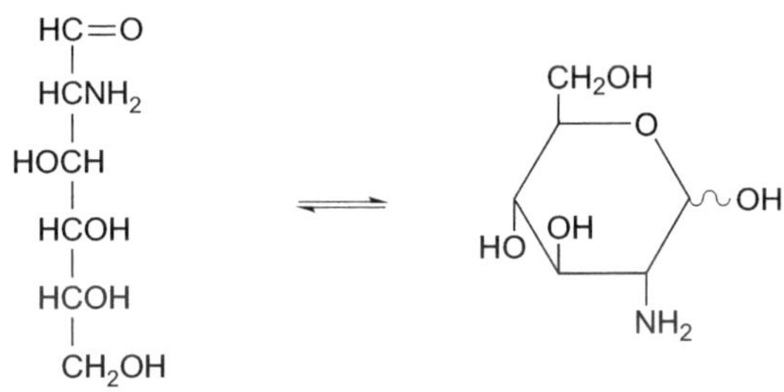

2-Amino-2-deoxy-D-glucose
(D-Glucosamine)

In some other naturally occurring sugars, the carbon atom on the opposite end of the carbon chain from the carbonyl group (that is, the one that is usually a $-CH_2OH$ group) is in the form of a carboxyl ($-CO_2H$) group. These sugars (known as uronic acids) are presented in Chapter 2.

Some functions of monosaccharides in foods

Monosaccharides found in foods are almost exclusively D-glucose and D-fructose and are almost exclusively present as added ingredients. D-Glucose and D-fructose are among the few carbohydrates used as ingredients to provide sweetness. While crystalline D-glucose (dextrose) or D-fructose may be used, they are primarily added together, whether in a natural product (honey) or a manufactured ingredient (invert sugar, high-fructose syrup) (Chapter 19).

The ability to bind water and to control the movement of water molecules and water activity is a basic and useful property of carbohydrates. Most carbohydrates attract and/or hold water, although there are differences in their ability to do this.

Carbohydrate molecules actually induce structure in the water molecules surrounding them, with the effectiveness of different carbohydrate molecules in doing this being different because of differences in orientation of their hydroxyl groups.

Carbohydrates are hydrophilic, water soluble, and in some cases **hygroscopic**[8] (making them useful as **humectants**[9]) because their hydroxyl groups form hydrogen bonds with water molecules. Hydrogen bonding to water molecules results in their hydration and dissolution. There are differences in degrees of hygroscopicity. For example, D-fructose is more hygroscopic than is D-glucose. After 9 days at 60% relative humidity, crystalline D-fructose will have absorbed 0.63% water, whereas crystalline D-glucose will have absorbed only 0.07% water. Amorphous sugars will absorb more water and absorb it at a faster rate than crystalline sugars, so noncrystalline materials such as glucose syrups (Chapter 7), high-fructose syrups (Chapter 7), and invert sugar syrups (Chapter 3) are better humectants than are crystalline sugars. This occurs because, when sugars crystallize, intermolecular hydrogen bonds form between the sugar molecules and, therefore, the hydroxyl groups are unavailable to hydrogen bond with water. In mixtures of sugars, each sugar acts as an impurity to the other(s) and prevents crystallization of individual sugars (that is, prevents formation of sugar molecule to sugar molecule hydrogen bonds), allowing sugar molecule to water molecule hydrogen bonds to form. In some foods, water holding is important; in others, moisture uptake from the air is undesirable. Not only is humectancy important with regards to texture, it is important in controlling water activity, which is related to microbial growth and spoilage.

Additional resources

Carbohydrate Nomenclature

Carbohydrate Research 297: 1-92; (1997) Advances in Carbohydrate Chemistry and Biochemistry 52: 43–177; http://www.chem.qmw.ac.uk/iupac/2carb/. (These documents are identical.).

Nomenclature of carbohydrates. Pure and Applied Chemistry 68, 1996, 1919–2008 (1997).

Monosaccharide Chemistry

Belitz, H.D., Grosch, W., Schieberle, P., 2004. Food Chemistry, third ed. Springer, Berlin.

Biliaderis, C.G., Izydorczyk, M.S., 2007. Functional Food Carbohydrates. CRC Press, Boca Raton.

Huber, K.C., BeMiller, J.N., 2017. Carbohydrates. Chap. 2. In: Damaodaran, S., Parkin, K.L. (Eds.), Fennema's Food Chemistry, fifth ed. CRC Press, Boca Raton.

Izydorczyk, M., 2005. Understanding the chemistry of food carbohydrates. In: Cui, S.W. (Ed.), Food Carbohydrates. CRC Press, Boca Raton, pp. 1–65.

[8] **Hygroscopic** refers to sorption of water vapor from the atmosphere, which effects a change in physical characteristics.

[9] A **humectant** is a substance that absorbs or retains moisture.

Jouppila, K., 2011. Mono- and disaccharides: selected physicochemical and functional aspects. In: Eliasson, A.-C. (Ed.), Carbohydrates in Food, fourth ed. CRC Press, Boca Raton, pp. 37–92.

Wrolstad, R.E., 2012. Food Carbohydrate Chemistry. Wiley-Blackwell, London.

Determination of Monosaccharides

BeMiller, J.N., 2017. Carbohydrate analysis. In: Nielsen, S.S. (Ed.), Food Analysis, fifth ed. Springer, New York, pp. 333–360.

Fructose

White, J.S., 2011. Crystalline fructose. In: O'Brien-Nabors, L. (Ed.), Alternative Sweeteners, fourth ed. CRC Press, Boca Raton, pp. 379–402.

Tagatose

Vastenavond, C.M., Bertelsen, H., Hansen, S.J., Laursen, R.,S., Saunders, J., Eriknauer, K., 2011. Tagatose (D-tagatose). In: O'Brien-Nabors, L. (Ed.), Alternative Sweeteners, fourth ed. CRC Press, Boca Raton, pp. 423–438.

Carbohydrate Reactions

2

Chapter Outline

Key information and skills that should be obtained from study of this chapter will enable you to

1. Define and/or identify

aldonic acid	xylitol
aldonate	meso compound
reducing sugars	erythritol
Somogyi-Nelson reagent	cyclitol
glucose oxidase	inositol
glucose oxidase/peroxidase/dye (GOPOD) method	phytic acid
lactone (1,4- and 1,5-)	phytate
glucono-delta-lactone (GDL)	phytin
alditol	aldaric acid

Continued

Carbohydrate Chemistry for Food Scientists. https://doi.org/10.1016/B978-0-12-812069-9.00002-9

polyol	aldobiouronic acid
polyhydroxy alcohol	anhydrosugar
sugar alcohol	sorbitan
D-glucitol	cyclic acetal
sorbitol	

2. Explain the principle of detection or measurement of reducing sugars with the Tollens, Fehling, Benedict, and Somogyi-Nelson reagents and with sodium 3,5-dinitrosalicylate (DNSA), hypoiodite, and hypobromite.
3. With equations, show the formation of D-glucono-1,5-lactone (GDL), including all reagents required.
4. Explain the principle of the GDL formation reaction.
5. Describe the uses and benefits of GDL in bakery, dairy, and processed meat products.
6. With equations, show how sorbitol (D-glucitol), mannitol, and xylitol are made, including all reagents required.
7. Explain the principle of the polyol formation reaction.
8. Describe the uses and benefits of sorbitol, mannitol, and/or xylitol in food products.
9. Using equations, show the effects of reaction with nitric acid, periodate anion, dinitrogen tetraoxide, and galactose oxidase on given carbohydrates.
10. When given the name of a monosaccharide acetate, phosphate, or methyl ether, give its chemical structure, and vice versa.

Introduction

All carbohydrate molecules have hydroxyl groups available for reaction. Simple monosaccharide and most oligosaccharide (Chapter 3) molecules also have carbonyl groups available for reaction. (Polysaccharide molecules have a maximum of one carbonyl group (at the reducing end [Chapter 4]), so the natural aldehydic or keto group in them is insignificant.) Reactions of the carbonyl and hydroxyl groups of carbohydrates are summarized in Table 2.1. Formation of pyranose and furanose rings (cyclic hemiacetals) and glycosides (acetals) of monosaccharides were covered in Chapter 1. Reactions of Maillard browning and related processes, while reactions of monosaccharides with an aldehydic carbonyl group, are covered separately in Chapter 18.

Oxidation of the aldehydic group and the anomeric hydroxyl group of aldopyranoses and aldofuranoses

Aldonic acids

The aldehydic group of aldoses is readily oxidized to a carboxyl/carboxylate ($-COOH/-COO^-$) group. The products of such an oxidation are carboxylic acids,

Table 2.1 Important reactions of carbohydrate molecules

Group Modified	**Reactions**
Carbonyl group (alone)	**1.** Oxidation to a carboxylic acid group
	2. Reduction to a hydroxyl group
	3. Additions of nucleophiles[a]
Hydroxyl groups	**1.** Ester formation
	2. Ether formation
	3. Cyclic acetal formation
	4. Oxidation to a carbonyl group
	5. Reduction to a deoxy carbon atom
	6. Replacement with amino, thiol, and halogeno groups
Both carbonyl and hydroxyl groups	**1.** Formation of cyclic hemiacetals: pyranose and furanose ring forms
	2. Formation of acetals (glycosides)
	3. Aldose ⇄ ketose isomerizations
Anomeric hydroxyl group	**1.** Oxidation to lactones
	2. Formation of glycosyl halides
	3. Formation of glycosylamines*

[a]Chapter 18.

which when formed from aldoses are called **aldonic acids** (that is, aldonic acids are monosaccharides in which C1 is a carboxyl group rather than an aldehydic group). The reaction is commonly used for quantitative determination of sugars and for the manufacture of acids, such as D-gluconic acid. Salts of aldonic acids are **aldonates**, so for example, the sodium salt of D-gluconic acid is sodium D-gluconate.

Aldoses are called **reducing sugars** because they effect reduction of reagents that will oxidize their aldehydic group to a carboxylate group (because in the process of the aldehydic group of an aldose being oxidized to the salt of a carboxylic acid group, the oxidizing agent is reduced). A keto group cannot be oxidized. However, ketoses are also called reducing sugars because, under the alkaline conditions of most reagent solutions used to detect reducing sugars, ketoses are isomerized to aldoses (Chapter 1), which are then oxidized.

A qualitative method for detecting the presence of aldoses (in fact, for detecting the presence of any aldehyde) is the Tollens' silver mirror test. The Tollens reagent is a basic solution of silver ammonia complex $[Ag(NH_3)_2^+]$. The oxidizing agent (Ag^+), which converts the aldehydic group to a carboxylic acid salt (aldonate), is reduced to silver metal (Ag^0), producing a silver mirror coating on the inside of a test tube.

$$2Ag(NH_3)_2^+ + R{-}\overset{\overset{\displaystyle O}{\|}}{C}{-}H + 3OH^- \longrightarrow 2\,Ag^0 + R{-}\overset{\overset{\displaystyle O}{\|}}{C}{-}O^- + 4NH_3 + 2H_2O$$

One of the earliest methods for detection and measurement of sugars employed the Fehling reagent, which is an alkaline solution of a copper(II) salt that oxidizes an aldose to an aldonate and in the process is reduced to copper(I), which precipitates as the brick-red oxide Cu_2O.

$$2\,Cu(OH)_2 + R{-}\overset{\overset{\displaystyle H}{|}}{C}{=}O \longrightarrow R{-}\overset{\overset{\displaystyle O}{\|}}{C}{-}OH + Cu_2O + 2H_2O$$

A variation of the Fehling reagent called the Benedict reagent contains copper citrate. It is less alkaline than the Fehling reagent and, as a consequence, is not as effective in isomerizing ketoses to aldoses (Chapter 1). Hence, the Benedict reagent can be used to detect aldoses in the presence of ketoses. Reducing sugars are also oxidized by sodium 3,5-dinitrosalicylate (DNSA) (yellow); the products are an aldonate and the reddish-brown reduction product (sodium 3-amino-5-nitrosalicylate), which can be measured spectrophotometrically. The reagent most used today to measure reducing sugars is a reagent called both the **Somogyi-Nelson reagent** and the Nelson-Somogyi reagent. It is an arsenomolybdate reagent that changes color on reduction so that the amount of reduced reagent (and hence oxidized sugar) can be measured spectrophotometrically.

In none of these methods is the amount of product formed related in an exact mole-to-mole ratio to the amount of the reducing sugar being measured. Therefore, each requires a standard curve and careful control to give quantitative results. A stoichiometric method (oxidation with hypoiodite anion [IO^-] at pH 9.5) is available, but seldom used. Hypochlorite and hypobromite similarly oxidize aldoses to aldonic acids but also can oxidize secondary hydroxyl groups.

Glucose oxidase

A simple and specific method for quantitative oxidation of D-glucose to D-gluconic acid uses the enzyme **glucose oxidase**, the initial product being the 1,5-lactone (delta-lactone, δ-lactone)[1] of the acid. The reaction is employed to determine the amount of D-glucose in foods and biological tissues, including the D-glucose concentration in blood and urine.

D-Gluconic acid and D-glucono-1,5-lactone (**D-glucano-delta-lactone [GDL]**) are natural constituents of fruit juices, honey, and wine and other fermented products. D-Gluconic acid contributes to the natural tangy flavor of these foods. The anion of

[1] A **lactone** is a cyclic ester. In the case of glucono-delta-lactone (GDL), the cyclic ester involves a carboxylic acid group (C1) and the hydroxyl group on C5. C2 is the alpha carbon atom. C3 is the beta carbon atom. C4 is the gamma carbon atom, and C5 is the delta carbon atom. Hence, the 1,5-lactone can also be called a delta lactone.

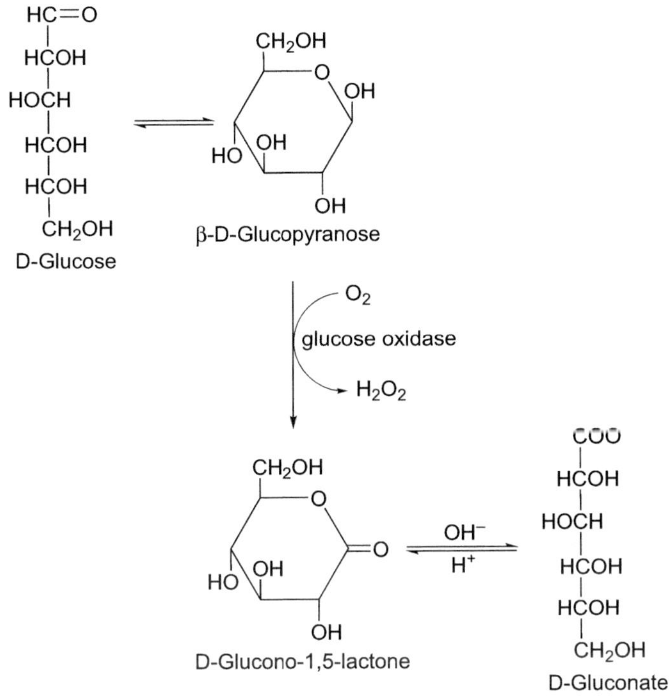

D-gluconic acid is the D-gluconate anion, so for example, (as already stated) the sodium salt form is sodium D-gluconate. Sodium D-gluconate and/or GDL (see below) are food-approved sequestrants.[2]

Calcium D-gluconate and mixtures of calcium gluconate and calcium lactate (which have a higher solubility than either salt individually) are used for calcium fortification of food because they are odorless and tasteless, nonirritating to the gastrointestinal tract, and highly bioavailable. These and other salts (copper gluconate, iron(I) gluconate, potassium gluconate, zinc gluconate) are used as nutritional supplements.

In the analytical procedure for D-glucose using glucose oxidase, the colorless form of a dye is added along with a second enzyme (peroxidase) that uses the hydrogen peroxide produced in the first enzyme-catalyzed reaction to oxidize the dye to a colored compound, the amount of which is determined spectrophotometrically. This method that uses two enzymes[3] and an oxidizable colorless compound is known as the glucose oxidase/peroxidase/dye method (**GOPOD method**)

A related enzyme (hexose oxidase) catalyzes the oxidation of a variety of mono- and disaccharides and is used to improve bread by catalyzing the oxidation of maltose

[2] A sequestrant is a substance that chelates di- and trivalent metal ions, some of which catalyze oxidations (primarily of lipids). Thus, sequestrants can act as preservatives.

[3] Such processes are known as coupled-enzyme reactions.

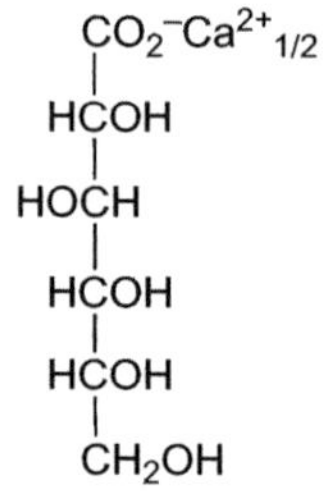

Calcium D-gluconate

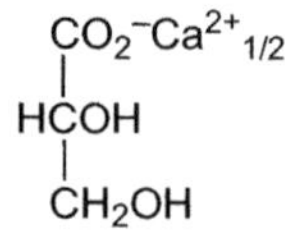

Calcium D-lactate

(a breakdown product of starch [Chapter 3]). Its use can eliminate the need for bread-making oxidants such as bromate and dehydroascorbic acid (section on Sorbitol in this chapter). The coproduct of the oxidation effected by the action of glucose oxidase on β-D-glucopyranose is hydrogen peroxide, which may be the key to the enzyme's effectiveness as a dough improver. It is also the key to use of glucose oxidase to determine amounts of D-glucose in foods and biological tissues and fluids..

Lactones[1]

Aldonic acids readily and spontaneously undergo intramolecular ester formation to produce lactone rings. Even drying down of an aqueous solution of an aldonic acid yields the lactone. While the six-membered **1,5-lactone** ring can be formed, the five-membered **1,4-lactone** ring (gamma-lactone ring) is usually more stable and often the only product isolated. Because of rapid equilibrium between an aldehydic sugar and its pyranose and furanose ring forms and equilibrium between an aldonic acid and its 1,4- and 1,5-lactone ring forms, oxidation of a formerly crystalline sugar in solution gives the same mixture of lactones at equilibrium.

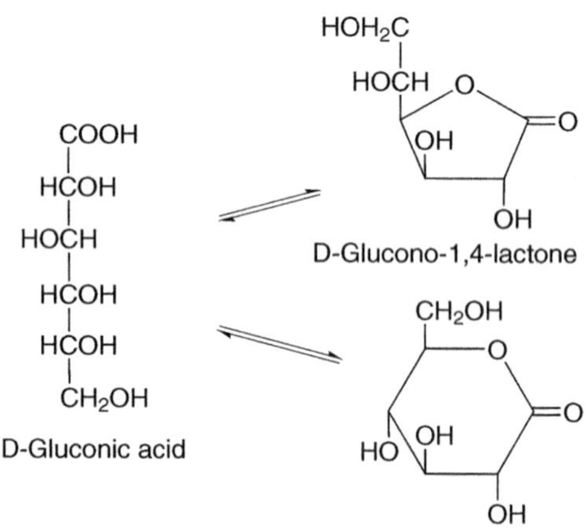

D-Gluconic acid

D-Glucono-1,4-lactone

D-Glucono-1,5-lactone

D-Glucono-delta-lactone (GDL) (properly D-glucono-1,5-lactone and often written as D-glucono-δ-lactone) is produced commercially by oxidation of β-D-glucopyranose using glucose oxidase as the catalyst. GDL undergoes hydrolysis to the open-chain carboxylic acid (D-gluconic acid) in water as shown above. The rate of hydrolysis of GDL to D-gluconic acid is a function of temperature. At room temperature, equilibrium pH (2.5–2.6) is reached in 3–5 h. During the slow hydrolysis, the initial sweet taste of the solution gradually changes to a mild, slightly acidic or sour taste.

GDL is used in the baking industry as an ingredient in chemical leavening agents and as a preservative. As a leavening agent, it is used to neutralize a bicarbonate or carbonate salt and release carbon dioxide. Because normal bakers' yeast does not tolerate cold temperatures well, GDL plus a bicarbonate or carbonate salt is often the leavening agent used in refrigerated and frozen dough products. The very slow hydrolysis of GDL at refrigerator temperatures releases carbon dioxide and often pressurizes the container/package containing the dough. GDL also increases the shelf life of refrigerated dough by preventing its turning gray and/or black spot formation. In doughs that are not refrigerated, very little acid is released during preparation of the dough, but acid is then released as the temperature of the dough rises during baking. GDL is also used to enhance the antimicrobial effect of benzoate, propionate, and sorbate salts (via a lowering of the pH) in bakery fillings and icings.

Addition of GDL (and other ingredients) to the cooking water of pasta and rice improves their color and texture and reduces the amount of dissolved carbohydrate, thus improving yield and reducing the BOD[4] of the waste water. Its addition to the cooking water also extends shelf life. GDL is also added to the flour mixture in preparation of certain noodles to extend shelf life.

Many cheese and tofu manufacturing processes require a slow lowering of the pH. Traditionally, pH lowering is effected by fermentation brought about by lactic acid bacteria. However, GDL will accomplish the same slow drop in pH and may be used in the manufacture of products such as cottage cheese, feta cheese, tofu, and yogurt. Eliminating the need for starter cultures reduces production time and makes the process easier to control and the end product more uniform. The result is higher yields (due to optimized process parameters), more constant quality, and longer shelf life of the product.

In fish cake and surimi, GDL acts as a preservative by lowering the pH and enhancing the antimicrobial effect of benzoate, propionate, and sorbate salts (without resulting in an acidic flavor) and aids in preventing graying. In meat products, GDL reduces the amount of nitrite required, accelerates the curing process, and lengthens shelf life by inhibiting the growth of lactic acid bacteria.

GDL maintains color and firmness in canned and frozen vegetables. Because GDL lowers pH, its addition means that canned fruits and vegetables can be processed at lower temperatures and shorter times. GDL chelates the metal ions that catalyze enzymic (enzymatic) browning and reduces such browning when used together with an antioxidant in foods such as sliced apples, peaches, and potatoes. GDL also

[4] BOD = biological oxygen demand.

functions as a pickling agent. In all of these applications, its slow hydrolysis (acidification) and mild flavor make GDL a desirable and unique acidulant.

Reduction of carbonyl groups

Reduction of the carbonyl group of an aldose or a ketose is accomplished via hydrogenation. (Hydrogenation is the addition of hydrogen to a double bond.) When applied to carbohydrates, it entails addition of hydrogen (H_2) to the double bond between the oxygen atom and the carbon atom of the carbonyl group of an aldehyde or ketone. Hydrogenation of mono- and oligosaccharides (Chapter 3) is accomplished commercially with hydrogen gas under pressure (30–100 atm, temperature 100–150°C [212–300°F]) in the presence of a catalyst, which often is a nickel- or ruthenium-based catalyst. In the laboratory, reduction of an aldehydic or keto group of a carbohydrate can be accomplished using sodium borohydride.

The product of reduction, a compound that has a hydroxyl group on every carbon atom (that is, has no carbonyl group), is an **alditol** (by systematic nomenclature). However, carbohydrates in this class are often called **polyols** or, sometimes, polyhydric or **polyhydroxy alcohols** or **sugar alcohols** in food ingredient literature. Members of this class of compounds are named by adding an -itol suffix to the root name of the sugar. Thus, the alditol produced by reduction of D-mannose is D-mannitol.

Because none of the compounds in this class of compounds contains a carbonyl group and is, therefore, a reducing sugar, none of them can participate in the Maillard nonenzymic (nonenzymatic) browning reaction (Chapter 18). Alditols are also resistant to both acids and alkalies and are nonfermentable by many microorganisms. In general, alditols are reduced-calorie, noncariogenic[5] sweeteners that are used to replace sucrose (Chapter 19) in sugarless products. They are also used as humectants (that is, to hold water and retain the moistness of a product), to depress the freezing point of a product, or to provide the desired texture without making the product overly sweet (as might happen if sucrose [Chapter 3] were used). The caloric values of alditols are not values that are agreed on by regulatory agencies because energy is obtained from polyols in two ways: (1) partial absorption from the small intestine and subsequent catabolism and (2) fermentation in the large intestine (colon) (Chapter 17), and because the distribution between the two pathways is dependent on the quantity of the polyol consumed and the individual. Relative caloric values and other properties of alditols/polyols used in foods are presented and discussed more extensively in Chapter 17.

Sorbitol (D-Glucitol)

When D-glucose is reduced, the product (**D-glucitol**) is obtained in an almost 100% yield. In food ingredient literature, D-glucitol is almost always referred to by its

[5] Noncariogenic is an adjective signifying that the substance does not cause dental caries (tooth decay). Alditols/polyols are noncariogenic because they are poor substrates for oral bacteria, which as a result produce little acid and the polysaccharides that form dental plaque.

common name (**sorbitol**). Because it is made from a hexose, D-glucitol/sorbitol is a hexitol. While sorbitol is widely distributed throughout the plant world (ranging from algae to higher plants, where it is found in fruits and berries), the amounts present in nature are generally small. Sorbitol was first discovered in the berries of the European mountain ash tree (also known as the Rowan tree), where its concentration is about 8.5% on a fresh weight basis. This tree belongs to the genus *Sorbus*, from which the compound gets its common name. Pears contain about 2.1% and cherries about 2.0% sorbitol on a fresh weight basis. Sorbitol is quite soluble and is available both as a liquid (syrup) and as crystals.

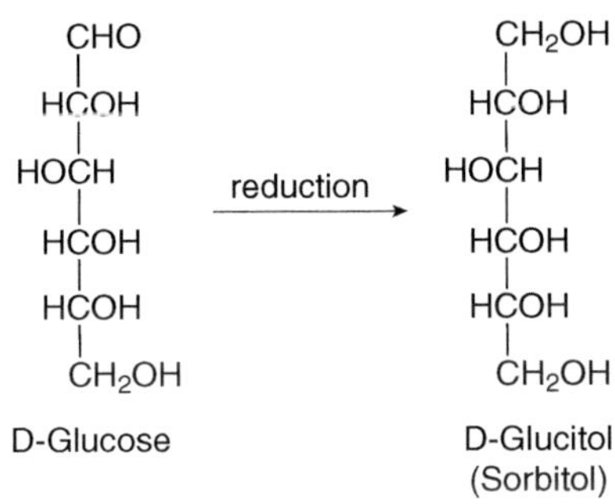

Sorbitol is the polyol that is produced and used in the greatest quantities. Sorbitol has a sweet taste (Chapter 19), resulting in its major food use being in candies, cough drops, mints, and sugarless chewing gums, where it functions as a noncariogenic[5] sweetener and imparts a cooling sensation because of its negative heat of solution. Because it is hygroscopic, it is also widely used as a general humectant. A large quantity of sorbitol is used in toothpaste where it acts as a noncariogenic humectant and a plasticizer and imparts a cool, sweet taste. Sorbitol is the starting material for preparation of sorbitan esters (section on Ethers in this chapter), which are useful as nonionic food emulsifiers.

Sorbitol replaces sugar (reduces calories) and controls humectancy in sugar-free bakery products, cake mixes, fillings, frostings, and icings. It replaces sugar (reduces calories), inhibits crystallization, and depresses the freezing point in sugar-free ice cream and frozen desserts. It replaces sugar (reduces calories) and reduces color formation in sugar-free pancake syrup and no-sugar added jams and jellies. It functions as a cryoprotectant[6] that cannot participate in the Maillard (nonenzymic) browning reaction (Chapter 18) and provides sweetness in surimi and frozen meat products, fruits, and vegetables. It controls humectancy and provides sweetness in dried fruit, and it replaces sugar (reduces calories) and controls humectancy in granola bars.

[6] A cryoprotectant is a substance that protects against freezing damage.

Sorbitol is the starting material in the chemical synthesis of L-ascorbic acid (vitamin C). First, sorbitol is oxidized at C5 by the microorganism *Acetobacter suboxydans* to produce L-sorbose (a ketohexose). L-Sorbose is then converted in three steps into 2-keto-L-gulonic acid. Heating 2-keto-L-gulonic acid under acidic conditions causes it to lactonize. Then the lactone is converted into L-ascorbic acid (also a lactone). High yields at each step in the synthesis allow vitamin C to be produced at a low cost.

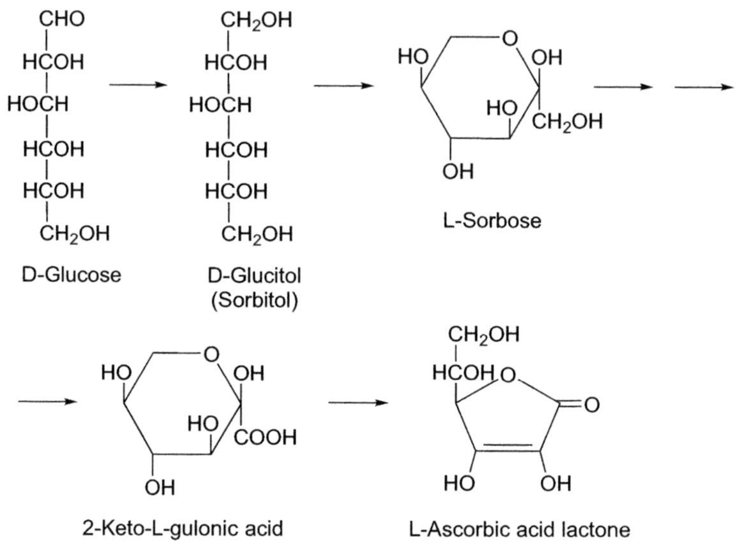

L-Ascorbic acid is widely distributed in plants and animals. Humans, other primates, guinea pigs, bats, birds, and fish lack a liver enzyme (L-gulono-γ-lactone oxidase) necessary for synthesis of L-ascorbic acid and require an exogeneous source of the vitamin. L-Ascorbic acid is required for collagen formation, fatty acid metabolism, good brain function, and drug detoxification; it prevents scurvy and reduces infection and fatigue. In plants, L-ascorbic acid is involved in cellular respiration, growth, and maintenance of carbon balance. Natural L-ascorbic acid is isolated commercially in small quantities from rose hips, persimmon and citrus fruits, and other plant sources.

When a solution of calcium L-ascorbate is applied to the surface of freshly cut fruit, it acts as an antioxidant and prevents discoloration (enzymic browning). L-Ascorbic acid and esters of it improve both doughs and the quality of breads and increase loaf volume by means of dehydroascorbic acid (the oxidized form)—mediated cross-linking of flour proteins.

D-Mannitol

Commercially, D-mannitol is obtained along with D-glucitol (sorbitol) via hydrogenation of a high-fructose syrup (Chapter 7). It can also be obtained by hydrogenolysis[7] of sucrose (Chapter 3), by hydrogenation of D-mannose, and by fermentation. When an aldose is reduced, no new chiral carbon atom is formed. However, when a ketose (like D-fructose) is reduced, a new chiral carbon atom is formed (because the carbonyl group to be reduced is at the C-2 position). Therefore, both D-glucitol and D-mannitol are formed via reduction of D-fructose. Because D-mannitol is much less soluble than is sorbitol, the two products can be separated by crystallization of mannitol.

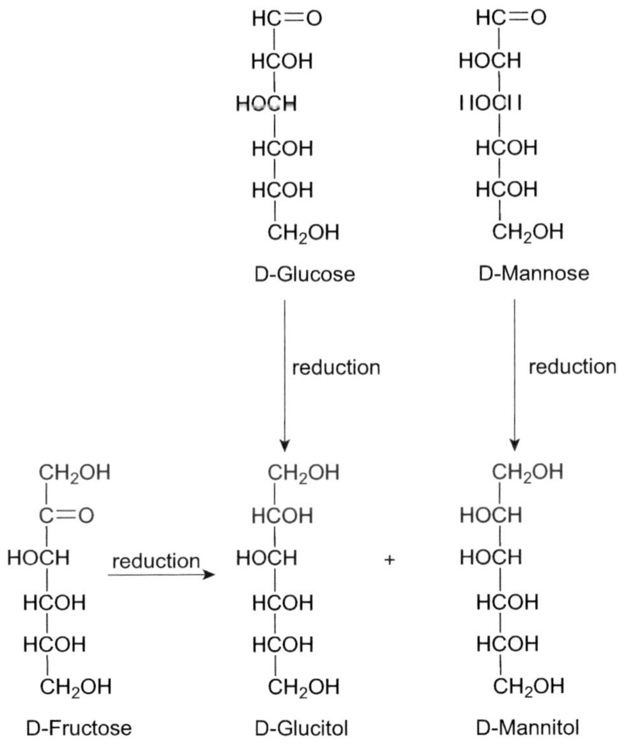

Mannitol occurs in nature as manna on the manna ash tree. Brown seaweeds may contain as much as 25% mannitol on a dry weight basis (both free and in combined forms). It is also a constituent of saw palmetto and plane trees.

Unlike sorbitol, mannitol is not a humectant, but like sorbitol, it produces a cooling effect in the mouth because of its negative heat of solution. It crystallizes easily and is only slightly soluble. It is used as an anticaking agent and for dusting confectionery

[7] The term hydrogenolysis denotes cleavage of a chemical bond (in this case a carbon—oxygen bond of the glycosidic linkage) with concurrent addition of H_2.

products. It is about 70% as sweet as sucrose (Chapter 19) and is used in sugar-free chocolates, hard and soft candies, and pressed mints.

Xylitol

Currently, **xylitol** is produced by hydrogenation of the pentose D-xylose, which is obtained by hydrolysis of hemicelluloses (Chapter 17) rich in D-xylose, especially those from birch trees. No D or L designation is needed for xylitol or erythritol (below) because they are not chiral compounds. Both have a plane of symmetry (that is, the upper half of the molecule is the mirror image of the lower half).[8] In the case of xylitol, the plane of symmetry goes through carbon atom 3 (C3). Compounds containing a plane of symmetry are known as **meso compounds**. Because they are not chiral compounds, neither xylitol nor erythritol (below) will rotate plane-polarized light.

Xylitol has a high negative heat of solution, so crystals of it produce a cool sensation when they dissolve in the mouth. It is used where a cooling sensation is desired (for example, in mint-flavored lozenges and hard candies and in sugarless chewing gum). Xylitol has about the same degree of sweetness as sucrose (Chapter 19). It is only slightly hygroscopic and quite soluble in water.

When xylitol is used in place of sucrose, there is a reduction in dental caries (tooth decay) because xylitol is not metabolized by the microflora of the mouth that produce dental plaques. Dental caries are primarily produced by the microorganism *Staphylococcus mutans,* which uses the D-fructose portion of the sucrose molecule as an energy source and polymerizes the D-glucose portion to a polysaccharide (Chapter 4) called dextran that surrounds the colony of bacteria. Oxygen is restricted beneath the plaque, so acidic metabolic products (such as lactic acid) accumulate, demineralize tooth enamel, and cause tooth decay. Because xylitol is not catabolized by the bacterial cells in the mouth, the polysaccharides that allow the cells to attach to teeth and the acids the bacteria produce (which cause tooth decay) are not made. Prevention of dental caries by replacing sucrose with another sweetener is therefore desirable, but xylitol is unique in that it not only reduces caries but also prevents tooth decay.

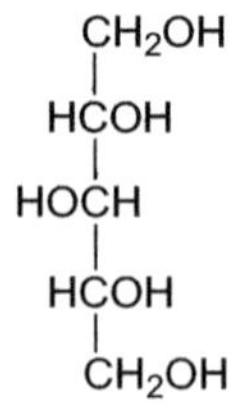

Xylitol

[8] It can be easily determined that these two alditols (and galactitol) are not optically active by writing their mirror-image structures. Then rotate one of the two structures 180 degrees and you will see that the structures are identical (that is, superimposable).

Erythritol

Currently, **erythritol**, the reduction product of the tetrose D-erythrose (Chapter 1), is produced by fermentation rather than by hydrogenation.[9] Erythritol is noncariogenic. It is almost noncaloric (0.2 Kcal/g; less than 5% of the caloric value of sucrose). Crystalline erythritol has a strong cooling effect (−23.3 Kcal/g heat of solution) (Chapter 17). It is readily soluble in water; a saturated solution contains about 60% erythritol. The sweetness of erythritol is about 60%−70% that of sucrose. It is not hygroscopic and improves the sweet taste (making it more like that of sucrose) when used in combination with several high-intensity sweeteners (Chapter 19). It produces few gastroenterological problems. Erythritol is suggested for use as a softener for chewing gum and in sugar-free hard candies, but by far its largest use is in soft drinks where it improves the taste of drinks sweetened with stevia and stabilizes and improves the taste of drinks sweetened with aspartame (Chapter 19). Erythritol occurs naturally in wine, soy sauce, melons, and fruits.

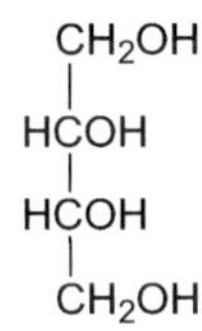

Erythritol

Use of alditols in carbohydrate analysis

Gas-liquid chromatography (GLC) is often used to determine monosaccharides—both qualitatively and quantitatively. To determine the monosaccharides present in a food product, they are first extracted (with water) from the product. To determine the monosaccharide composition of an oligo- or polysaccharide, the saccharide is hydrolyzed with aqueous acid to release the monosaccharides. The monosaccharides in the mixture of sugars are then reduced to alditols using sodium or potassium borohydride. The next step is complete acetylation of the resulting mixture of alditols with acetic anhydride to make what are called peracetate esters of the alditols. The alditol peracetates are volatile enough, that mixtures of them can be analyzed by gas-liquid chromatography (GLC). (Unsubstituted carbohydrates are nonvolatile.) This process is discussed more thoroughly in the section on Esters in this chapter.

[9] A factory to produce erythritol from glucose via an electrochemical process is under construction.

Cyclitols

Cyclitols, like alditols, are carbohydrates that contain only hydroxyl groups (that is, contain no carbonyl group). The difference is that cyclitols have cyclic (ring) structures. One member of this class of compounds is known as myo-inositol (or meso-inositol because it has a plane of symmetry and, hence, no optical rotation). Like the aldo- and ketohexoses, myo-inositol has the general formula $C_6H_{12}O_6$.

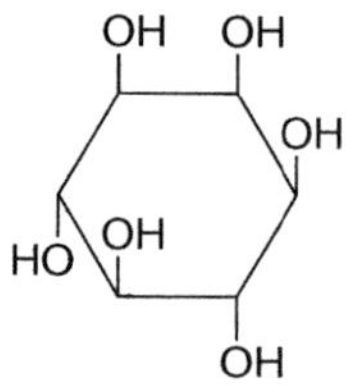

myo-Inositol

Myo-inositol occurs as a free compound in virtually all plants. Isomers of myo-inositol (that is, cyclitols with configurations of the chiral carbon atoms different from those of myo-inositol) also exist in nature. Likewise, cyclitols containing groups in addition to hydroxyl groups are present in nature. Myo-inositol (the most common isomer and, therefore, often simply called **inositol**) most often occurs as a hexaphosphate ester known as **phytic acid**. (There is controversy about the precise structure of phytic acid. It may be a mixture of isomeric, fully phosphorylated inositols, but it is primarily myo-inositol hexaphosphate.) Salts of phytic acid are **phytates**. Phytates are ubiquitous in plants; 1%–5% of the dry weights of the seeds of cereal grains and legumes (beans) may be phytates. Naturally occurring phytates are mainly mixed potassium–magnesium salts. **Phytin** is a complex salt of phytic acid, inorganic cations, and protein and is the form in which most phytic acid occurs in plants. As much as 90% of the total phosphorus of cereal grain and legume seeds may be present as phytin.

The effects of phytic acid (phytates) in human nutrition are unclear. Historically, phytic acid has been considered to be an antinutritional component of cereal grains and legumes because phytate binds cations and reduces mineral bioavailability. (This effect can be reduced by treatment of the foodstuff with the enzyme phytase.) However, research has revealed that myo-inositol and/or phytic acid may be beneficial to human health. Reported potential health benefits include a reduction in digestion of starch (which is especially beneficial to diabetics), reduction in blood cholesterol (and as a result a reduction in cardiovascular disease), prevention of kidney stones, removal of lead and other heavy metal ions, and anticancer activity.

Oxidation of nonanomeric hydroxyl groups

Boiling 30% nitric acid oxidizes both the aldehydic group and the primary alcohol group of aldoses, forming dicarboxylic acids that belong to the class of carbohydrates known as **aldaric acids**. In this way, D-glucose is converted to D-glucaric acid (commonly called saccharic acid) in a yield of 50%–65%. D-Galactose is converted to galactaric acid[10] (commonly called mucic acid) in a yield of about 75%. Galactaric acid is so insoluble that its formation was once used to measure the amount of galactose in a product.

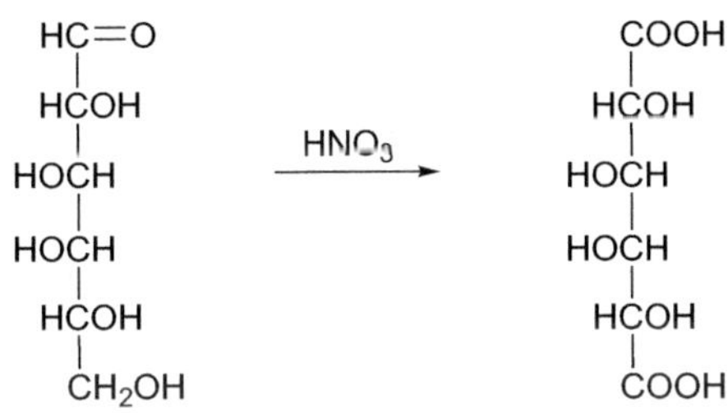

D-Galactose

Galactaric acid
(Mucic acid)

Periodate anion (IO_4^-) is a specific oxidant for adjacent (vicinal) hydroxyl groups and is convenient for measuring the number of such groups in a molecule. With each oxidative cleavage of the molecular chain, one periodate ion is consumed/reduced and the carbon atoms containing adjacent secondary hydroxyl groups are converted into aldehydic groups. When three secondary carbon atoms containing hydroxyl groups adjoin each other, the central one is oxidized twice, resulting in its transformation into formic acid. The most rapid oxidation occurs at pH 3–5. In the past, periodate oxidation was used in determinations of the structures of polysaccharides.

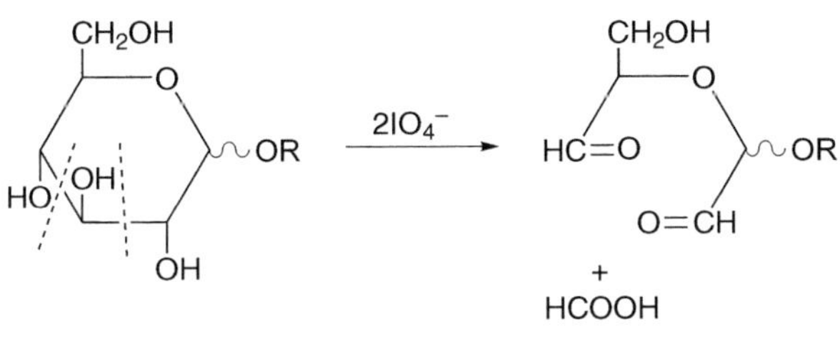

[10] Galactaric acid is a meso compound, so no D or L designation is used.

Dinitrogen tetraoxide (N_2O_4) is a fairly selective oxidant for converting primary hydroxyl groups to carboxyl groups. It is most often used experimentally to oxidize polysaccharides, but is also applicable to glycosides. (The aldehydic group must be protected. If it is not, an aldaric acid is formed.) The product of oxidation of a glycoside or a polysaccharide with dinitrogen tetraoxide is a uronic acid unit (Chapter 1). For example, a glycoside of D-glucose is converted into a glycoside of D-glucuronic acid.

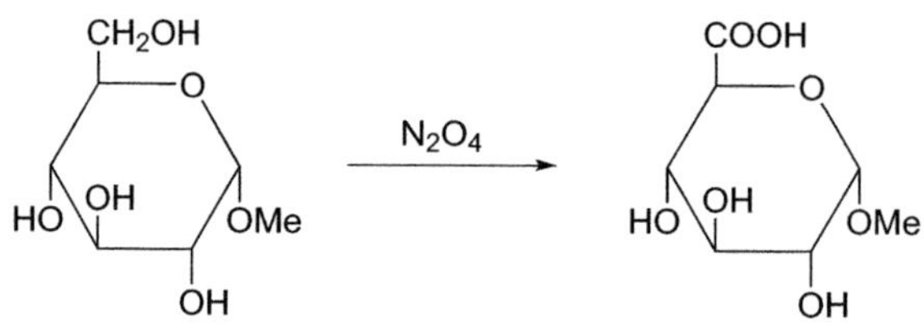

Methyl α-D-glucopyranoside Methyl α-D-glucopyranosiduronic acid

If a uronic unit occurs as part of an oligo- or polysaccharide, its glycosidic linkage is rather resistant to hydrolysis, and hydrolysis of the polymer containing it gives a high yield of a disaccharide called an **aldobiouronic acid.**[11]

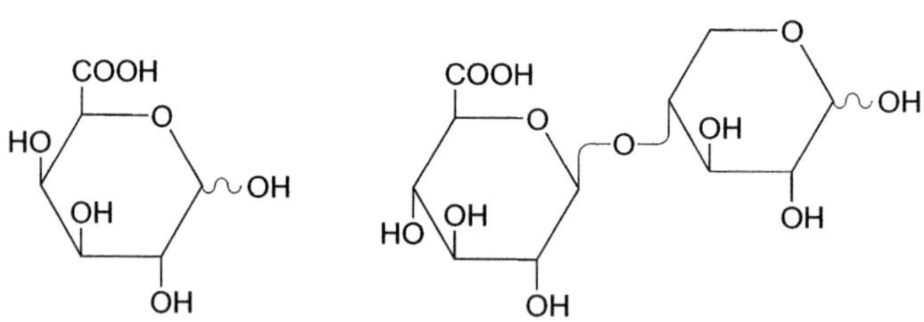

D-Galacturonic acid

An aldobiouronic acid (4-O-β-D-Glucuronopyranosyl-D-xylose)

Selective oxidation of primary hydroxyl groups in aqueous solution using oxygen and a platinum or palladium catalyst or the TEMPO reagent[12] is used for laboratory preparation of individual uronic acids and of polysaccharides containing uronic acid units.

[11] An aldobiouronic acid is a disaccharide (Chapter 3) with a uronic acid unit at its nonreducing end.

[12] 2,2,6,6-Tetramethylpiperidine-1-oxyl.

The enzyme galactose oxidase catalyzes oxidation of the primary hydroxyl group (the C6 position) of D-galactopyranosides to produce an aldehydic group. Methyl lactoside (the methyl glycoside of the disaccharide lactose [Chapter 3]), for example, can be oxidized to produce an aldehydic group from the hydroxymethyl group (C6) of the β-D-galactopyranosyl unit as shown below. If the aldehydic group is further oxidized, it becomes a carboxyl group and the product is an aldobiouronic acid glycoside. ([O] is a symbol signifying an oxidation.) Its methyl group aglycon remains easily hydrolyzable, whereas the D-galactouranosyl linkage becomes resistant to hydrolysis.

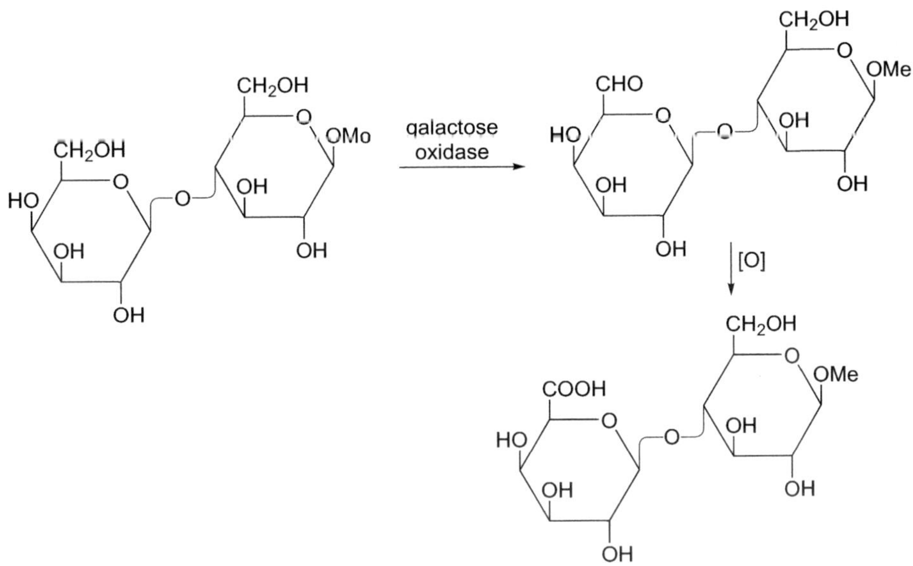

The following scheme depicts the relation of a hexose (specifically D-glucose) to the three acids that can be formed from it by oxidation of the aldehydic group (an aldonic acid), the carbon atom with the primary hydroxyl group (a uronic acid), and both ends of the molecule (an aldaric acid).

Hydrogen peroxide is a nonspecific oxidant that has been used to depolymerize oligo- or polysaccharides. It acts via a free-radical mechanism in a reaction catalyzed by Fe(II) ions which donate electrons to hydrogen peroxide, resulting in its splitting into hydroxide ions and hydroxyl radicals (as shown in the reaction below). The hydroxyl radicals attack the hydroxyl groups of carbohydrates, leading to the formation of carbonyl groups. These reactions are a reminder that carbohydrates are susceptible to a variety of oxidants.

$$Fe^{2+} + HO{-}OH \rightarrow Fe^{3+} + HO^{\cdot} + OH^{-}$$

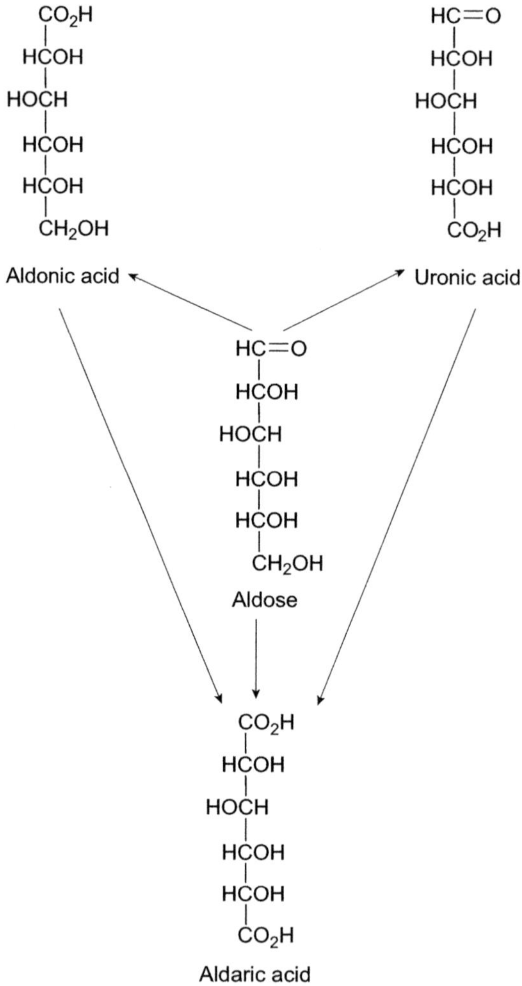

Esters

The hydroxyl groups of carbohydrates (like the hydroxyl groups of simple alcohols) can form esters with organic and some inorganic acids. Reaction of hydroxyl groups with an activated form of a carboxylic acid (namely, a carboxylic acid anhydride or chloride [an acyl chloride]) in the presence of a suitable base produces an ester according to the following reactions in which ROH is the carbohydrate molecule.

Already mentioned in the section on alditols was a method for both analysis of the sugar composition of food products and for determination of the monosaccharide compositions of polysaccharides (Chapter 4). Monosaccharides that occur naturally, that have been added to a food product, or which have been released by acid-catalyzed

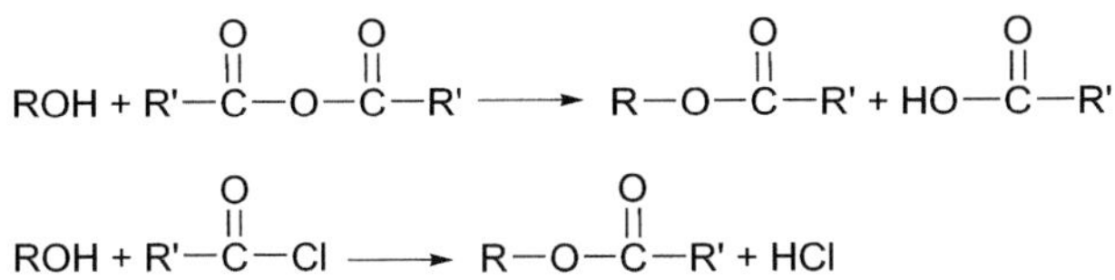

hydrolysis (Chapter 4) of oligo- and polysaccharides are first reduced to their corresponding alditols.[13] The alditols are then treated with acetic anhydride in pyridine to produce the fully acetylated derivatives (peracetylated alditols, alditol peracetates). (For example, D-glucose is reduced to D-glucitol, which in turn is acetylated to form D-glucitol 1,2,3,4,5,6-hexaacetate.) The alditol peracetates are then separated, identified, and measured quantitatively by gas-liquid chromatography (GLC). The value in this reaction sequence lies in the fact that each aldose gives a single, volatile and thermo stable alditol acetate.

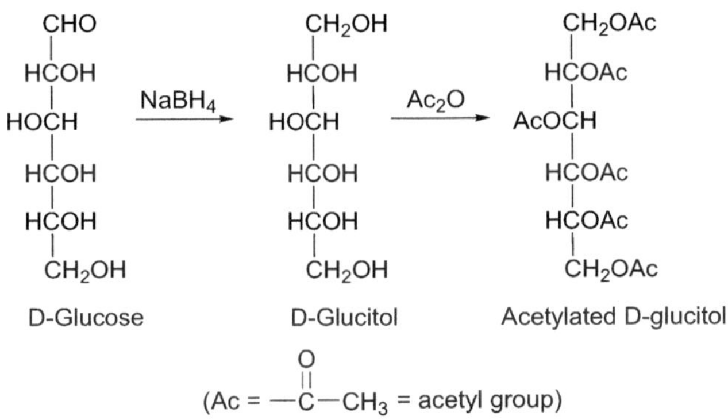

Acetates, succinate half-esters, and other carboxylic acid esters of carbohydrates occur in nature (in polysaccharides [Chapter 4]). For example, the polysaccharide xanthan (Chapter 11) contains a 6-*O*-acetyl-α-D-mannopyranosyl unit.[14]

Sugar phosphates (that is, the phosphate esters of sugars) are common metabolic intermediates. Examples of such compounds are D-glucose 6-phosphate and D-fructose 1,6-bisphosphate, which like their nonphosphorylated counterparts occur primarily in ring forms.

Monoesters of phosphoric acid are also constituents of polysaccharides. Potato starch contains a small percentage of phosphate ester groups (Chapter 6). Small amounts of phosphate ester groups may also be present in other starches. Corn/maize

[13] In a research laboratory, the usual reducing agent is sodium borohydride ($NaBH_4$) in dilute ammonium hydroxide.

[14] 6-*O*-acetyl indicates that the acetyl group is on O6 (the oxygen atom on C6).

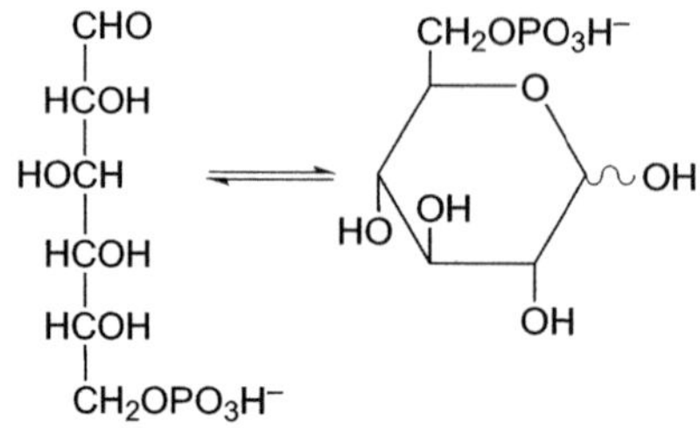

D-Glucose 6-phosphate

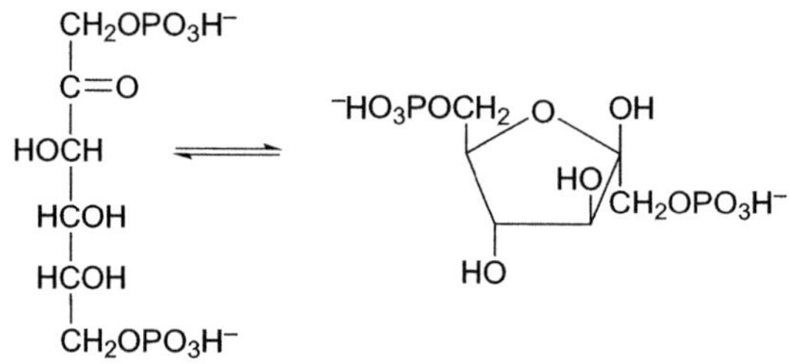

D-Fructose 1,6-bisphosphate

starch contains only traces. In producing modified food starch, corn starch may be derivatized by addition of mono- and/or distarch ester groups (Chapter 7,).

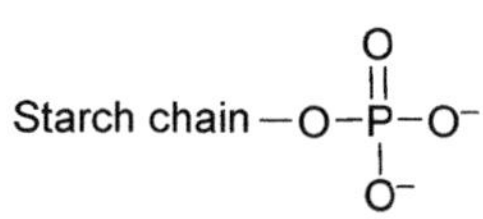

Monostarch phosphate

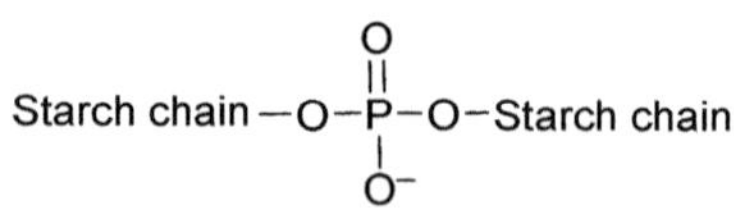

Distarch phosphate
(Cross-linked starch)

Other esters of starch (most notably, the acetate, succinate half-ester, substituted succinate half-ester, and distarch adipates [diester]) are produced and sold as modified food starches (Chapter 7). Sucrose fatty acid esters (Chapter 3) are produced and used commercially, primarily as water-in-oil emulsifiers. Polysaccharides of the family of red seaweed polysaccharides, which include the carrageenans (Chapter 13), contain sulfate groups (half-esters of sulfuric acid; $R{-}OSO_3^-$).

Ethers

The hydroxyl groups of carbohydrates, like the hydroxyl groups of simple alcohols, can also participate in the formation of ethers (R-O-R′). Ethers of carbohydrates are not as common in nature as are esters. One of only a few examples of carbohydrates containing an ether group is 4-*O*-methyl-D-glucuronic acid,[15] a common constituent of hemicelluloses (Chapter 17) and exudate gums (Chapter 16,).

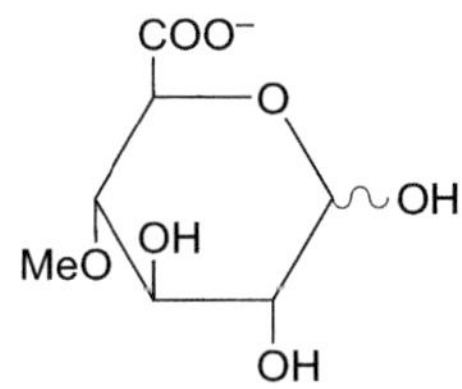

4-O-Methyl-D-glucuronate

Polysaccharides are etherified commercially to modify their properties and make them more useful. Examples are the production of methyl ($-CH_3$), sodium carboxymethyl ($-CH_2-CO_2{}^-Na^+$), and hydroxypropyl ($-CH_2-CHOH-CH_3$) ethers of cellulose (Chapter 8) and hydroxypropyl ethers of starch (Chapter 7), all of which are approved for food use.

A special type of ether (an internal ether formed between carbon atoms three and six of a D-galactosyl unit) is found in red seaweed polysaccharides (specifically in agar, furcellaran, kappa-carrageenan, and iota-carrageenan [Chapter 13]). Such an internal ether is known as a 3,6-anhydro ring (Fig. 2.1), the name of which derives from the fact that two −OH groups (on C3 and C6) have been replaced with one ether (−O−) linkage and, therefore, the end result is in essence removal of one molecule of water (HOH) from the two hydroxyl groups. Monosaccharides such as the parent sugar of the unit shown in Fig. 2.1 are known as **anhydrosugars.**

Nonionic surfactants based on sorbitol (D-glucitol) are used in foods as water-in-oil emulsifiers and as defoamers. They are produced during esterification of sorbitol with

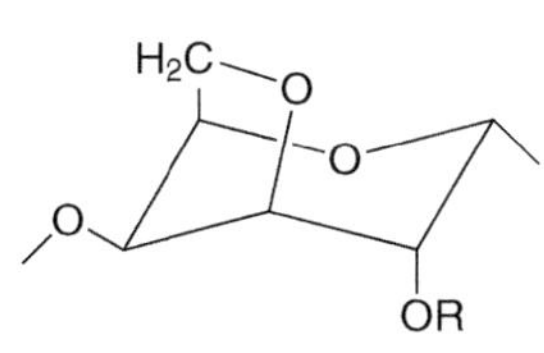

Figure 2.1 A 3,6-anhydro-α-D-galactopyranosyl unit of kappa-carrageenan (R = H) and iota-carrageenan ($R = SO_3{}^-$, a sulfate half-ester group).

[15] 4-*O*-methyl indicates that the methyl group is on O4.

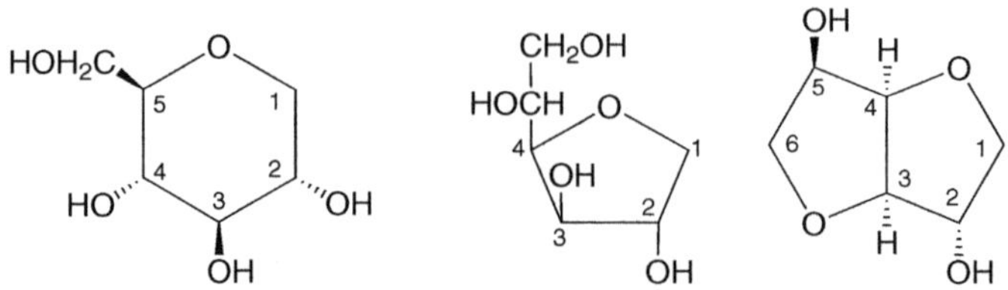

Figure 2.2 Anhydro-D-glucitols (sorbitans). Numbering refers to the carbon atoms of the original molecule of D-glucose (and of sorbitol).

fatty acids. Cyclic dehydration accompanies esterification when sorbitol is heated with a fatty acid (primarily at a primary hydroxyl group [that is, a hydroxyl group at C1 or C6]), so that the carbohydrate (hydrophilic) portion is not only sorbitol but also its mono- and dianhydrides (cyclic ethers). The products are known as **sorbitan** esters (Spans). The product called sorbitan monostearate is actually a mixture of compounds formed by partial esterification (with perhaps a mixture of stearic [C_{18}] and palmitic [C_{16}] acids) of the mixture of sorbitol (D-glucitol), 1,5-anhydro-D-glucitol (1,5-sorbitan; the left-hand structure in Fig. 2.2), 1,4-anhydro-D-glucitol (1,4-sorbitan; the middle structure) (both internal [cyclic] ethers), and 1,4:3,6-dianhydro-D-glucitol (isosorbide, the right-hand structure) (an internal dicyclic ether). (The designation mono-, di-, and tri-simply indicates the ratio of fatty acid ester groups to each molecule in the mixture of compounds called sorbitan.) Therefore, sorbitan monostearate is a mixture of partial stearic and palmitic acid esters of the four polyols (sorbitol and three anhydrosugars derived from sorbitol) in the mixture with a molar ratio of fatty acid to the original sorbitol being 1:1. Some sorbitan fatty acid esters (such as sorbitan monostearate, sorbitan monolaurate, and sorbitan monooleate) are also modified by reaction with ethylene oxide to produce so-called ethoxylated sorbitan esters, which also have short poly(ethylene glycol) chains attached to the four polyols and which are more hydrophilic than are the unmodified sorbitan esters. Ethoxylated sorbitan esters are nonionic detergents (called Tweens) approved by the US Food and Drug Administration for food use (as are Spans).

Cyclic acetals

Cyclic acetals are formed when the two hydroxyl groups that react with a carbonyl group to form an acetal are on the same molecule. They have the structure shown below. Cyclic acetals occur only rarely in nature. One structure in which a cyclic acetal is found is that of the polysaccharide xanthan (Chapter 11), which contains a cyclic acetal[16] formed from pyruvic acid ($CH_3{-}CO{-}CO_2H$) (a compound containing a keto group) on one of the monomer units in its repeating unit structure (Fig. 2.3).

[16] Sometimes called a ketal as it is formed from a ketone.

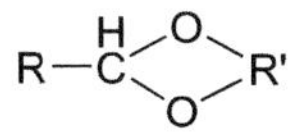

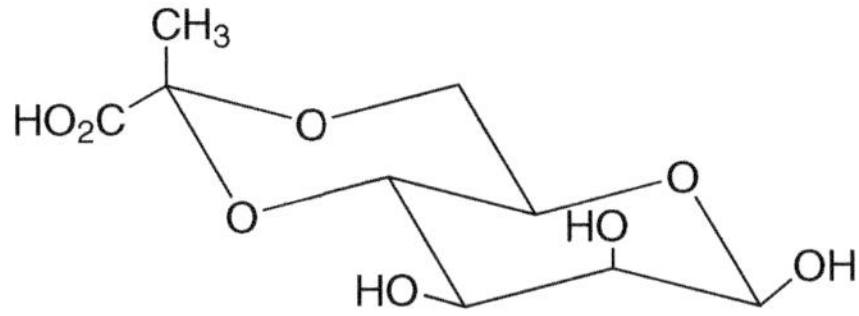

Figure 2.3 The 4,6-O-(1-carboxyethylidene)-β-D-mannopyranosyl unit of the polysaccharide xanthan. The upper left ring is a cyclic acetal of pyruvic acid.

Additional resources

General (Reactions of Sugar Carbonyl and Hydroxyl Groups)

Folkes, D.J., Jordan, M.A., 2006. Mono- and disaccharides: analytical aspects. Chap. 2. In: Eliasson, A.-C. (Ed.), Carbohydrates in Food, third ed. CRC Press, Boca Raton.

Tomasik, P. (Ed.), 2004. Chemical and Functional Properties of Food Saccharides. CRC Press, Boca Raton.

Determination of Monosaccharides

BeMiller, J.N., 2017. Carbohydrate analysis. In: Nielsen, S.S. (Ed.), Food Analysis, fifth ed. Springer, New York, pp. 333–360.

Brummer, Y., Cui, S., 2005. Understanding carbohydrate analysis. Chap. 2. In: Cui, S.W. (Ed.), Food Carbohydrates: Chemistry, Physical Properties, and Applications. CRC Press, Boca Raton.

Polyols (see also Chapter 19)

de Cock, P., 2011. Erythritol. In: O'Brien-Nabors, L. (Ed.), Alternative Sweeteners, fourth ed. CRC Press, Boca Raton, pp. 249–264.

de Silva, S.S., Chandel, A.K. (Eds.), 2012. Xylitol: Fermentative Production, Application and Commercialization. Springer, New York.

Deis, R.C., 2011. Maltitol syrups and polyglycitols. In: O'Brien-Nabors, L. (Ed.), Alternative Sweeteners, fourth ed. CRC Press, Boca Raton, pp. 265–274.

Jamieson, P.R., Lee, A.S., Mulderrig, K.B., 2011. Sorbitol and mannitol. In: O'Brien-Nabors, L. (Ed.), Alternative Sweeteners, fourth ed. CRC Press, Boca Raton, pp. 333–348.

Kearsley, M.W., Boghani, N., 2011. Maltitol. In: O'Brien-Nabors, L. (Ed.), Alternative Sweeteners, fourth ed. CRC Press, Boca Raton, pp. 333–348.

Sentko, A., Bernard, J., 2011. Isomalt. In: O'Brien-Nabors, L. (Ed.), Alternative Sweeteners, fourth ed. CRC Press, Boca Raton, pp. 275–298.

Sentko, A., Bernard, J., 2011. Isomaltulose. In: O'Brien-Nabors, L. (Ed.), Alternative Sweeteners, fourth ed. CRC Press, Boca Raton, pp. 423–438.

Zacharis, C., Stowell, J., Olinger, P.M., Pepper, T., 2011. Xylitol. In: O'Brien- Nabors, L. (Ed.), Alternative Sweeteners, fourth ed. CRC Press, Boca Raton, pp. 349–378.

Zacharis, C., Stowell, J.W., Mesters, P.H.J., Velthuijsen, J.A., Brokx, S., 2011. Lactitol. In: O'Brien-Nabors, L. (Ed.), Alternative Sweeteners, fourth ed. CRC Press, Boca Raton, pp. 315–332.

Ascorbic Acid Use in Foods

Every, D., Simmons, L., Ross, M., Wilson, P.E., Schofield, J.D., Bollecker, S.S.J., Dobraszczyk, B., 2000. Mechanism of the ascorbic acid improver effect on baking. In: Shewry, P.R., Tatham, A.S. (Eds.), Wheat Gluten. The Royal Society of Chemistry, London, pp. 277–282. Special Publication 261.

Cyclitols/Inositols

Minihane, A.M., Rimbach, G., 2002. Iron absorption and the iron binding and anti-oxidant properties of phytic Acid. International Journal of Food Science and Technology 37, 741–748.

Oatway, L., Vasanthan, T., Helm, J.H., 2001. Phytic acid. Food Reviews International 17, 419–431.

Plaami, S., 1997. Myoinositol phosphates: analysis, content in foods and effects in nutrition. Food Science & Technology (London) 30, 633–647.

Oligosaccharides

3

Chapter Outline

Carbohydrate Chemistry for Food Scientists. https://doi.org/10.1016/B978-0-12-812069-9.00003-0

Key information and skills that can be obtained from the study of this chapter will enable you to do the following:

1. Define and/or identify:

oligosaccharide	sucrose
di-, tri, tetra-, penta-, hexa-, hepta-, octa-, nona-, and decasaccharide	invert sugar invertase
glycosidic linkage/bond	molasses
reducing end	raffinose
nonreducing end	stachyose
maltose	cryoprotectant
reversion	olestra
polydextrose	sucralose
maltitol	isomaltulose
lactose	isomaltitol
lactase	isomalt
lactose intolerance	fructooligosaccharide (FOS)
lactitol	kestose
lactulose	α,α-Trehalose

2. Give chemical structures for maltose, lactose, sucrose, and α,α-trehalose.
3. Give shorthand structures for the disaccharides listed in objective 2 plus raffinose and stachyose.
4. Make interconversions between names and Haworth and/or conformational structures, e.g., 4-*O*-(α-L-arabinofuranosyl)-D-xylopyranose (structure below).

αLAra*f* (1 → 4) Xyl*p* ↔

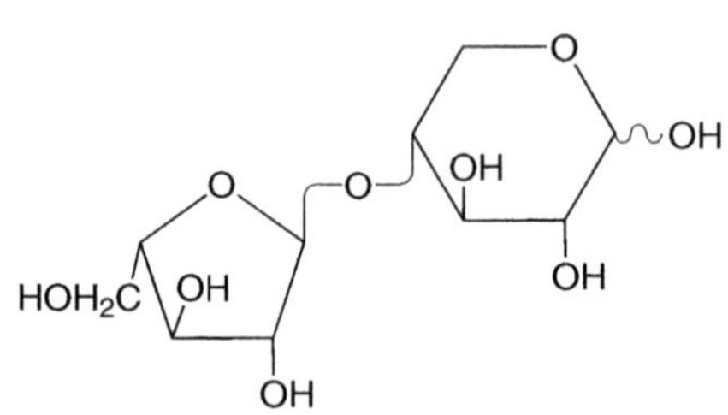

5. Describe (a) the cause and (b) the effects of lactose intolerance.
6. Describe how substances such as sucrose and sorbitol act as cryoprotectants.

Introduction

Short chains of monomer[1] units are known as oligomers. *Oligo* means few in Greek. The suffix *mer* designates a structure composed of parts (*meros* is the Greek word for part). So an **oligosaccharide** is a molecule that consists of a few monosaccharide units joined together. There is no exact specification of the size range of oligosaccharides. They are most often defined as molecules containing 2 to 10 sugar units joined by glycosidic linkages (see below). (The generic prefix indicating a carbohydrate-containing compound is *glyco-*, the generic term for any monosaccharide is glycose, and the generic term for a monosaccharide unit of an oligo- or polysaccharide [Chapter 4] is glycosyl unit.) However, since the smallest known naturally occurring polysaccharide contains about 35 glycosyl units, the intermediate group (containing between 10 and 35 monosaccharide units) is unnamed. The author of this book prefers to designate the demarcation between oligosaccharides and polysaccharides (Chapter 4) as 20 glycosyl units. A depiction of a homologous series of oligosaccharides is given in Fig. 3.1.

Disaccharides are glycosides in which the aglycon is another monosaccharide unit (that is, they are composed of two monosaccharide units joined together). Addition of a third monosaccharide unit forms a **trisaccharide** (a three-monosaccharide unit structure). Progressive addition of glycosyl units to form larger and larger molecules, results in **tetra-, penta-, hexa-, hepta-, octa-, nona-,** and **decasaccharides**, containing 4, 5, 6, 7, 8, 9, and 10 glycosyl units, respectively. Larger oligosaccharides are named similarly.

The glycosyl unit-to-glycosyl unit bond joining any two monosaccharide units is known as a **glycosidic linkage** or **bond**. As described in Chapter 1, glycosidic linkages are part of an acetal structure and, as such, can be cleaved by hydrolysis (that is, hydrolyzed) when the oligo- or polysaccharide or other glycoside is subjected to an enzyme specific for it or to acidic conditions. (Heat is usually also required in the latter case.)

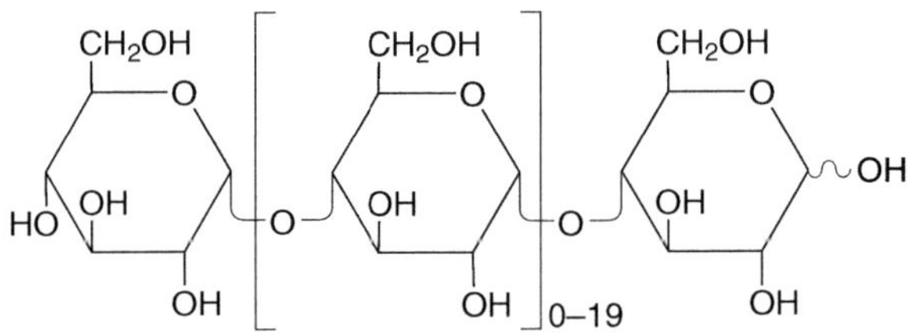

Figure 3.1 An example of homologous oligosaccharides: n = 0–19. In this example, each monosaccharide unit is an α-D-glucopyranosyl unit, and each is joined to the O4 position of the unit to the right of it by a glycosidic bond. Such molecules are obtained from starches (Chapter 6) and are known as maltooligosaccharides.

[1] Monomers are basic building blocks that can be joined together in a chainlike fashion to form large molecules (macromolecules). In carbohydrate chemistry, the monomers of oligo- and polysaccharides are monosaccharide aldoses or ketoses.

Oligo- and polysaccharides (Chapter 4) are composed of glycosyl (saccharide) units joined together in a head-to-tail fashion by glycosidic linkages. As a result, each oligo- and polysaccharide molecule has one reducing end and at least one nonreducing end (Fig. 3.2). The **reducing end** is so named because it has (or potentially has) a free aldehydic group that can act as a reducing agent (Fig. 3.3). Because glycosyl units have multiple hydroxyl groups that can be involved with the anomeric carbon atom of another glycosyl unit in a glycosidic bond, more than one unit can be joined to a monosaccharide unit, and therefore, branched structures are possible (Fig. 3.2). All oligo- and polysaccharides have one, and only one, reducing end, although it may not have a free carbonyl group. All linear oligo- and polysaccharides have one **nonreducing end** (that is, an end that does not have a free (or potentially free) carbonyl group. Branched oligo- and polysaccharides have more than one nonreducing end (one additional one for each branch).

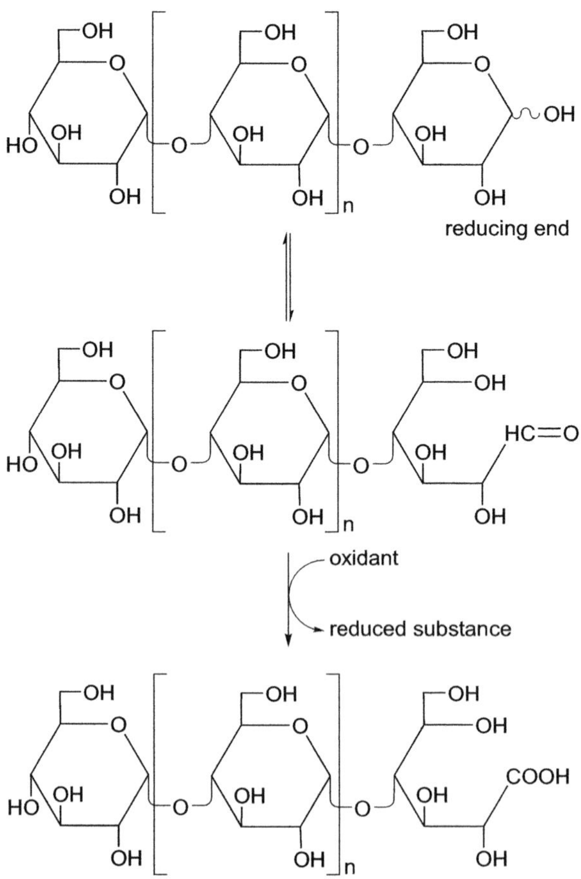

Figure 3.2 A depiction of how the reducing-end unit of an oligo- or polysaccharide could act as a reducing agent, and therefore, an indication of the origin of the term. The reducing end (usually the right-hand end when the structure is written or printed) may not actually have an aldehydic group, but the term indicates the orientation of the molecule.

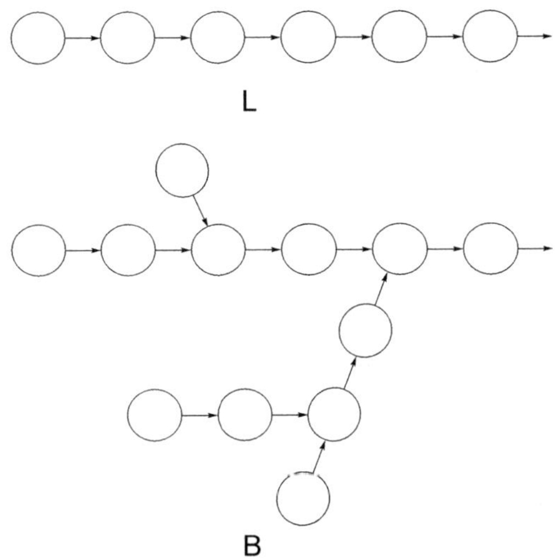

Figure 3.3 Schematic representations of linear (L) and branched (B) oligosaccharides. *Arrows* represent the linkages joining the anomeric carbon atom of one glycosyl unit and a hydroxyl group of another. Thus, the end with a free arrow is the end with a free aldehydic (or keto) group and, therefore, the reducing end. Ends that are glycosyl units without another unit attached to one of their hydroxyl groups by a glycosidic bond (represented by *circles* only) are nonreducing ends. Thus, the specific branch-on-branch structure presented (B) has one reducing end and four nonreducing ends.

Maltose (obtained by hydrolysis of starch) is a good example of a disaccharide (section Maltose). The precise chemical name of maltose is that of a derivative of D-glucose, which is the aglycon (Chapter 1; the unit on the right as customarily written) and is derivatized by the attachment of an α-D-glucopyranosyl unit at the oxygen atom on carbon atom 4 (O4). It is, therefore, 4-*O*-(α-D-glucopyranosyl)-D-glucose. Because the reducing end has a potentially free aldehydic group, it may form α and β six-membered rings (Chapter 1). Therefore, the nature of the reducing end group (ring size and anomeric configuration) cannot be specified, except for molecules in a specific crystalline state. (In the case of maltose, because O4 is blocked by attachment of the second D-glucopyranosyl unit, a furanose ring cannot be formed.) Maltose is a reducing sugar because its aldehydic group is free to react with oxidants, such as the Cu(II) ions in Fehling solution. In fact, maltose and similar oligosaccharides will undergo essentially all reactions of aldoses.

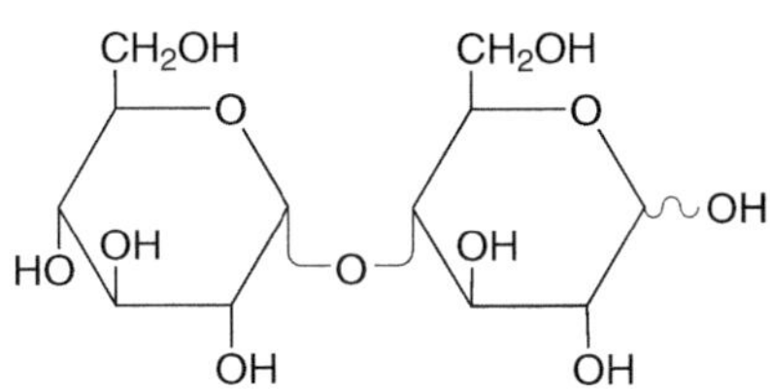

Maltose

Origins of oligosaccharides

Only a few oligosaccharides occur in nature. Most are obtained by hydrolysis of polysaccharides into smaller (oligosaccharide- size) units by either acid- or enzyme-catalyzed hydrolysis. They, in turn, can be converted by hydrolysis into smaller and smaller units (eventually completely into monosaccharides). (An abbreviated mechanism for acid-catalyzed hydrolysis [that is, cleavage in the presence of aqueous acid and heat] is shown below for a segment of a starch molecule).

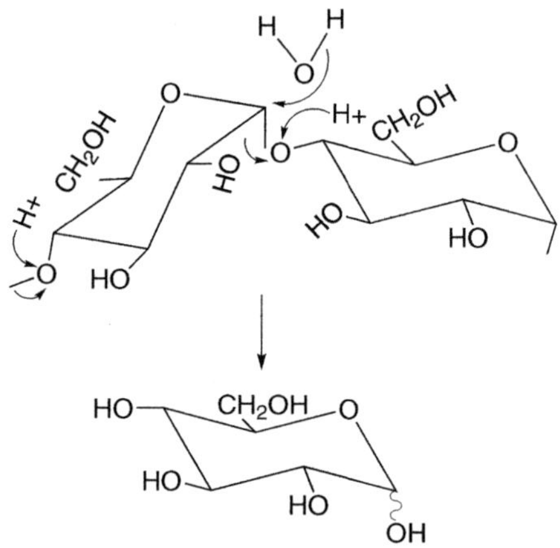

The hydrolysis reaction is reversible, so sugars can react with each other when there is a deficiency of water and low pH; but often only traces of water are required to effect hydrolysis, so it is almost always the predominate reaction. As presented in Chapter 2, sugar acetals (or ketals) are formed by the reaction of the aldehydic (or keto) group of a sugar with two hydroxyl groups. Because the stable form of sugars is a hemiacetal ring (involving one of the sugar's own hydroxyl groups), only the second reaction is necessary to make a complete acetal or ketal, called a glycoside. When the second reacting hydroxyl group is part of another monosaccharide molecule, the product, a combination of two sugar units, is a disaccharide. And somewhat larger structures can be formed. But this is not the way that either disaccharides or larger oligosaccharides are produced (either in nature or commercially) for the most part.

The acid-catalyzed combination of two monosaccharide molecules is called **reversion**. The joining of one monosaccharide molecule to another can occur at any hydroxyl position (for example, at O2, O3, O4, or O6 of a hexopyranose) and in either anomeric configuration (α or β). Thus, it is theoretically possible to form all possible disaccharides containing only one type of sugar from that sugar alone. For example, condensation of D-glucose will produce some disaccharide molecules containing a glycosidic union with a configuration opposite to that of maltose, that is 4-*O*-(β-D-glucopyranosyl)-D-glucopyranose (cellobiose), which has the glycosidic linkage in the β-D or

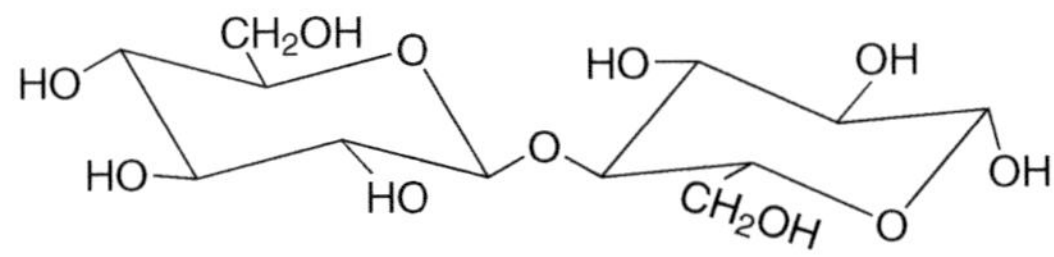

β-Cellobiose

equatorial position. (In the structure given for cellobiose, the reducing end is portrayed in the most common β-D pyranose ring).

The more or less random reaction between sugar molecules produces a complex mixture of disaccharides, with the amount of each depending on its comparative stability. In addition, glycoside-forming reactions can continue, especially under extremely dry conditions, to produce trisaccharides, tetrasaccharides, and larger oligosaccharides. Such mixtures produced by acid catalysis contain a variety of glycosidic linkages formed in somewhat random fashion. They are rather highly branched because a second sugar molecule can react with the hydroxyl groups on a sugar unit that is already derivatized with a sugar unit attached to hydroxyl group. One such randomly branched structure called **polydextrose** is made commercially. Polydextrose (Chapter 17) is made from D-glucose (dextrose), sorbitol (Chapter 2), and citric acid (as a catalyst but also incorporated into the final product via ester formation). The mixture of molecules formed contains molecules containing from 1 to 20 glucosyl units. Polydextrose is essentially nondigestible and noncaloric because the structures formed are unnatural and not recognized by human digestive enzymes.

Other oligosaccharides are made by rearrangements of the structures of existing oligosaccharides (usually in reactions catalyzed by enzymes).

Shorthand designations

In shorthand notations of oligo- and polysaccharides, the glycosyl units are designated by the first three letters of their names with the first letter being capitalized—except for glucose, whose shorthand notation is Glc. If the monosaccharide unit is that of a D sugar, the D is usually omitted; only L sugars must be so designated (for example, LAra). The size of the ring is designated by an italicized *p* for pyranosyl or *f* for furanosyl. The anomeric configuration is designated with α or β. Thus, an α-D-glucopyranosyl unit is indicated as αGlc*p*. Uronic acids are designated with a capital A. Thus, an L-gulopyranosyluronic acid unit is indicated as LGul*p*A. The positions of linkages are designated either as, for example, 1 → 3 or 1,3, the latter designation being more commonly used by biochemists and the former designation more commonly used by carbohydrate chemists. Using this shorthand notation, the structure of maltose is represented as αGlc*p*(1 → 4)Glc or αGlc*p*1,4Glc. Hyphens may be used, with the structure of maltose represented as αGlc*p*-(1 → 4)-Glc or α-Glc*p*-1,4-Glc, for example. (Note that the reducing end is not designated as α or β or as pyranose, except when a specific crystal structure is being described, because in solution, the reducing-end ring can open and reclose in different ways. Thus, in solutions of mono-, oligo- and polysaccharides, reducing-end units of the molecules occur as mixtures of α- and β-ring forms and the acyclic form, with rapid interconversion between them.) The scheme below indicates the relationships between names, shorthand designations, and structures.

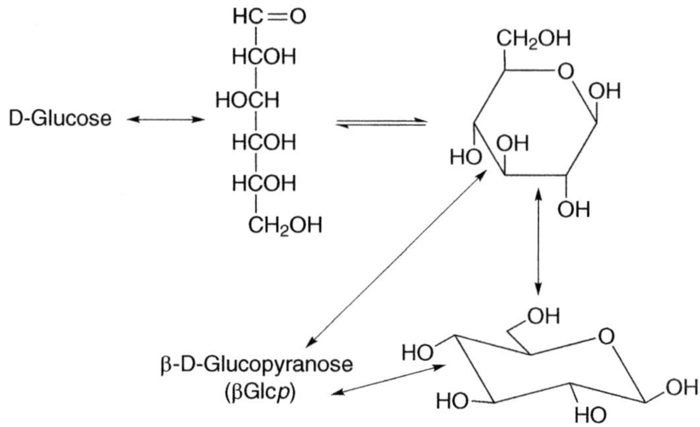

Maltose

Maltose (which was introduced and its structure given earlier in this chapter) is produced by hydrolysis of starch using the enzyme β-amylase (Chapter 7), which specifically releases disaccharide (maltose) units sequentially from the nonreducing ends of the starch polysaccharides (Chapter 6). Maltose occurs only rarely in nature and only in plants, mostly in germinating seeds as a result of partial hydrolysis of starch by β-amylase. It is present in considerable amounts in malt.[2] Commercially, β-amylase from *Bacillus* species is used for the production of maltose. However, β-amylase from any source (including that from barley seed, soybeans, and sweet potatoes) produces maltose in yields greater than 80%. Higher yields are obtained by using a mixture of β-amylase and a starch debranching enzyme (Chapter 6). Maltose is easily crystallized from aqueous solutions as α-maltose monohydrate. A more conformationally correct structure for maltose is given in Figure 3.4.

The reducing end of a maltose molecule can be reduced, converting it into an alditol (D-glucitol/sorbitol; Chapter 2) unit. The reduced disaccharide is called **maltitol**. Commercially, reduction is done by catalytic hydrogenation. Maltitol is available as a crystalline powder. Maltitol is relatively stable, both chemically and thermally, and like other polyols, does not participate in nonenzymic (Maillard) browning

[2] Malt is germinated and dried cereal grain. During germination, enzymes required to break down starch into maltooligosaccharides and D-glucose (Chapter 7) are produced. The process of producing malt is called malting.

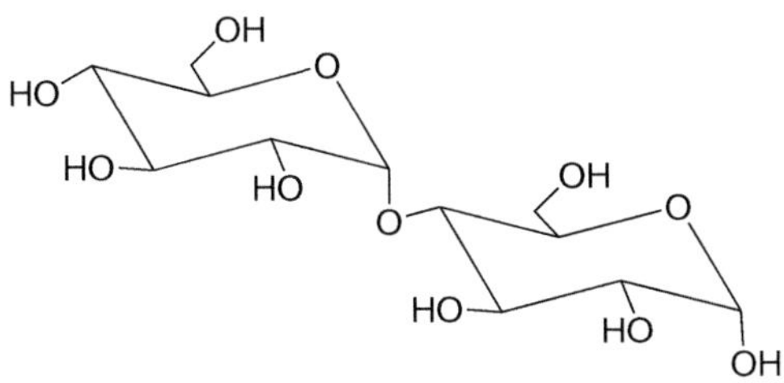

Figure 3.4 A more conformationally correct structure of α-maltose than the Haworth representation given earlier in this chapter.

reactions (Chapter 18). Its sweetness is about 90% that of sucrose, and it has a similar taste. It has very low hygroscopicity, absorbing no moisture until the relative humidity reaches 89%. Because only monosaccharides can be absorbed from the small intestine, maltitol is not absorbed, but it is partially hydrolyzed there into D-glucose and sorbitol. The glucose is absorbed from the small intestine; sorbitol is partially absorbed. Unhydrolyzed maltitol and the nonabsorbed sorbitol pass into the large intestine (colon) where they are fermented to short-chain fatty acids that are absorbed and used as an energy source, giving maltitol an overall caloric value of 2.4–3.0 Kcal/g (a range because of differences in distribution between the two pathways in different individuals). Like other polyols, maltitol is nonacidogenic[3] and noncariogenic.[4] The principal use of maltitol is in the manufacture of sugarless chocolate (Chapter 19). (See Chapter 19 for more about maltitol).

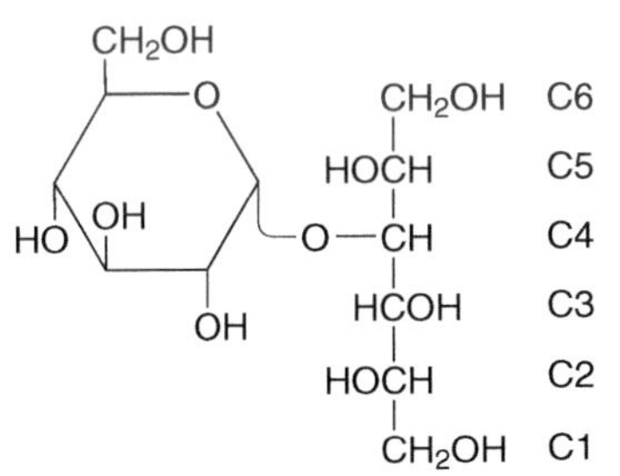

Maltitol

Lactose

The disaccharide **lactose** occurs in milk, mainly free, but to a small extent as a component of higher oligosaccharides. Lactose is 4-*O*-(β-D-galactopyranosyl)-D-

[3] Acidogenic means acid-producing.

[4] A noncariogenic substance does not promote dental caries (tooth decay).

glucopyranose (βGal*p*(1 → 4)Glc or βGal*p*1,4Glc). Like maltose, it is a reducing sugar with a potentially free aldehydic group. It undergoes hydrolysis in hot, acidic solutions or via the action of the enzyme **lactase** (below) to produce one molecule of D-galactose and one molecule of D-glucose. Like maltose and other reducing sugars, it exists in aqueous solution as an equilibrium mixture of its anomeric forms.

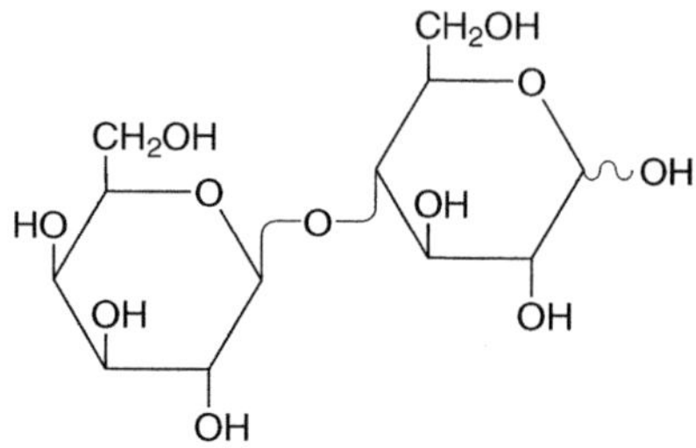

Lactose

The concentration of lactose in milk varies with the mammalian source from 2.0% to 8.5%. Cow and goat milks contain 4.5%–4.8% lactose; human milk contains about 7%. Lactose is the primary carbohydrate source for developing mammals. In humans, lactose supplies about 40% of the energy consumed during nursing. Utilization of lactose for energy must be preceded by its hydrolysis to its constituent monosaccharides (D-glucose and D-galactose) because only monosaccharides can be absorbed from the small intestine. Milk also contains 0.3%–0.6% of larger lactose-containing oligosaccharides, many of which are important as energy sources for growth of a specific variant of *Lactobacillus bifidus*, which as a result is the predominant microorganism of the intestinal flora of breast-fed infants.

Production and uses of lactose

Lactose (still sometimes referred to as milk sugar) is produced commercially from cows' milk that has had the casein coagulated by heat after adjustment of the milk to casein's isoelectric pH of 4.5–4.7 or by use of the enzyme chymosin (rennet). The insoluble casein (curds) is separated from the sweet whey, which is then subjected to ultrafiltration to remove remaining proteins. Then, minerals are removed by ion-exchange, and the solution is concentrated to 50%–65% solids to allow lactose to crystallize or to be precipitated. The lactose is redissolved; the solution is decolorized with carbon, and lactose is recrystallized. For every pound (454 g) of cheese produced, about 9 pounds (4 kg) of whey is recovered. Since whey contains about 4.7% lactose, about 0.4 lb (200 g) of lactose is potentially available as a by-product from the production of each pound of cheese; and since whey from cheese production amounts to millions of metric tons per year, there is an enormous potential source of lactose. However, little commercial use is made of it.

The most common form of lactose is α-lactose monohydrate, which crystallizes from supersaturated solutions at temperatures below 93.5°C (200°F). This form dissolves slowly, which means that it cannot be used in instant food products. There are three forms of anhydrous α-lactose. A stable form is prepared by heating α-lactose • H_2O in air at 130°C. β-Lactose is more soluble than α-lactose. It is prepared by heating α-lactose • H_2O in an alcohol in the presence of a base. β-Lactose is sweeter than α-lactose, but each anomer will form the same equilibrium mixture in solution. Overall lactose is 15%–30% as sweet as a sucrose solution of the same concentration. Crystalline lactose has adsorptive properties and can be used as a carrier for aromas and flavors.

Lactose is used to a small extent in confections, ice creams, icings, pie fillings, and toppings. It contributes body to foods and enhances colors and flavors. Amorphous lactose is used as an excipient[5] in pharmaceutical tablets and caplets because it is cost-effective, has a bland taste, has low hygroscopicity,[6] is compatible with active ingredients, is water soluble, and provides rapid dissolution.

Digestion of lactose and lactose intolerance and related conditions

Lactose is ingested in milk and other unfermented dairy products (such as ice cream). Cheese contains less lactose because most of it is removed in the whey. Fermented dairy products (such as yogurt made with active cultures) also contain less lactose because, during fermentation, some of the lactose is converted into lactic acid. Lactose stimulates intestinal adsorption and retention of calcium ions.

Lactose is not digested until it reaches the small intestine, where the hydrolytic enzyme lactase is found. **Lactase** (a β-galactosidase) is a membrane-bound enzyme located in the brush border epithelial cells of the small intestine. It catalyzes the hydrolysis of lactose into its constituent monosaccharides, D-glucose, and D-galactose. Both D-glucose and D-galactose are rapidly absorbed and enter the blood stream.

$$\text{Lactose} \xrightarrow{\text{lactase}} \text{D} - \text{glucose} + \text{D} - \text{galactose} \tag{3.1}$$

If for some reason the ingested lactose is only partially hydrolyzed, or is not hydrolyzed at all, a clinical syndrome called **lactose intolerance** results. The symptoms of this syndrome are abdominal distention, cramps, diarrhea, and flatulence (commonly referred to as "passing gas"). These symptoms are produced because, if a person has a deficiency of lactase, some lactose remains in the lumen of the small intestine. The presence of lactose draws fluid into the lumen because of the increase in osmotic pressure. It is this fluid that (in part) leads to abdominal distention and cramps. From the small intestine, the lactose passes into the large intestine (colon), where anaerobic

[5] An excipient is an inactive substance that serves as a vehicle or medium for drug delivery (for example, a material that enables tablet formation and subsequent dissolution in the body).

[6] Hygroscopicity is the property of absorbing water/moisture from the air.

bacteria ferment it to lactic acid (largely present as the lactate anion) and other short-chain acids. The increase in the concentration of molecules increases the osmolality of the intestinal fluid and results in still greater retention of water. The acidic products of fermentation lower the pH and irritate the lining of the colon, leading to increased movement of the contents. Watery stools result from the retention of fluid and an increased movement of the intestinal contents. The gaseous products of fermentation (carbon dioxide, hydrogen, and methane) cause bloating and flatulence.

Lactose intolerance is not usually seen in children until after about 3–5 years of age. At this point, the percentage of lactose-intolerant individuals begins to rise and

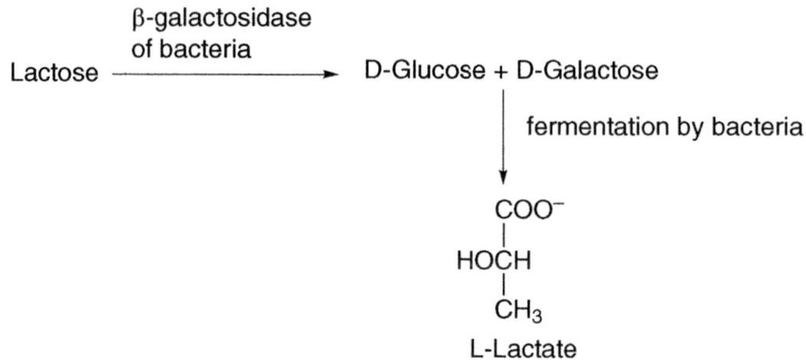

increases throughout the life span, with the greatest incidence in the elderly; a majority of elderly people around the world have a 90%–95% loss of lactase activity.

There are varying degrees of lactose intolerance. It is estimated[7] that as much as 70% of adults in the world are lactose intolerant to some extent. The incidence of lactose intolerance varies among ethnic groups; as much as 90% of the adult population of Chinese and Japanese ancestry may be lactose intolerant to some extent. In the United States, about one-third of the adults have problems with lactose consumption. By 12 years of age, 45% of African Americans develop the symptoms of lactose intolerance; among teenagers, the incidence climbs to 70%; and by adulthood, 80% of the African American population shows symptoms of lactose intolerance. Lactose intolerance is also high among Asian Americans (65%–100% incidence). Among Caucasians of Western European ancestry, the peak incidence in adulthood is 6%–25%. The incidence among Native Americans is 50%–75%; 47%–74% of Mexican Americans are estimated to exhibit symptoms. Lactose intolerance is also high in persons of Jewish descent. Overall, this information indicates that the presence or absence of lactase is determined by the genetic makeup and the age of the person. Lactose intolerance often results in avoidance of dairy products, leading to insufficient intake of calcium and a variety of health problems, such as osteoporosis.

[7] All numbers in this paragraph are estimates because most cases are unreported.

There are three ways to overcome the effects of lactase deficiency. One is to reduce the amount of lactose in the food product by fermentation. Fermentation of milk produces yogurt[8] and buttermilk products. Another way to reduce the amount of lactose in the diet is to add lactase to milk. However, both products of hydrolysis (D-glucose and D-galactose) are sweeter than lactose, and at about 80% hydrolysis, the taste change becomes evident. Therefore, most of these products have the lactose reduced as close as possible to the 70% government-mandated minimum for reduced-lactose milk. In an alternative technology, live yogurt cultures are added to refrigerated milk. The bacteria remain dormant in the cold and do not change the flavor of the milk, but upon reaching the small intestine, they release lactase. The third way to overcome the effects of lactase deficiency for the lactase-deficient person is to consume lactase, along with the dairy product. Lactase is available in caplet form.

Likewise, other carbohydrates that are not completely broken down into monosaccharides by intestinal enzymes are not absorbed and passed into the colon. There, they also are metabolized by microorganisms producing lactic acid and gas. Watery stools, bloating, and cramping result. This problem can occur from eating beans because beans contain a trisaccharide (raffinose) and a tetrasaccharide (stachyose) (section Sucrose) that are not hydrolyzed to monosaccharides by intestinal enzymes and, thus, pass into the colon where they are fermented. Some such compounds are called prebiotics and are beneficial to human health (Chapters 10 and 17).

Derivatives of lactose

The chemistry of lactose has been rather extensively examined, in part because of an interest in converting the rather abundantly available lactose into a more useful product. Like maltose, lactose can be reduced to a polyol, namely **lactitol** [4-*O*-(β-D-galactopyranosyl)-D-glucitol] by hydrogenation. Lactitol is a nonnutritive sweetener (Chapter 19) with a sweetness of about 40% that of sucrose. When used, it is most often used together with a high-intensity sweetener. Lactitol can be crystallized as either a monohydrate or a dihydrate, both of which are nonhygroscopic. It, therefore, can be used in the manufacture of products, such as chocolates, which require no moisture pickup during processing, and bakery products that should remain crisp. Lactitol provides foods with a bulk and texture similar to that provided by sucrose. Since it is a nonreducing disaccharide, it cannot be a reactant in nonenzymic browning. Like maltitol, lactitol is not absorbed from the small intestine and is metabolized by microorganisms of the large intestine to produce lactic and other low-molecular-weight

[8] If yogurt is made without added milk solids, the concentration of lactose is reduced as the microorganisms in the culture convert lactose into lactic acid. However, some yogurt producers add considerable amounts of milk solids before culturing the milk, or start entirely with nonfat dry milk solids, which are dispersed in water at a higher solids content than is found in milk, increasing the amount of lactose in the formulation. In those yogurts, the amount of lactose present after fermentation may be 10–15 g per cup (about the same as that of milk). However, even so, this yogurt is usually tolerated better than milk in lactose-sensitive (lactose-intolerant) persons when it contains live bacteria because some organisms survive the stomach and help digest lactose in the small intestine.

acids that, by their hydrophilic and water-binding characteristics, facilitate bowel movement. In excess amounts, they produce diarrhea, but lactitol can be used in controlled amounts as a stool softener (see Chapters 17 and 19 for more about lactitol).

A product with similar physiological action is made from lactose by isomerizing it with an alkaline substance to the keto form (**lactulose** [4-O-(β-D-galactopyranosyl)-D-fructose]), a molecule in which the D-glucose unit at the reducing end has been converted to D-fructose. Lactulose is not acted on by human digestive enzymes but is readily fermented by colonic microorganisms, which accounts for its major use (that is, in treatment of chronic constipation). Lactulose is a bifidus factor (that is, it is effective in promoting the proliferation of Bifidobacteria in the intestine, increasing the level of the organisms in the feces of bottle-fed infants to that in breast-fed infants) (Chapter 17).

Sucrose

High concentrations of **sucrose** (commonly called sugar) are found in only a few plant tissues. Early humans used fruits and honey as a source of sweetness. Many thousands of years passed before another source of sweetness was found. Sugarcane was first cultivated along the Ganges River in what is now the state of Bihar in India. Cultivation spread slowly, reaching China by about the 1st century B.C.E. Sugarcane growing spread to the Middle East and was brought to Europe by the Crusaders who, like Pliny the Elder, much earlier, called its juice "a kind of honey made from reeds." For several centuries after sucrose was first brought to Europe, it was a delicacy enjoyed only by the nobility. Not until Spanish and Portuguese travelers brought sugarcane to the Americas, where it was grown on large plantations, did the price come down to a level where the general public could afford it.

The structure of sucrose (Fig. 3.5), a disaccharide, is an exception to the general rule for structures of oligo- and polysaccharides. In sucrose, the constituent glycosyl units (an α-D-glucopyranosyl unit and a β-D-fructofuranosyl unit) are linked head-to-head, that is, reducing end to reducing end (anomeric carbon atom to anomeric carbon atom) (Fig. 3.6), rather than the very much more common head-to-tail (anomeric carbon atom to a carbon atom containing an alcoholic hydroxyl group) type of linkage. Since both the aldehydic group of the D-glucosyl unit and the keto group of the D-fructosyl unit are covalently bound in a mutual glycosidic bond, sucrose has no reducing end. Therefore, it is classified as a nonreducing sugar and does not react with Fehling solution or Tollens' reagent in tests for detection of aldehydes or ketones (Chapter 2).

A majority of sucrose molecules in solution have the structure illustrated below, where the primary hydroxyl group at C1 of the D-fructofuranosyl unit is hydrogen bonded to the C2 hydroxyl group of the D-glucopyranosyl moiety to the extent that it is much less reactive than the remaining primary hydroxyl groups. In crystalline sucrose, there is an additional hydrogen bond between the primary hydroxyl group at C6′ and the ring oxygen atom of the D-glucopyranosyl unit.

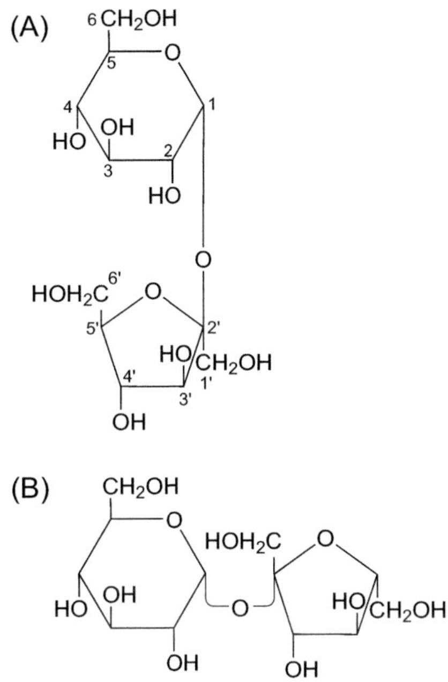

Figure 3.5 Two depictions of the structure of sucrose (Haworth representation) with the carbon atoms numbered in one.

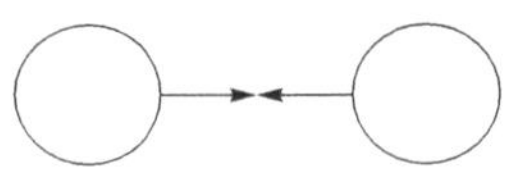

Figure 3.6 Depiction of the head-to-head linkage found in sucrose.

The glycosidic bond between the two sugar rings is a high-energy and unstable bond due in part to the strained fructofuranosyl ring. As a result, sucrose is very easily hydrolyzed (even in the presence of very dilute acid) to give an equimolar mixture of D-glucose and D-fructose, termed **invert sugar**. Sucrose is so acid labile that a very weak acidic solution, such as dilute vinegar (dilute acetic acid), at the boiling

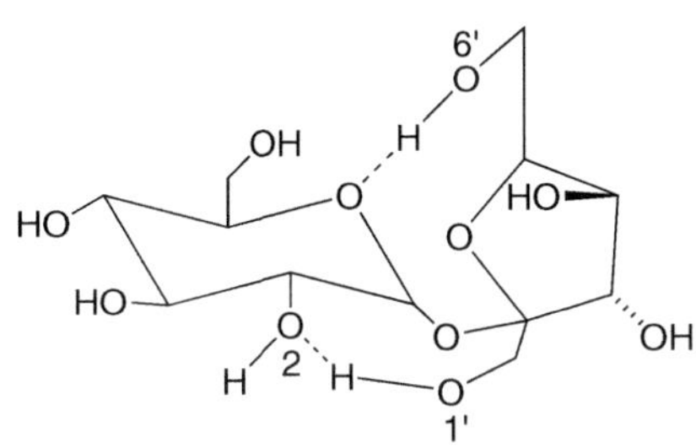

temperature of a sugar solution is sufficient to effect some hydrolysis. The name invert sugar, and therefore the name of the enzyme that catalyzes hydrolysis of sucrose, namely **invertase**, originates from the fact that early investigators noticed that the specific optical rotation ([α]D) of sucrose (+66.5 degrees) changed to −33.3 degrees, the [α]D of an equimolar mixture of the two constituent sugars. As a result of the change from a positive (dextrorotatory) rotation to a negative (levorotatory) rotation, they called the process an inversion and the product, invert sugar. Invertases are, therefore, β-D-fructofuranosidases.

Other enzymes also catalyze the hydrolysis of sucrose into D-glucose and D-fructose. Sucrase-isomaltase (Chapter 17) of the human intestinal tract is one such enzyme, making sucrose one of the oligosaccharides humans can digest and utilize for energy (the other digestible carbohydrates being the disaccharide lactose and oligosaccharides derived from starch [Chapter 6]). Sucrase-isomaltase is an α-D-glucosidase because it recognizes and binds the α-D-glucopyranosyl unit in its active site and catalyzes cleavage between the oxygen atom of the glycosidic bond and the α-D glucopyranosyl unit. Invertases of yeast and bacteria are sucrose hydrolases that catalyze cleavage on the other side of the glycosidic bond (that is, they recognize and bind the β-D-fructofuranosyl unit in their active sites and leave the oxygen atom with the D-glucosyl unit). Sucrose is relatively stable under alkaline conditions because it has no free carbonyl group and acetals and ketals are stable to alkali.

$$\text{Sucrose} \xrightarrow{\text{intestinal sucrase } \textit{or} \text{ bacterial invertase}} \text{D-glucose} + \text{D-fructose} \qquad (3.2)$$

Sucrose is formed in the process of photosynthesis. It is transported from leaves and other photosynthetic tissues to other parts of the plant to supply energy and to provide the carbon atoms (through numerous metabolic routes) for synthesis of all molecules and structures in the plant. Biosynthesis of sucrose occurs by reaction of uridine diphosphate D-glucose (UDPGlc) with D-fructose 6-phosphate (Fru-6-P), which is produced early in the photosynthetic carbon-fixation pathway. These two components are combined through the action of the enzyme sucrose phosphate uridylyl transferase to produce sucrose phosphate. The phosphate ester group is then removed by sucrose phosphatase to yield sucrose.

$$\text{UDPGlc} + \text{Fru-6-P} \rightarrow \text{sucrose phosphate} \rightarrow \text{sucrose} \qquad (3.3)$$

Sources of sucrose

Cane sugar

There are two principal sources of commercial sucrose: sugarcane and sugar beets. To obtain sugar from sugarcane, 12- to 18-month-old cane is crushed between rolls. A small spray of water helps remove the juice, which contains about 16% sugar. The juice is made slightly alkaline by the addition of calcium hydroxide (lime) to prevent hydrolysis of the acid-labile glycosidic linkage, and the mixture is heated to coagulate proteins, including the hydrolyzing enzyme invertase. The scum containing a variety of compounds produced in this process is removed by filtration. The filtrate is concentrated under reduced pressure at a carefully controlled temperature to about 50% solids. The crystals that form are removed by centrifugation and washed. The mother liquor is further concentrated to obtain another crop of crystals. Such concentration and crystallization is continued until impurities build up to the point where the remaining sucrose will not crystallize. This usually occurs after two or three crops of crystals are obtained. The final mother liquor, termed black strap **molasses**, is a dark, heavy, bitter-flavored syrup, high in inorganic salts (ash).

Cane must be brought to the mill and crushed as soon as possible to reduce exposure to microorganisms that would quickly begin to grow on it. Microorganisms hydrolyze sugar by means of the enzyme invertase. Some organisms convert a portion of the sugar to dextran (Chapter 12), a soluble polysaccharide that thickens the sugar solution, causing clogging of filters and other mechanical problems and loss of yield.

Raw sugar, the product of the process described above, is light brown and is shipped to a refinery for purification. There, it is dissolved. Calcium hydroxide (lime) and a phosphate salt are added to the raw sugar solution to further flocculate and precipitate impurities. The mixture is clarified by centrifugation and by filtration through diatomaceous earth. The resulting solution is decolorized with carbon (charcoal). Crystallization under reduced pressure yields pure, white table sugar.

Beet sugar

Beet sugar is obtained by countercurrent extraction of sugar beet slices (called cossettes) until the liquor contains about 12% sugar. The liquor is agitated with calcium hydroxide (lime) for several hours. Carbon dioxide is then bubbled through the mixture to neutralize the alkaline solution and form calcium carbonate, and the mixture is filtered. The solution is decolorized with sulfur dioxide. Sucrose is crystallized by concentration of the solution under reduced pressure. After removal of the crystals by centrifugation, the mother liquor is again concentrated, and more sugar crystals are collected. This sequence is repeated until crystals no longer form. Additional quantities of sucrose can be obtained from the mother liquor by diluting it to about 7% sugar, cooling it to 12°C (54°F), and adding lime. A compound of lime and sucrose (termed tricalcium saccharate) forms and can be isolated by centrifugation. Calcium ions are removed as insoluble calcium carbonate after bubbling in carbon dioxide.

Also present in sugar beet extracts are a trisaccharide, **raffinose**, which has a single D-galactopyranosyl unit attached to sucrose and a tetrasaccharide **stachyose**, which contains a second D-galactosyl unit. These oligosaccharides are also found in beans, are nondigestible, and are the source of the flatulence associated with eating beans (Chapter 17).

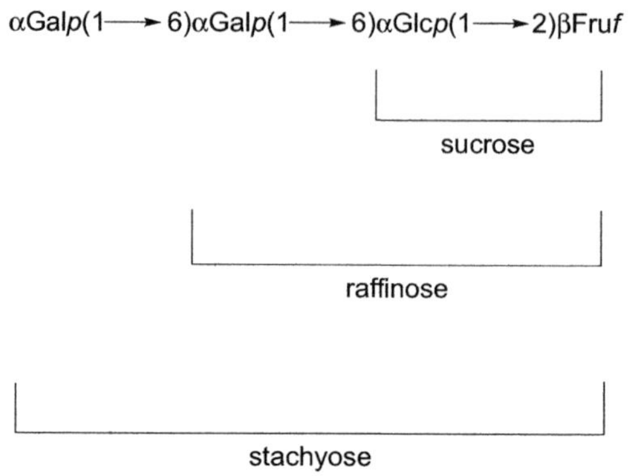

Sugar products

White granulated and pulverized products

There are no agreed-upon industry designations for types of sugars. Granulated sugar is subdivided into products that may be designated (in decreasing order of crystal size) as coarse, sanding, fine or extra fine, fruit, Bakers' special, superfine/ultrafine or bar, and confectioners' or powdered (6X, 10X) sugar. However, the crystal size range in a given type may vary from producer to producer; not all producers make the same range of types; different producers use different names for their products, and not all products are listed above.

Brown sugars

Brown sugars are most often made by coating white sugar with a specific molasses syrup. Grades range from light brown to dark brown. Brown sugars contain moisture and ash, have some molasses flavor, and are usually somewhat sticky, although free-flowing brown sugars are known. Brown sugars are known as "soft sugars" in the sugar industry. The darker the color (that is, the more molasses is present), the stronger the flavor and the more moisture present. Turbinado sugar is a raw sugar that is light brown in color and mild in flavor.

Liquid sugar

For many food and beverage industry applications, sucrose is not crystallized, rather it is shipped as a refined aqueous solution known as liquid sugar.

Some properties and functionalities of sucrose[9]

Sucrose performs many functions in foods other than providing sweetness (Chapter 19). Sucrose is widely used to control water activity and thereby acts as a preservative (for example, in full-sugar jams and jellies). It is also widely used to add body, provide a carbon source for fermentation (for example, in breadmaking so that the dough will rise), provide crispness, modify flavor, give surface frost or glaze, form a glass, impart desirable texture, and create viscosity. In nonsweet foods, such as salad dressings and sauces, sucrose balances the acidity from added vinegar and tomato products.

Sucrose and most other low-molecular-weight carbohydrates (for example, monosaccharides, alditols, disaccharides, and mixtures of low-molecular-weight oligosaccharides) will form highly concentrated solutions of high osmolality because of their great hydrophilicity and solubility. Examples of rather concentrated solutions of mono- and oligosaccharides are pancake and waffle syrups and honey. Such solutions need no preservatives (because of their high osmotic strength and low water activity) and can be used not only as sweeteners but also as preservatives and humectants. A sucrose syrup must have a concentration of more than 75% (a water activity of more than 0.6) for microbiological stability. Sucrose syrups generally have a concentration of about 66%; (corn/glucose) syrup solids (Chapter 7), glucose syrups (Chapter 7), or invert sugar are added to prevent crystallization of the sucrose.

A portion of the water in any carbohydrate solution is nonfreezable. When the freezable water crystallizes (that is, forms ice), the concentration of solute in the remaining liquid phase and the viscosity of the remaining solution increase, and the freezing point decreases. Eventually, the liquid phase solidifies as a glass in which the mobility of all molecules becomes restricted and diffusion-dependent events become very slow. Because of their restricted motion, water molecules cannot form crystals (that is, they become nonfreezable). Water molecules may also become nonfreezable by an increase in the solution concentration to the point that the freezing point is lowered to a temperature below the temperature of the freezer. In this way, sucrose, sorbitol, and other carbohydrates function as **cryoprotectants** (that is, they protect against the destruction of structure and texture caused by freezing).

Chocolate contains stable sugar particles (partly microcrystals) dispersed in a partially crystalline fat phase. The sugar particles impart non-Newtonian rheology (Chapter 5) to the chocolate. The particle size distribution is critical to the texture perceived in the mouth. In the United States, the average size of sugar particles in chocolate is 30–33 μm, with a maximum size of 50 μm. In Europe, the average size is 20–23 μm, with a maximum size of 40 μm, giving that chocolate a slightly more slippery texture. In both cases, some sugar particles are less than 1 μm in size, but the human mouth does not detect particles less than 12–15 μm. Small particles

[9] The functionalities of sucrose in foods, especially in bakery products, are discussed further in Chapter 19.

require more fat for lubrication and to achieve the required viscosity because of their large surface area.

Some crystalline sugar is converted into amorphous (that is, noncrystalline) sugar during the refining of chocolate. Amorphous sucrose has properties different from that of crystalline sugar. It can be obtained by (1) melting, followed by rapid solidification (such as in making cotton candy), (2) rapid drying of solutions (such as in spray drying), and (3) milling. One property of amorphous sugar is its ability to absorb flavor components. This same property is held by other amorphous carbohydrates (for example, lactose). Amorphous sugar not only absorbs flavors but also plays a key role in the development of caramel flavor. As will be described in Chapter 18, the Maillard reaction (which can occur only after hydrolysis of some of the sucrose) is important in producing the flavors of milk chocolate, caramel, toffees, and fudges.

Sucrose has the ability to form supersaturated solutions and glasses. A hard candy may contain about 60% sugar, about 39% corn/glucose syrup solids (Chapter 7) to prevent crystallization, and 1%–3% water. In other confectionary products, sugar forms microcrystals (as it does in chocolate). In still others, sugar crystallization is inhibited by forming invert sugar from some of the sucrose (primarily via addition of citric acid to bring about some hydrolysis), that is, by introducing D-glucose and D-fructose, which act as contaminants and prevent crystallization of sucrose.

Derivatives of sucrose

Sucrose esters

Sucrose has eight free hydroxyl groups that are potential sites for esterification. Esterification from 1 to 3 of the 8 hydroxyl groups with a fatty acid (such as stearic acid) produces surfactants. These fatty acid esters[10] of sucrose are approved for use as food-grade emulsifiers. They also have antimicrobial properties (that is, they can inhibit the growth of certain bacteria, inactivate bacterial spores, and reduce mold growth). Fully acetylated sucrose (sucrose octaacetate) is extremely bitter and has been used as a denaturant for ethanol.

Sucrose derivatives that are more highly esterified with fatty acids (such as stearic, palmitic, and oleic acids) have been employed as noncaloric fat substitutes. One such product (**olestra**) is a mixture of molecules in which six, seven, or eight fatty acid molecules are esterified to the eight hydroxyl groups of a sucrose molecule. Different types of olestra can be made by changing the fatty acid moieties (different chain lengths and degrees of saturation) and the degree of esterification. The

[10] The products are mixtures of mono-, di-, and triesters of sucrose because it is not practical to make a defined product with only certain hydroxyl groups esterified.

different types can be made for different applications (fried foods, cooking oils, margarine, shortening, spreads, and bakery products, for example). Olestra is not absorbed by the body and, therefore, not metabolized. (Sucrose esters can be utilized by humans only when there are four or less ester groups per sucrose molecule.) Olestra products provide the same taste and cooking properties as fats and oils and, because they can withstand high temperatures, can be used in frying. They have been determined to be neither toxic, carcinogenic,[11] genotoxic,[12] nor teratogenic.[13] Because olestras are neither absorbed from the intestinal tract nor broken down in the small intestine into fatty acids and monosaccharides, they are noncaloric. Products containing an olestra must have the following statement on the label (in the United States): "This product contains Olestra. Olestra may cause abdominal cramping and loose stools. Olestra inhibits the absorption of some vitamins and other nutrients. Vitamins A, D, E, and K have been added." Olestra has been approved as a calorie-free replacement for up to 100% of the conventional fats and oils used in the preparation of savory snacks (such as flavored and unflavored potato chips, extruded smacks, and crackers). Other uses include substitution for all or part of the frying fat as well as use in dough conditioners, oil sprays, flavors, and fillings. However, because olestra does cause abdominal cramping and loose stools in some persons, it is not widely used.

Sucralose[14]

A chlorinated product having a derivatized D-galactopyranosyl unit in place of the normal D-glucopyranosyl unit of sucrose (called **sucralose**) is 600 times sweeter than sucrose. Sucralose has a taste profile much like that of sugar. It has a clean, quickly perceptible sweet taste without an unpleasant aftertaste. It has good water solubility and is not hydrolyzed by intestinal invertase. It can be hydrolyzed with acid, but because of the high electron-withdrawing power of the chlorine atom, it is much more stable to low pH and heat than sucrose. Because it is not hydrolyzed in either the small or large intestine, sucralose is noncaloric and passes through the body. It has been approved by the US Food and Drug Administration as a general purpose sweetener for foods and beverages. It can be used in any application where sugar is used (baked goods, beverages, confectionaries, desserts, etc.), but like other high-intensity sweeteners, it contributes none of the body or other attributes contributed by sugar because it is used in such small amounts.

[11] A carcinogenic substance is one that has the potential of inducing cancer.

[12] A mutagenic substance is one that changes the DNA of an organism and, thus, increases the frequency of mutations (over the natural background level).

[13] A teratogenic substance is one that is capable of interfering with the development of a fetus and, as a result, causes birth defects.

[14] See Chapter 19 for additional information.

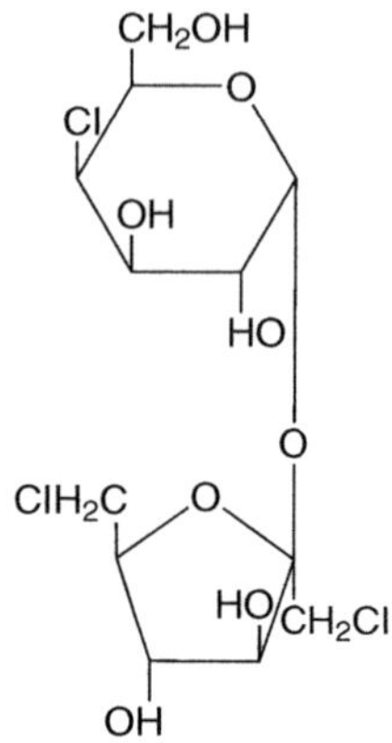

Sucralose

Oligosaccharides related to sucrose

Isomaltulose and isomalt

Isomaltulose (6-*O*-(α-D-glucopyranosyl)-D-fructose) is derived from sucrose by a specific enzyme-catalyzed transfer of the D-glucopyranosyl group from the O2 position to the O6 position of the α-D-fructofuranosyl unit. Isomaltulose is crystalline and has a clean sweet taste of about half the intensity of sucrose. It is used in Europe for making noncariogenic candies, specialty chocolates, chewing gums, and cookies (biscuits). Hydrogenation of isomaltulose produces **isomaltitol** (commonly called **isomalt**), which is an equimolar mixture of disaccharide polyols that is about 45% as sweet as sucrose and is used as a substitute for sucrose in making chewing gum, chocolate, and marzipan.

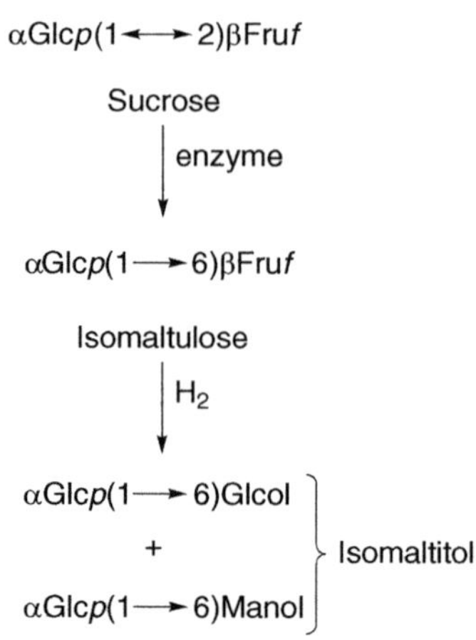

Leucrose

Leucrose (5-*O*-(α-D-glucopyranosyl)-D-fructose) is a reducing disaccharide, produced from sucrose in another specific enzyme-catalyzed reaction. It has about half the sweetness of sucrose.

Lactosucrose

Lactosucrose (4-galactosylsucrose; β-D-fructofuranosyl 4-*O*-β-D-galactopyranosyl-α-D-glucopyranoside) is made by an enzyme-catalyzed transfer of a β-D-galactopyranosyl unit from lactose to sucrose. It occurs in yogurt produced by adding sucrose before fermentation.

Fructooligosaccharides

Fructooligosaccharides (FOS) are basically oligomers of β-D-fructofuranosyl units linked (2 → 1). The action of invertase or a specific fungal transferase on a concentrated solution of sucrose effects cleavage of some of the glycosidic linkages with transfer of some D-fructofuranosyl units to O1 of the D-fructofuranosyl unit of another sucrose molecule in a β-D-linkage. The trisaccharide formed is a **kestose**, the structure of which is sometimes abbreviated as GF_2—an indication of a disaccharide of two fructofuranosyl units attached to a glucosyl unit (Fig. 3.7). Transfer of another D-fructofuranosyl group to the O1 position of the added D-fructofuranosyl unit extends the chain another unit, forming the tetrasaccharide GF_3. This lengthening of the chain produces a mixture of oligosaccharides which is about half as sweet as sucrose and is said to be noncariogenic and noncaloric. (When a fructooligosaccharide mixture is made from sucrose by enzyme-catalyzed transglycosylation, the individual molecules are terminated at their reducing end with a sucrosyl unit [Fig. 3.7]).

The main commercial source of FOS is hydrolysis of inulin (Chapter 10), which results in products similar to those described above. Many native inulin molecules are terminated at their reducing end with a sucrosyl unit. Therefore, some, but not many, of the FOS molecules made by hydrolysis of inulin will also be terminated at their reducing end with a sucrosyl unit. The FOS made by partial hydrolysis of inulin can also be called inulodextrins.

The most important attribute of fructooligosaccharide mixtures, whether from inulin or by action of an enzyme on sucrose, is their function as prebiotics (a functionality discussed in Chapters 10 and 17).

Trehalose

Trehalose, like sucrose, is a naturally occurring, nonreducing disaccharide. The commercial form is **α,α-trehalose** (α-D-glucopyranosyl α-D-glucopyranoside). In it, two α-D-glucopyranosyl units are linked head-to-head (that is, anomeric carbon atom to anomeric carbon atom) by a glycosidic bond (Fig. 3.6).

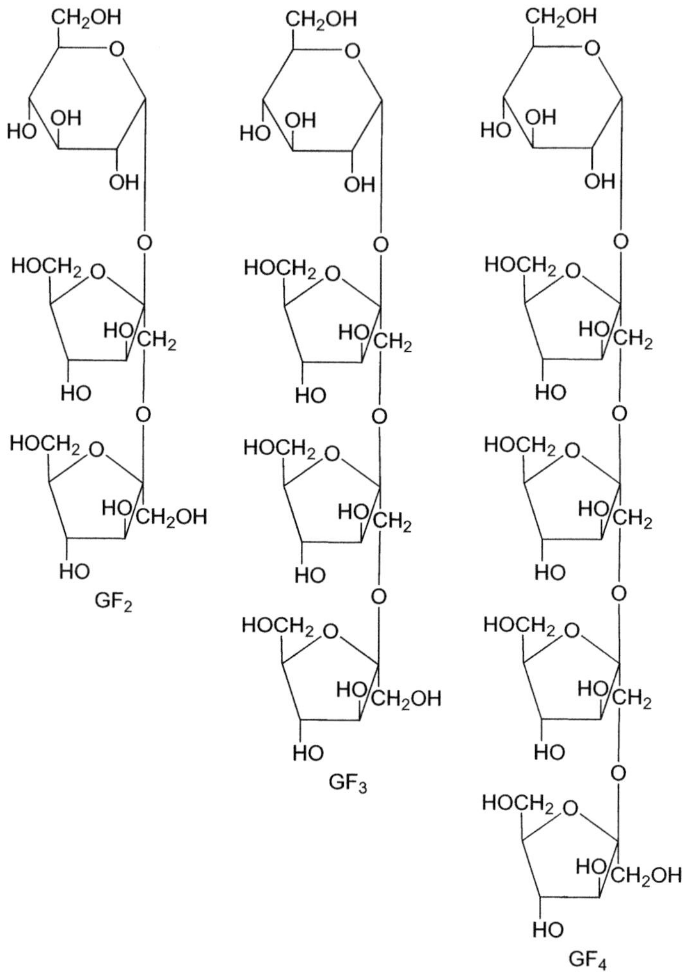

Figure 3.7 Fructooligosaccharides (FOS) terminated at their reducing ends with an α-D-glucopyranosyl unit. They can also be described as FOS terminated at their reducing ends with a sucrosyl unit. GF_2 is the abbreviation for a trisaccharide whose structure is that of a sucrose molecule with an additional fructofuranosyl unit attached to it via a 2 → 1 linkage to O1 of the fructofuranosyl unit of the sucrose moiety. GF_3 indicates a tetrasaccharide that contains an additional fructofuranosyl unit linked 2 → 1 to O1 of the nonreducing end of the new fructofuranosyl unit. GF_4 is the pentasaccharide containing a chain of three fructofuranosyl units attached to a sucrose molecule.

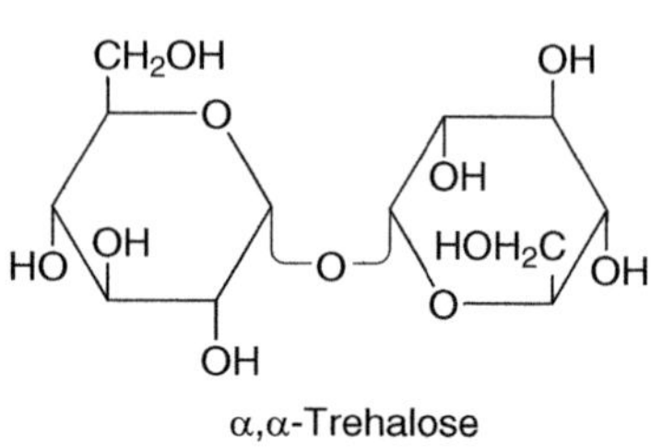

α,α-Trehalose

α,α-Trehalose (αGlc*p*(1 ↔ 1)αGlc*p*) occurs widely in nature. For example, it is found in small amounts in honey, lobster, mushrooms, certain seaweeds, shrimp, and foods produced using yeast. Commercially, trehalose is made from starch by an enzyme-catalyzed reaction. It is claimed that α,α-trehalose is able to stabilize proteins and protect them from the detrimental effects of freezing and drying. It is used to maintain the color, flavor, and texture of frozen and dehydrated foods, especially in Japan. It is also reported to reduce retrogradation of starch (Chapter 6) and to preserve cell structure. Trehalose has low hygroscopicity and about 45% of the sweetness of sucrose. It does not participate in nonenzymic (Maillard) browning reactions (Chapter 18). Compared with other nutritive sweeteners, trehalose effects less of an insulin response. It is an ingredient in a few sports drinks and nutrition bars.

Oligosaccharides related to starch

Starches are depolymerized using both acids and enzymes to produce a variety of lower-molecular-weight products, including oligosaccharide products. These products and the processes by which they are made are described in Chapter 7.

Oligosaccharides from other sources

Oligosaccharides produced by partial hydrolysis of the polysaccharide of guar gum are presented in Chapters 9 and 17.

Oligosaccharides of β-D-galactopyranosyl units with D-glucose at the reducing end are prepared from lactose by treating it with a β-galactosidase, which, in addition to being a hydrolase, also transfers D-galactopyranosyl units to the nonreducing ends of lactose and growing oligosaccharide chains produced by adding β-D-galactopyranosyl units to the lactose. The transgalactosylating activity of the enzyme produces a family of oligosaccharides of variable chain length containing a mixture of (1 → 4) and (1 → 6) linkages. Galactooligosaccharides are added to some functional foods as prebiotics (Chapter 17).

Glucosylsucrose oligomers (gentiooligosaccharides; (1 → 6)-linked β-D-glucopyranosyl units), xylooligosaccharides ((1 → 4)-linked β-D-xylopyranosyl units), and soybean oligosaccharides also function as prebiotics.

Additional resources

General

Jouppila, K., 2016. Mono- and disaccharides: selected physicochemical and functional aspects. In: Eliasson, A.-C. (Ed.), Carbohydrates in Food, third ed. CRC Press, Boca Raton, pp. 37–92.

BeMiller, J.N., 2001. Oligosaccharides: occurrence and significance in nature. In: Fraser-Reid, B.O., Tatsuta, K., Thiem, J. (Eds.), Glycoscience: Chemistry and Chemical Biology, Vol. II. Springer-Verlag, Berlin, pp. 1435–1444.

Crittenden, R.G., Playne, M.J., 1996. Production, properties and applications of food-grade oligosaccharides. Trends in Food Science & Technology 7, 353–361.

Lactose

Suarez, F.L., Savaiano, D.A., 1997. Diet, genetics, and lactose intolerance. Food Technology 51, 74–76.

Sucrose

Mathouthi, M., Reiser, P., 1995. In: Sucrose. Properties and Applications. Blackie Academic and Professional, London.

Pennington, N.L., Baker, C.W. (Eds.), 1990. Sugar. A User's Guide to Sucrose. AVI Publishing, Roslyn, New York.

Sucralose

Grotz, V.L., Molinary, S., Peterson, R.C., Quinlan, M.E., Reo, R., 2011. Sucralose. In: O'Brien Nabors, L. (Ed.), Alternative Sweeteners, fourth ed. CRC Press, Boca Raton, pp. 181–196.

Oligosaccharides Related to Sucrose

Sentko, A., Bernard, J., 2011. Isomalt. In: O'Brien Nabors, L. (Ed.), Alternative Sweeteners, fourth ed. CRC Press, Boca Raton, pp. 275–297.

Sentko, A., Bernard, J., 2011. Isomaltulose. In: O'Brien Nabors, L. (Ed.), Alternative Sweeteners, fourth ed. CRC Press, Boca Raton, pp. 423–438.

Trehalose

Richards, A.B., Dexter, L.B., 2011. Trehalose. In: O'Brien Nabors, L. (Ed.), Alternative Sweeteners, fourth ed. CRC Press, Boca Raton, pp. 439–470.

Reduced Oligosaccharides

Zacharis, C., Stowell, J., 2011. Lactitol. In: O'Brien Nabors, L. (Ed.), Alternative Sweeteners, fourth ed. CRC Press, Boca Raton, pp. 315–332.

Kearsley, M.W., Boghani, N., 2011. Maltitol. In: O'Brien Nabors, L. (Ed.), Alternative Sweeteners, fourth ed. CRC Press, Boca Raton, pp. 299–313.

Deis, R.C., 2011. Maltitol syrups and polyglycitols. In: O'Brien Nabors, L. (Ed.), Alternative Sweeteners, fourth ed. CRC Press, Boca Raton, pp. 265–273.

Polydextrose

Auerbach, M.H., Mitchell, H., Moppett, F.K., 2011. Polydextrose. In: O'Brien Nabors, L. (Ed.), Alternative Sweeteners, fourth ed. CRC Press, Boca Raton, pp. 489–505.

Polysaccharides: Occurrence, Structures, and Chemistry

4

Chapter Outline

Key information and skills that can be obtained from study of this chapter will enable you to

1. Define and/or describe

hydrocolloid	polymolecular, polymolecularity
polysaccharide	number average molecular weight (M_n)
degree of polymerization (DP)	weight average molecular weight (M_w)
glycan	polydispersity index (PI)
homoglycan	degree of substitution (DS)
heteroglycan	molar substitution/moles of substitution (MS)
polydisperse, polydispersity	polysaccharidase

2. Describe and/or discuss the polydisperse and/or the polymolecular nature of polysaccharides with respect to their purification, fractionation, homogeneity, and/or structural analysis.
3. Describe the quantitative determination of polydispersity.
4. Correctly use the several official names for polysaccharides.

Carbohydrate Chemistry for Food Scientists. https://doi.org/10.1016/B978-0-12-812069-9.00004-2

5. Describe the compositions and structures of polysaccharides based on names, and vice versa. For example,

galactomannan
D-galacto-D-mannan
poly(D-galactose-D-mannose)

6. Make interconversions between abbreviated names such as αLAra*f*(1→4)Xyl*p* and Haworth and/or conformational structures for polysaccharide structures or segments of polysaccharide structures containing D-glucose, D-galactose, D-mannose, D-glucuronic acid, D-galacturonic acid, D-mannuronic acid, L-guluronic acid, D-xylose, and/or L-arabinose.

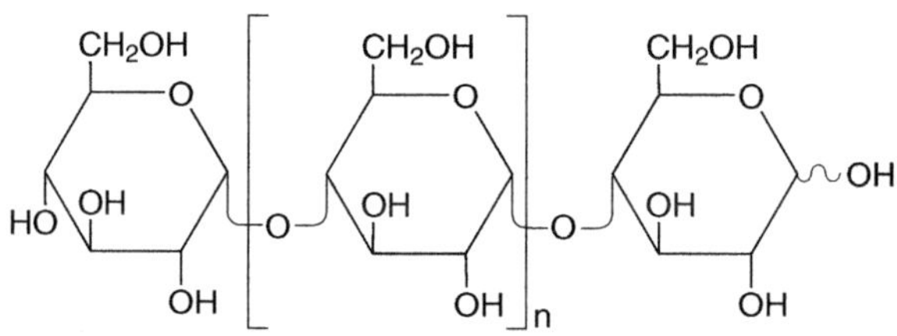

Amylose

7. Outline the general steps in the determination of chemical structures of polysaccharides.
8. Give the general procedures for determination of (1) monosaccharide compositions, (2) linkage positions, (3) linkage configurations, (4) functional groups, and/or (5) chemical fine structures of polysaccharides.
9. Deduce the chemical structure of an oligo- or polysaccharide from methylation analysis and other data on that oligo- or polysaccharide, and discuss any limitations of the data with regards to a description of its complete chemical structure.
10. Describe and discuss the stability of polysaccharides to heat, acidic conditions, basic conditions, and/or oxidants.
11. Describe depolymerization of polysaccharides via acid-catalyzed hydrolysis, discussing the variables involved (acid concentration, temperature, time, monomer unit type, ring form, linkages).
12. Calculate the DS or MS as appropriate from analytical information about a modified polysaccharide; or calculate the number of substituent groups per polysaccharide unit of given length from the DS value.
13. Describe the depolymerization of polysaccharides in alkaline solution in the presence of air.
14. Describe the depolymerization of polysaccharides by mechanical forces.

Chemical structures and names of polysaccharides

Polysaccharides are the largest component of biomass. It is estimated that more than 90% of the carbohydrate mass in nature is in the form of polysaccharides. Starches and **hydrocolloids**[1] (other than gelatin) are polysaccharides. **Polysaccharides** are polymers[2] of monosaccharides. (Long chains of structural units are called polymers [*poly* means many in Greek].) Thus, polysaccharides are high-molecular weight carbohydrate molecules that contain many monosaccharide units. Most polysaccharides are much larger than the 20-unit limit of oligosaccharides. The number of monosaccharide units in a polysaccharide, which is termed its **degree of polymerization (DP),** varies with polysaccharide type. Only a few naturally occurring polysaccharides have DPs less than 100; most have DPs in the range 200–3000. The larger polysaccharides, like cellulose (Chapter 8), have DPs of 7000–15,000. The amylopectin component of starch (Chapter 6) is even larger, having average DPs of more than 100,000 (average molecular weights of more than 10^7). The general scientific term for a polysaccharide is **glycan,** a word derived from *glyc*-for sweet or sugar and *-an* for polymer.

As in most oligosaccharides, the monosaccharide units in polysaccharides are joined together in a head-to-tail fashion by glycosidic linkages. Also like oligosaccharides, polysaccharide molecules can be either linear or branched. All polysaccharides, therefore, have one, and only one, reducing end. Branched polysaccharides have multiple nonreducing ends (Fig. 4.1).

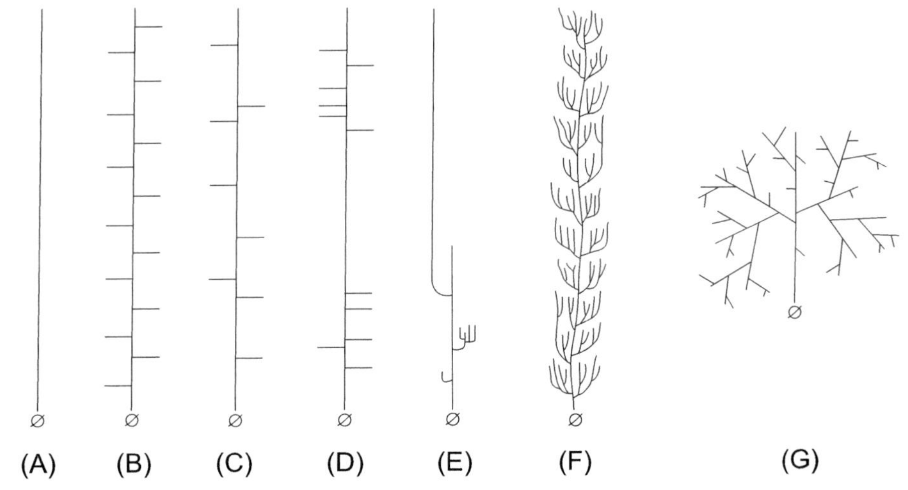

Figure 4.1 Schematic structural representations of small segments of polysaccharide molecules. ∅ = reducing end. A = unbranched molecule; B–D = molecules with short branches of mono-, di-, or trisaccharide units which are evenly spaced (B), randomly spaced (C), or clustered (D) along the backbone chain; F= the cluster type of branching found in amylopectin (Chapter 6); G = a branch-on-branch, bushlike structure such as that of gum arabic (Chapter 16). The latter structures also contain short branches on the backbone structure. While molecules with structures B–E are technically branched, they behave as linear polymers (Chapter 5).

[1] A **hydrocolloid** is a water-soluble, native or modified, nonstarch polysaccharide used in foods. Hydrocolloids are also known as food gums.

[2] Polymers are sometimes referred to as macromolecules.

If all the glycosyl units are of the same sugar type, the polysaccharide is homogeneous as to monomer units and is a classified as a **homoglycan.** Homoglycans can be linear or branched (Table 4.1). Examples of homoglycans are cellulose and the amylose component of starch, which are linear, and the amylopectin component of starch, which is branched; each of these polysaccharides is composed only of D-glucopyranosyl units.

When a polysaccharide is composed of two or more different monosaccharide units, it is classified as a **heteroglycan.** A polysaccharide that contains two different monosaccharide units is a diheteroglycan; a polysaccharide that contains three different monosaccharide units is a triheteroglycan, and so on. Diheteroglycans are generally, but not always, either linear polymers of blocks of similar glycosyl units alternating along the chain or consist of a linear chain of one type of glycosyl unit with a second present as single-unit branches. Examples of the former type are algins (Chapter 14) and of the latter type are guaran (which has a main chain composed of D-mannopyranosyl units to which are attached single-unit side chains of D-galactopyranosyl units) and locust bean gum (Chapter 9).

Whenever three or more types of monosaccharide units (Fig. 4.2) occur in plant polysaccharides, such as in exudate gums (Chapter 16), the polymers usually have branch-on-branch structures (Fig. 4.1G). Even in such branched structures, simplified arrangements of glycosyl units occur, with one type as the main chain and other units in different short branches or mixed in short branches. On the other hand, triheteroglycans from bacteria, such as xanthan (Chapter 11) and gellan (Chapter 12), are usually linear or essentially linear molecules. No glycans with more than seven different basic sugar units are known to be present in foods.

Linear glycans are the most abundant polysaccharides in nature (in terms of quantity) because of the enormous quantity of cellulose existing as the main structural component of the cell walls of higher land plants. However, branched polysaccharides are by far the most numerous, occurring in an immense variety of branched forms and with a variety of sugars in their structures (Table 4.1).

Polysaccharides found in food products come from a variety of sources—from the farm, the forest, the ocean, fermentation vats, and via chemical modification of natural polysaccharides, especially cellulose and starch (Table 4.2).

It is estimated that more than 90% of the carbohydrate mass on earth is in the form of polysaccharides, and as carbohydrates comprise more than 90% of the dry matter of plants of all types, polysaccharides constitute more than 80% of all plant material (dry weight basis). Polysaccharides have important roles in living organisms (Table 4.3). The greatest amounts are structural components of plant cell walls (for example, cellulose) and next comes plant food reserve materials (for example, starch). However, polysaccharides have a variety of other essential roles in plants and in animals.

Molecules in a preparation of a specific polysaccharide (in contrast to molecules of a protein and molecules of a nucleic acid) contain different numbers of monosaccharide units. Thus, polysaccharide preparations contain molecules of the same polysaccharide with a range of degrees of polymerization and, hence, molecular weights. Preparations which contain molecules of the same substance but of different molecular

Table 4.1 Classification of selected food polysaccharides by structure and composition

Classification schemes	Examples
By shape[a]	
Linear	Agar (agaran, agarose), algins, amyloses, carrageenans, cellulose, curdlan, furcellaran, gellan, inulin, pectic acids/pectates, pectins, carboxymethylcelluloses, hydroxypropylcelluloses, hydroxypropylmethylcelluloses, methylcelluloses, pullulan
Branched	
Short branches on an essentially linear backbone	Arabinans,[b] arabinogalactans, arabinoxylans, curdlan, galactomannans (guar gum, locust bean gum, tara gum), konjac glucomannan, psyllium seed gum, xanthan, xylans, xyloglucans, tamarind seed polysaccharide
Branch-on-branch structures	Amylopectins, gum arabics, gum gum ghatti, gum karaya, gum tragacanth(tragacanthin), okra gum
By monomeric units[c]	
Homoglycans	Amylopectins, amyloses, arabinans, cellulose, dextrans, fructans/levans
Diheteroglycans	Algins, arabinogalactans, carrageenans, furcellarans, galactomannans, glucomannans, konjac glucomannan, pectic acids, pectins, xylans
Triheteroglucans	Arabinoxylans, gellan, gum karaya, xanthan
Tetraheteroglycans	Gum arabics, okra gum, psyllium seed gum, xyloglucans
Pentaheteroglycans	Gum ghatti, gum tragacanth (tragacanthin)
By charge	
Neutral	Agar (agaran, agarose), amylopectins, amyloses, arabinans, arabinogalactans, beta-glucans, cellulose, curdlan, dextrans, galactomannans, glucomannans, inulin, konjac glucomannan, pullulan, xyloglucans, hydroxypropylcelluloses, hydroxypropylmethylcelluloses, methylcelluloses, tamarind seed polysaccharide
Anionic (acidic)[d]	Algins, arabinoxylans, carrageenans, furcellarans, gellans, gum arabics, gum ghatti, gum karaya, gum tragacanth(tragacanthin), okra gum, pectic acids/pectates, pectins, psyllium seed gum, xanthan, xylans, carboxymethylcelluloses

[a]Primary examples. For example, arabinoxylans occur in different architectures, compositions, and charges.
[b]The predominant structure.
[c]Considers only the basic monosaccharide units. A derivatized monosaccharide unit, such as D-galactopyranosyl 6-sulfate unit, is not considered as a unit separate from a D-galactopyranosyl unit, for example.
[d]From the presence of uronic acid (Chapter 2), sulfate half-ester (Chapter 13), pyruvyl cyclic acetal (Chapters 2 and 11), or succinate half-ester groups (Fig. 4.3).

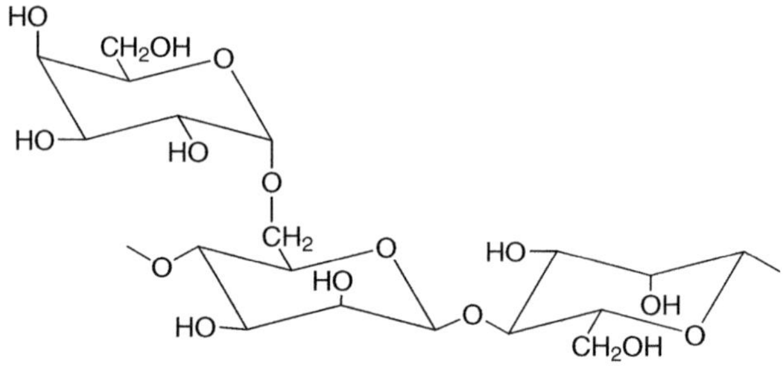

Idealized structure of guaran

weights are said to be **polydisperse**, so polysaccharide preparations are polydisperse. The molecular weight range may be narrow or broad.

Bacterial polysaccharides, such as xanthan (Chapter 11) and gellan (Chapter 12), and some plant polysaccharides, such as cellulose (Chapter 8), are chemically homogeneous with regard to monosaccharide units and their sequence (while being polydisperse), but bacterial polysaccharides can vary with respect to attached noncarbohydrate substituent groups (Fig. 4.3). However, the structures of most plant polysaccharides vary in linkage types, branching frequency (if branched), and/or in proportions of monosaccharide constituents from one molecule to another. A polysaccharide whose individual molecules have different fine structures is said to be **polymolecular**. The difference in fine structure from molecule to molecule in a preparation of a single kind of polysaccharide is known as microheterogeneity. Essentially, all naturally occurring plant polysaccharides (with the exception of cellulose) are polymoelcular. However, polymolecularity (microheterogeneity) is introduced in cellulose and other polysaccharides when they are chemically modified because the modification takes place at different hydroxyl groups of a glycosyl unit and/or at different locations along chains.

With only a few rare exceptions, polysaccharides in foods are both polydisperse and polymolecular. (All polysaccharides are polydisperse; most are also polymolecular.) In other words, all polysaccharides contain a heterogeneous population of molecules that vary in chemical structure and/or molecular size. (Bacterial polysaccharides such as gellan [Chapter 12], curdlan [Chapter 12], and xanthan [Chapter 11] have regular repeating unit structures, with regard to the sequence of monosaccharide units, but gellan and xanthan are polymolecular because of variations in the amounts of ester groups and/or the pyruvyl cyclic acetal group [in the case of xanthan].) Therefore, a description of most polysaccharides is a statistically most probable structure from a population of molecules, and the reported molecular weight is one of the several types of averages (see Section Molecular weights) that can be calculated for polymeric molecules, rather than being absolute structures and molecular weights.

Preparation of a polysaccharide, whether in the laboratory for characterization or in commercial production, begins with extraction from the source (in the case of a plant

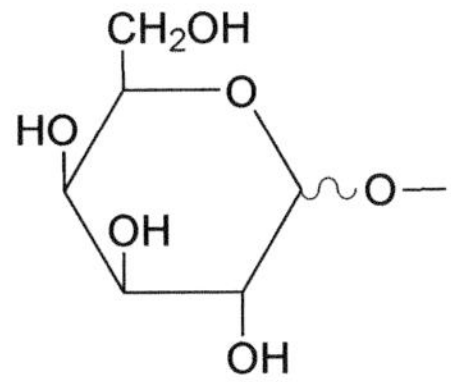

α-and β-D-Galactopyranosyl
(α,βGal*p*)

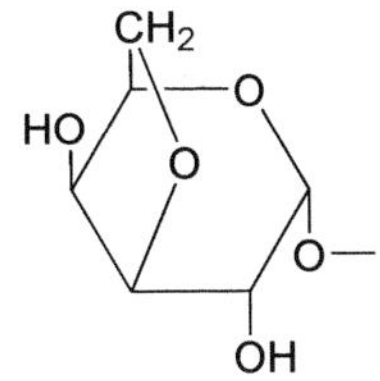

3,6-Anhydro α-D-Galactopyranosyl
(αGal*p*3,6An)

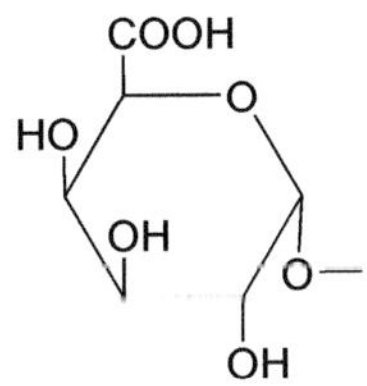

α-D-Galactopyranosyluronic acid
(αGal*p*A)

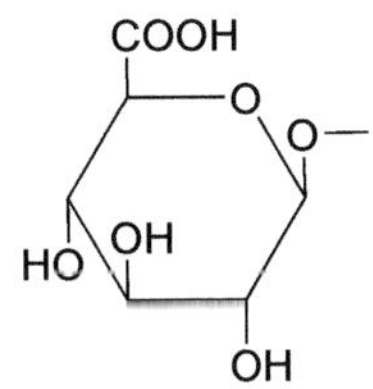

β-D-Glucopyranosyluronic acid
(βGlc*p*A)

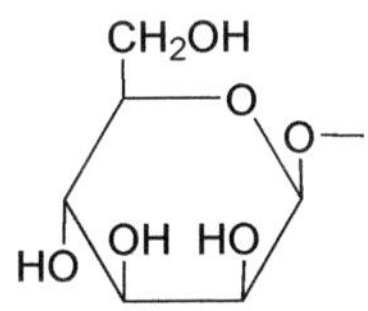

β-D-Mannopyranosyl
(βMan*p*)

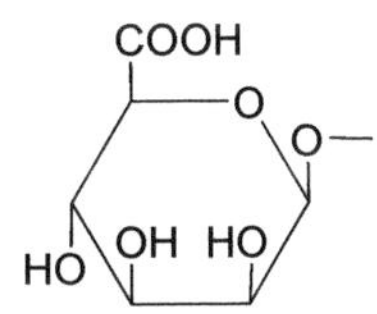

β-D-Mannopyranosyluronic acid
(βMan*p*A)

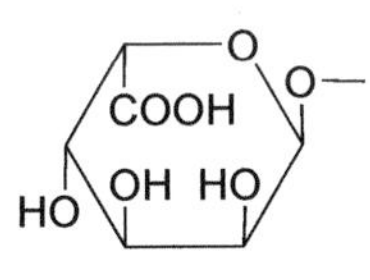

α-L-Gulopyranosyluronic acid
(αLGul*p*A)

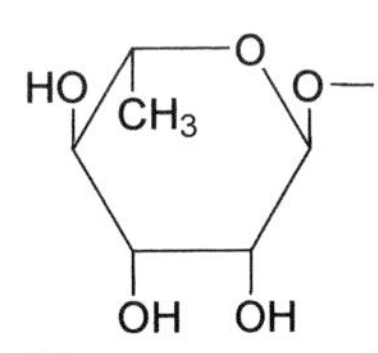

α-L-Rhamnopyranosyl
(αLRha*p*)

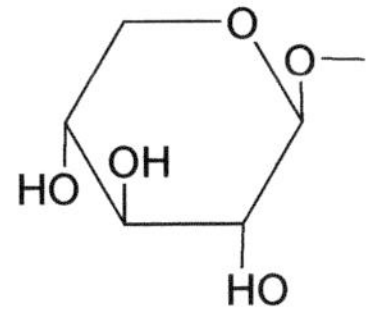

β-D-Xylopyranosyl
(βXyl*p*)

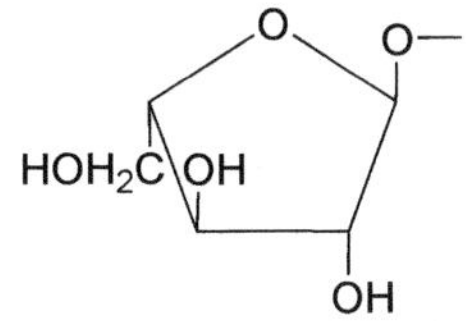

α-L-Arabinofuranosyl
(αLAra*f*)

Figure 4.2 Monomer units, other than the very common α- and β-D-glucopyranosyl units whose structures were given in Chapter 1, that are found in the food polysaccharides covered in this book. Additional monomer units are found in polysaccharides other than the common food polysaccharides.

Table 4.2 Classification of selected polysaccharides in foods by source

Class	Examples
Algal (seaweed extracts)	Agars, algins, carrageenans, furcellaran
Higher plants	
Insoluble	Cellulose
Extract of fruits	Pectins
Seeds	Corn starches, rice starches, wheat starches, beta-glucans, guar gum, locust bean gum, tara gum, psyllium seed gum, tamarind seed polysaccharide
Tubers and roots	Potato starches, tapioca (cassava) starches, konjac glucomannan
Exudates	Gum arabics, gum karaya, gum tragacanth
Microorganisms (fermentation gums)	Xanthans, gellans, curdlan, pullulan, dextrans
Derived	
From cellulose	Carboxymethylcelluloses, hydroxypropylcelluloses, hydroxypropylmethylcelluloses, methylcelluloses
From starch	Starch acetates, starch adipates, starch 1-octenylsuccinates, starch phosphates, starch succinates, hydroxypropylstarches, dextrins
Synthetic	Polydextrose

polysaccharide) or isolation from a fermentation culture medium (in the case of a bacterial polysaccharide). In laboratory preparations, extractions from a plant tissue are usually preceded by removal of interfering substances, such as lipids and lignin. Extraction may be done with water in a few cases but most often involves an alkaline solution. Both extraction and recovery from a fermentation medium are followed by purification to separate the desired polysaccharide from noncarbohydrate materials (such as proteins) and fractionation to separate the desired polysaccharide from other polysaccharides. Purification most often involves precipitation, sometimes fractional precipitation. Precipitation is usually achieved by addition of a water-soluble alcohol such as ethanol (in the laboratory) or 2-propanol (isopropanol) (industrially). Precipitation can sometimes be effected by addition of a complexing agent for the polysaccharide or by changing the pH of the solution. In the laboratory, size-exclusion and/or ion-exchange chromatographic techniques may be used to obtain reasonably homogeneous preparations.

Polysaccharides have a great variety of structures (Fig. 4.1, Table 4.1), the only common feature being that each is composed of monosaccharide units (in some cases,

Table 4.3 Biological functions of selected polysaccharides found either naturally or as additives in foods

Higher land plants
Structural (cell wall) components
Cellulose
Hemicelluloses
Annual plants
Arabinoxylans
Xylans
Xyloglucans (dicotyledons)
Pectic polysaccharides
Arabinans
Arabinogalactans
Homogalacturonan
Rhamnogalacturonan I
Rhamnogalacturonan II (same in all plants)
Storage (food reserve) materials
Fructans
Galactomannans
Glucomannans
Starches
Exudates
Gum arabic
Gum ghatti
Gum karaya
Gum tragacanth
Marine algae
Structual (cell wall) components
Agar
Alginates
Carrageenans
Cellulose
Furcellaran
Fungi
Structural (cell wall) components
Cellulose
Chitin
Crustaceans
Structural components
Chitin

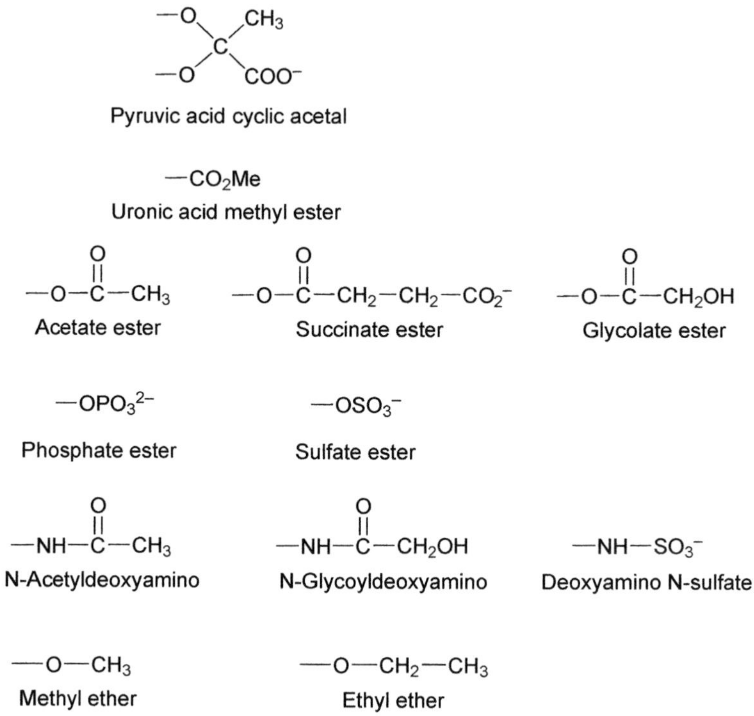

Figure 4.3 Noncarbohydrate substituent groups found on naturally occurring polysaccharides. Which polysaccharides contain which groups is found in Chapters 10–16.

esterified, etherified, or otherwise derivatized monosaccharide units). As with other biopolymers, structures of polysaccharides vary from species to species and from variety to variety within a species. The variability is greater in polysaccharides than it is in other biopolymers because there are not only differences in structure due to genetic differences in different species and varieties but also differences that arise from variations in growth environments of the plant or microorganism that makes them.

Structural analysis of a polysaccharide may be undertaken once it is obtained in an acceptable degree of purity. Structural characterization involves determination of (1) monosaccharide composition, (2) linkage types, (3) anomeric configurations, (4) presence and location of noncarbohydrate substituent groups, and (5) average DP (average molecular weight). Because there is such variability in structures, there is some variability in methods used to determine their structures, but some generalizations can be described.

Names of polysaccharides

Polysaccharides (glycans) can be named in several different ways. One way is by using a common name, such as amylose or pectin. (Only polysaccharides that were named before there was a nomenclature system for them are named in this way. Many of these older names end in either -ose or -in.) In systematic nomenclature, a polysaccharide is named by adding the suffix *-an* to the name(s) of the principal sugar(s) in its structure. For example, a polysaccharide composed of D-xylopyranosyl units is a xylan. A polysaccharide composed of D-galactopyranosyl and D-mannopyranosyl units, such as guaran[3] and locust bean gum (Chapter 9), is a galactomannan or a D-galacto-D-mannan. (The organic chemical name poly(D-galactose-D-mannose) is almost never used.) A polysaccharide composed of α-D-galacturonopyranosyl units (pectin, Chapter 15) can be named a galacturonoglycan, a galacturonan, or poly(α-D-galacturonic acid). When a backbone chain is composed of units of a single, or primarily a single, monosaccharide, the name of that monosaccharide is placed at the end of the name with the -an suffix (such as xylan). In the case of the galactomannans, the structures of which are a main chain of β-D-mannopyranosyl units with α-D-galactopyranosyl units attached to that main chain (Chapter 9), the mannan name comes last because the polysaccharide is a substituted (with D-galactosyl units). However, one cannot tell from the name galactomannan what specific type of structure it is, that has to be known from experience. The glucomannan obtained from konjac flour (Chapter 10) contains both D-glucopyranosyl and D-mannopyranosyl units in its slightly branched polymer chains. Here, the two units are simply listed alphabetically. So although the names galactomannan and glucomannan are similar, the general structures of the two classes of polysaccharides are quite different. Likewise, if there are two or more types of sugars attached to a main chain, they are listed in alphabetical order, as for example, in arabinoglucuronoxylan, which contains both L-arabinosyl and D-glucuronosyl units attached to a xylan backbone. (The arabino and glucrono parts of the name are incorporated alphabetically.)

Determination of structures

Determination of the monosaccharide composition of a polysaccharide begins with acid-catalyzed hydrolysis under conditions that give maximum depolymerization and minimum destruction of the released sugars in the hot acid used. Released monosaccharides are then determined both qualitatively and quantitatively, not only by high-performance liquid chromatography, but also by gas-liquid chromatography (GLC) after their conversion to volatile, thermostable derivatives—usually alditol peracetates (Chapter 2).

Linkages may be determined by methylation analysis. Methylation analysis, which reveals the linkage position, ring size, and the nature of the monosaccharide, is outlined in Fig. 4.4. All exposed hydroxyl groups of the polysaccharide are converted

[3] Guaran is the polysaccharide component of guar gum (Chapter 9).

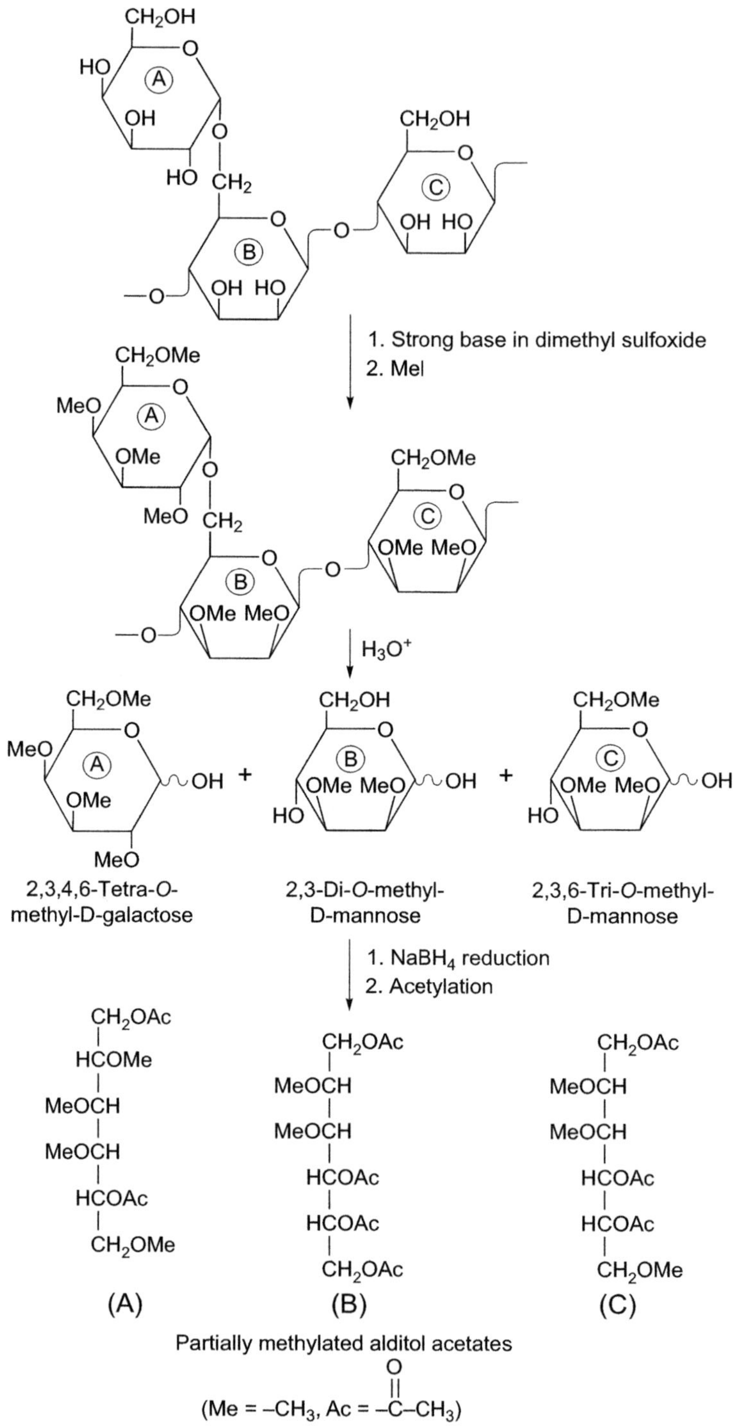

Figure 4.4 Methylation analysis of an idealized trisaccharide repeating unit of guaran, the polysaccharide of guar gum (Chapter 9). A, a D-galactopyranosyl unit; B, a 4,6-di-O-substituted D-mannopyranosyl unit; C, a 4-O-substituted D-mannopyranosyl unit. (Me = methyl ether group. Ac = acetyl ester group.)

into methyl ethers by treating the polysaccharide (dissolved in dimethyl sulfoxide) with a strong base and reacting it with methyl iodide (MeI/CH_3I). Hydrolysis of the completely methylated polysaccharide then exposes the hydroxyl groups that were protected from methylation by being involved in glycosidic linkages, thus revealing the locations of the linkages. (Of course, the anomeric hydroxyl group of each unit will be involved in a glycosidic bond, so only the other hydroxyl groups are significant.) Units that are nonreducing end units are completely methylated, except for the anomeric hydroxyl group and the hydroxyl group that is involved in ring formation. The released monosaccharides are then reduced to the corresponding alditols, which releases the hydroxyl group involved in ring formation (the one on C5 if the unit was originally an aldopyranosyl unit and the one on C4 if the unit was an originally an aldofuranosyl unit). For separation by GLC, the alditols are acetylated to form partially methylated alditol acetates.

Methylation analysis pinpoints the position of linkages to each monomer unit. It does not reveal the sequence. For example, the products produced by methylation analysis of the trisaccharide sequence given in Fig. 4.4 would also be produced by three other sequences (Fig. 4.5).

To determine the monosaccharide sequence, the polysaccharide is partially depolymerized (using enzyme- and/or acid-catalyzed hydrolysis) to oligosaccharides, whose structures are then determined. A polysaccharide with a repeating unit structure such as that in Fig. 4.4, which is an idealized structure of guaran (Chapter 9), will produce the following trisaccharide fragments.

The anomeric configuration of the glycosidic linkage also needs to be determined. This can be done on either the intact polysaccharide or fragments derived from it. Most

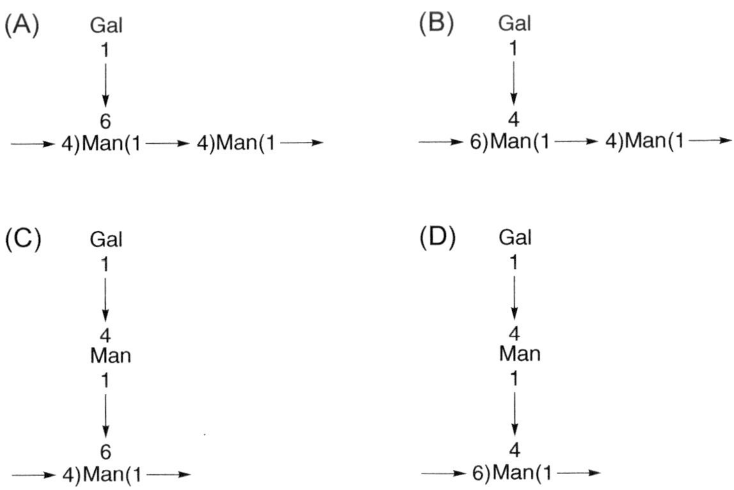

Figure 4.5 The four structures that will give the three partially methylated alditol acetates shown in Fig. 4.4. Structure A is the same as the structure in Fig. 4.4.

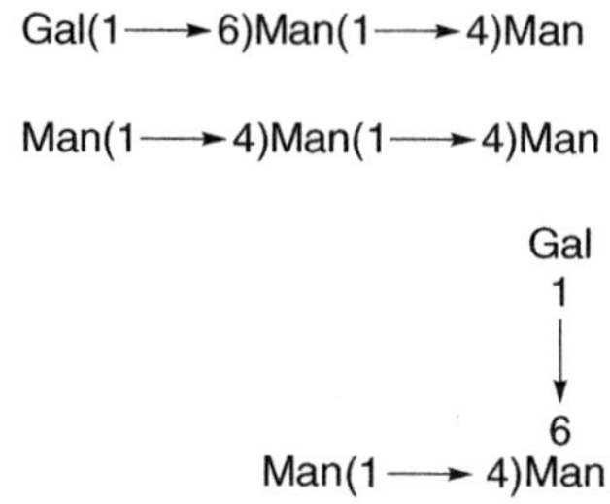

Structure A is obviously the only one that could give rise to these products.

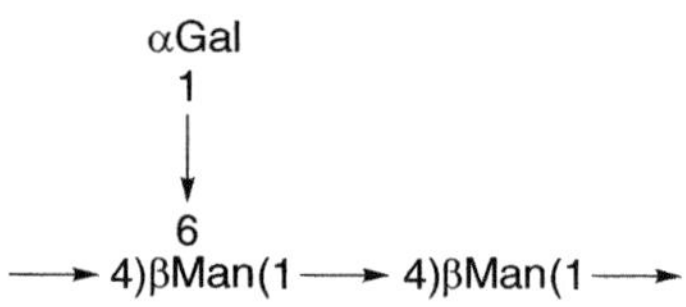

commonly, the anomeric configuration is determined using nuclear magnetic resonance (NMR), but it can also be determined by testing with enzymes that are specific for a particular kind of linkage. For example, evidence that a polysaccharide contains β-D-galactopyranosyl units as nonreducing end units is obtained if the polysaccharide is acted on by a β-galactosidase, an enzyme that catalyzes only the hydrolysis of such units. After the anomeric configuration is determined, the structure becomes the following.

For complete characterization of the polysaccharide, information is needed on the type and location of any substituent groups (Fig. 4.3) and the polysaccharide's average molecular weight. Typical noncarbohydrate substituent groups are acetate ester groups (for example, xanthan [Chapter 11]), sulfate half-ester groups (carrageenans [Chapter 13]), phosphate ester groups (potato starch [Chapter 6]), and pyruvyl cyclic acetal groups (xanthan [Chapter 11]).

Average molecular weights

Because all polysaccharide preparations are polydisperse, only average molecular weights for them can be determined. There are several ways to determine average molecular weights and several molecular weight averages of polymers; four of these (particularly two) are commonly used with polysaccharides. The formulas below give the mathematical basis for each type of average; in all formulas, N_i is the number of molecules that have a molecular weight of M_i and w_i is the weight of the molecules with a molecular weight of M_i.

The number average molecular weight (M_n) is the total weight of the sample (w) divided by the number of molecules in the sample ($\Sigma_i N_i$). It is obtained experimentally by summing the number of molecules of each molecular weight multiplied by the molecular weight of that species ($\Sigma_i N_i M_i$), which is the mass of the sample (w), divided by the total number of molecules. It is the simple average (mean) that is used with all kinds of data. For many simple Gaussian-type molecular weight distributions, M_n is near the peak molecular weight.

$$M_n = \frac{\sum_i N_i M_i}{\sum_i N_i} = \frac{w}{\sum_i N_i}$$

There are two important ways to determine the number average molecular weight: (1) End-group analysis: Because, as has been emphasized, essentially each polysaccharide molecule has one, and only one, reducing end (an aldehydic group), its measurement by a quantitative reducing sugar method is a measurement of the total number of molecules. However, for molecular weights above about 25,000 (2.5×10^4) (DP about 150, quite small for a polysaccharide), most reducing end-unit methods become insensitive because the end-group is present in a too low concentration. Amylopectins (Chapter 6), for example, probably have average molecular weights of more than 10^7. (2) Because colligative properties (boiling point elevation, freezing point depression, osmotic pressure, etc.) are a function of the number of molecules in solution, they can be used as a measure of the number of molecules. Vapor phase osmometry may be used for this purpose. Number average molecular weights are strongly influenced by the smaller molecules.

Another average is the weight average molecular weight (M_w). The M_w is obtained by summing the weight of all polymer molecules with molecular weight M_i each multiplied by this molecular weight and dividing this number by the sum of the number of molecules each multiplied by its molecular weight. This is the same as summing the number of molecules of each molecular weight multiplied by the molecular weight of that species squared divided by the mass of the sample. The principle method of determining M_w is light scattering. Because this molecular weight average involves the square of the molecular weight (equation below), it is strongly influenced by the larger molecules and is close to the molecular weight of the largest molecules in a polysaccharide preparation.

$$M_w = \frac{\sum_i N_i M_i^2}{\sum_i N_i M_i} = \frac{\sum_i N_i M_i^2}{w}$$

The **polydispersity index (PI)** (M_w/M_n) is a measure of the degree of polydispersity of the polysaccharide preparation. Because M_n is weighted toward the smallest molecules and M_w is weighted toward the largest molecules, M_w/M_n reflects the range of molecular weights in the preparation. This value can be illustrated with the following very simple illustration of a population of molecules in a preparation with a peak molecular weight of 200,000. (Of course, in a real preparation, there would be a continuous spectrum of molecular weight values rather than eight distinct populations.)

M_i	N_i
50,000	100
100,000	200
150,000	500
200,000	1000
250,000	600
300,000	300
350,000	200
400,000	100

Using the data to calculate M_n and M_w, they are found to be 217,000 and 244,000, respectively. As a result, the value for the PI (M_w/M_n) is 1.12. (The PI for a polysaccharide is usually a considerably larger number. A PI value of 1.12 indicates very little polydispersity, although the range in molecular weights is eightfold.)

A third method based on measurement of intrinsic viscosity, which gives a viscosity average molecular weight (a relative molecular weight value), is rapid and inexpensive.

A fourth commonly used method is that of size-exclusion chromatography (SEC). SEC can provide good separation of polymer molecules. Separation by SEC is based on the polymer molecules' radii of gyration (R_g)[4] rather than their molecular weights. To conduct SEC of a polysaccharide preparation, a column containing a packing material suitable for the molecular weight range of the polysaccharide to be characterized is selected. Dextran and pullulan standards of known average or peak molecular weights are used to calibrate the column, but unknowns are unlikely to have the same folding pattern as the standards and, thus, are unlikely to have the same radius of gyration for the same molecular weight. After plotting elution volumes against molecular weights for the standards to create a standard curve, the molecular weight of the unknown is determined by measuring its elution value and comparing it with the standard curve. The SEC method gives relative molecular weight values (that is, values relative to those of the standards used) and information about the broadness of the molecular weight distribution.

The chromatographic method gives considerably more information if a multiangle laser light scattering (MALLS) detector is used. Modern SEC employs a refractive

[4] The radius of gyration is the root mean square distance away from the center of gravity of the folded polysaccharide molecule. It is a measure of the volume swept out by the hydrated polymer molecule and, thus, is related to both the flexibility and the extent of folding of polymer molecules.

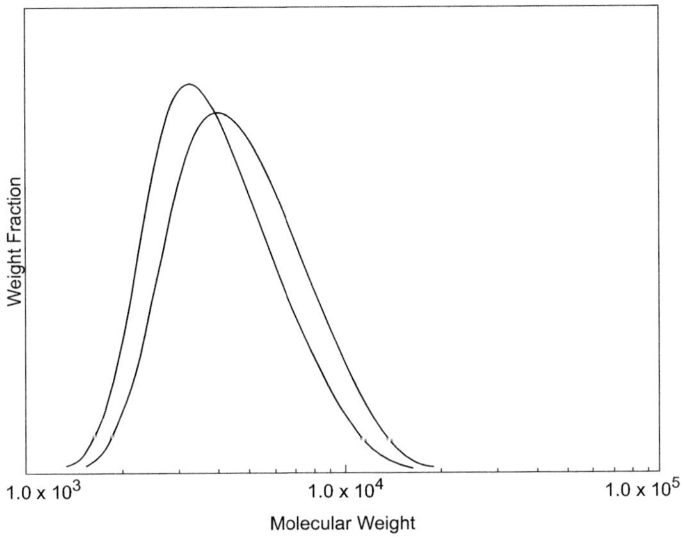

Figure 4.6 A hypothetical depiction of molar mass distributions of two preparations of the same polysaccharide.

index detector, which determines the amount (mass) of material eluted at a given time, and a MALLS detector, which determines the M_w of the same material. A plot of one value against the other gives a profile of the amounts of each molecular weight species (Fig. 4.6). With this method, column calibration is unnecessary. Software with the MALLS detector will also calculate R_g.

Structural modifications

A summary of ways in which modifications of the structures of polysaccharides can be achieved, either deliberately for preparation of a food ingredient or during raw material handling, food processing, or storage, are summarized in Table 4.4. Other modifications may be practiced for nonfood applications.

Derivatization

Hydroxyl groups of polysaccharide molecules can be etherified and esterified, just as hydroxyl groups of monosaccharides can be (Chapter 2). Both ethers and esters of polysaccharides are made commercially. Ethers are the more common derivatives used to modify polysaccharides for food use.

The average number of hydroxyl groups per glycosyl unit that have been derivatized by etherification or esterification is called the **degree of substitution (DS)**. A polysaccharide that contains only hexosyl units that are neither uronic acid nor deoxy-hexosyl units contains an average of three hydroxyl groups available for derivatization (substitution) per monosaccharide unit. (When a glycosyl unit is a

Table 4.4 Procedures that can be used to modify food polysaccharide structures

Chemical Reactions
Derivatizations
• Etherification of hydroxyl groups[a]
• Esterification of hydroxyl groups[b]
Oxidations of hydroxyl groups[a]
Depolymerizations
• Acid-catalyzed hydrolysis[c]
• Enzyme-catalyzed hydrolysis[d]
• Alkali-catalyzed beta-elimination[a]
• Enzyme-catalyzed beta-elimination[e]
Removal of specific units via *enzyme-catalyzed hydrolysis*[f]
Deacylations/deesterifications
• Of uronic acid esters[g]
• Of hydroxyl group esters[h]
Transglycosylation[b]
Physical treatments
Heat treatments[i]
• Pregelatinization
• Heat-moisture treatment
• Annealing
• Dry heating
Nonthermal treatments
• Ultrasonication
• Milling
• Homogenization
• Pulsed electric field
Biological approaches
By selection and breeding of cultivars of the producing plant
By mutant production and/or selection and breeding of the producing plant
By selection of strains of the producing microorganism[j]
Via changes in environment (growth conditions) of the producing plant[k] or microorganism[l]

[a]Starches and cellulose.
[b]Starches.
[c]May occur during processing.
[d]May occur during processing or storage of either raw material or finished product.
[e]Pectins, during fruit ripening or processing.
[f]Guar gum.
[g]Pectins.
[h]Gellan and konjac glucomannan.
[i]Practiced with starches (Chapter 7), but inadvertent overheating can affect the quality of hydrocolloids.
[j]Xanthan.
[k]Generally related to climate and cannot be controlled.
[l]Xanthan and gellan.

branch point, there is one less hydroxyl group for every branching unit attached to that unit. However, every branch, even if it is a single-unit branch, is terminated with a nonreducing end unit which contains four unsubstituted hydroxyl groups, thus balancing the one lost through branching. So, no matter how highly branched it is, a polysaccharide composed only of regular, neutral hexosyl units will have an average of three hydroxyl groups per glycosyl unit.) Thus, a maximum degree of hydroxyl group substitution of three (a maximum DS of 3.0) can be obtained for a neutral polysaccharide containing only hexosyl units, such as a cellulose (Chapter 7) or a starch polymer molecule (Chapter 6).

When hydroxypropyl (-CH_2-CHOH-CH_2OH) ether groups are added by reaction with propylene oxide, the substituent group itself contains a hydroxyl group that can react with another derivatizing reagent molecule (starch ethers [Chapter 6] and methylcelluloses and hydroxypropylmethylcelluloses and hydroxypropylcelluloses [Chapter 8]). As a result, chains of substituent groups can theoretically form. In this case, the term **molar substitution** or **moles of substitution (MS)** is used to denote the average number of moles of substituent added to a glycosyl unit. The actual distribution of derivatizing groups among the three hydroxyl groups available for reaction and along the polymer chain varies with reaction conditions, reagent type, and the extent of substitution. Of polysaccharide ethers, only hydroxypropylstarch (Chapter 7), hydroxypropylcellulose, and hydroxypropylmethylcellulose (Chapter 8) are approved for use in food products. Modified polysaccharides are characterized by molecular weight and physical properties as well as by DS or MS.

Any polysaccharide that is esterified or etherified and is intended for food use is only partially derivatized, so the unbalanced equation for the reaction in an alkaline system (which is always used) is the following (after neutralization):

$$\text{Polysaccharide(OH)}_x \xrightarrow{\text{RX}} \text{Polysaccharide(OR)}_y\text{(OH)}_z + X^-$$

in which $y + z = x$ and z is larger than y. In the remainder of this book, the reaction is simplified as

$$\text{Polysaccharide-OH} \xrightarrow{\text{RX}} \text{Polysaccharide-OR} + X^-$$

Depolymerization

Polysaccharides may, at times, undergo depolymerization under food processing or storage conditions. Depolymerization may be deliberately effected during the preparation of a modified food starch (Chapter 7) or a hydrocolloid[1] (Chapters 8–16). Depolymerization obviously reduces the average molecular weight. It also dramatically reduces the viscosity of a solution of the polysaccharide at a given concentration (Chapter 5). Often, polysaccharide ingredients that provide functional properties without producing highly viscous solutions are needed. Solutions of high solids content without excessive viscosity are desired, for example, when film formation or

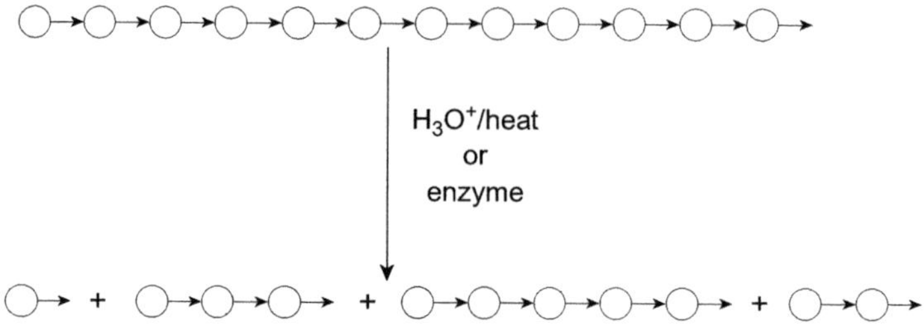

binding is needed or to provide texture,[5] particularly body,[6] such as in syrups. Two types of depolymerization via chemical reactions are employed: acid- or enzyme-catalyzed hydrolysis and alkali-induced beta-elimination following oxidation. Enzyme-catalyzed beta-eliminations may occur naturally with pectins (Chapter 15) during ripening processes. Extensive hydrolysis of a polysaccharide will result in, first, the production of oligosaccharides and, finally, in monosaccharides.

Hydrolysis

Polysaccharides are generally relatively more susceptible to hydrolytic breakdown than are proteins. Hydrolysis of glycosidic bonds joining monosaccharide (glycosyl) units in oligo- and polysaccharides may be catalyzed by either aqueous acids (H^+) plus heat or enzymes. In enzyme-catalyzed hydrolysis, the enzyme is the proton (H^+)-donating reagent. A simplified mechanism of the hydrolysis reaction was presented in Chapter 3.

The extent of depolymerization during thermal processing, which reduces viscosity at a given concentration, is determined by the following factors:

1. pH: The lower the pH, the faster the rate of hydrolysis/depolymerization at any given temperature.
2. Temperatures experienced by the polysaccharide in the process.
3. Times at each temperature.
4. Structure of the polysaccharide (monosaccharide units, ring size, anomeric configurations, linkage positions).

Because many foods are slightly acidic, depolymerization can occur during thermal processing and storage. Loss of viscosity during processing can usually be overcome by using more of the polysaccharide (starch or hydrocolloid) in the formulation to compensate for any loss of viscosity during processing by using a higher viscosity grade of the hydrocolloid (Chapter 5), or by using a more acid-stable starch or hydrocolloid. Polysaccharides can even undergo depolymerization while stored as a semidry

[5] Texture refers to the overall physical properties and palatability of a food product, which is a function of its constituents, their properties, and their arrangement and relation to each other in a food product.

[6] Body refers to the organoleptic attributes (often referred to as mouthfeel) due to the rheological properties (Chapter 5) of a food or beverage.

powder when hydrogen ions (H^+) are present. Of course, any depolymerization during storage affects shelf life.

Polysaccharides are also subject to enzyme-catalyzed hydrolysis. The rates and end products of breakdown by enzyme-catalyzed hydrolysis are also affected by several factors:

1. The specificity of the enzyme: Most hydrolytic enzymes are rather specific for a particular monosaccharide unit, its anomeric configuration, the position of the linkage [that is, $(1 \rightarrow 3)$, $(1 \rightarrow 4)$, $(1 \rightarrow 6)$, etc.], and its position within a polymer chain, including the natures of neighboring units.
2. pH: There is an optimum pH, usually below pH 7, for each **polysaccharidase** (polysaccharide-hydrolyzing enzyme).
3. Time and temperature: The rate of enzyme-catalyzed hydrolysis increases with increasing temperature up to the temperature of denaturation of the enzyme.
4. Other environmental factors: Enzyme activity and stability may be affected by factors such as salt concentration.

Polysaccharides, like other biological substances, are subject to microbial attack, a property which is related to their susceptibility to enzyme-catalyzed hydrolysis. Starch and hydrocolloid products are not delivered as sterile materials. Thus, food products containing a polysaccharide as an added ingredient should be sterilized. Preservatives, such as sodium benzoate, methyl *p*-hydroxybenzoate, propyl *p*-hydroxybenzoate, or sorbic acid, are often added to food products to prevent microbial growth.

Polysaccharides can also be depolymerized (reduced in molecular weight) by oxidation in an alkaline system (below) and by physical treatments (see Section Physical methods).

Oxidation–elimination

Oxidation of polysaccharides converts any hydroxyl groups into carbonyl (aldehydic or keto) groups. The reaction involves radical formation, generally utilizes oxygen, and involves catalysis by a transition metal ion. Intermediate radicals formed from other molecules, such as unsaturated fatty acids, may also be involved. These reactions may produce other even more undesirable defects such as off flavors and aromas.

An oxidized polysaccharide will undergo beta-elimination in an alkaline environment. Beta-elimination is a reaction (Fig. 4.7) that occurs under alkaline conditions, when there is a leaving group (OR in Fig. 4.7) on the second carbon atom (the beta carbon atom) away from a carbonyl group. Because of the electron-withdrawing power of the carbonyl group, the proton on the carbon atom adjacent to it (the alpha carbon atom) is acidic and is removed in alkaline systems. Its removal leaves an unshared pair of electrons on the alpha carbon atom. A carbon atom does not accommodate negative charge easily, so to stabilize the system, this pair of electrons moves between the alpha and beta carbon atoms, forming a double bond and eliminating the OR group as RO^-. (An oxygen atom is much more accommodating to the extra pair of electrons [that is, the negative charge] than is a carbon atom.) In the case of polysaccharides, R is a chain

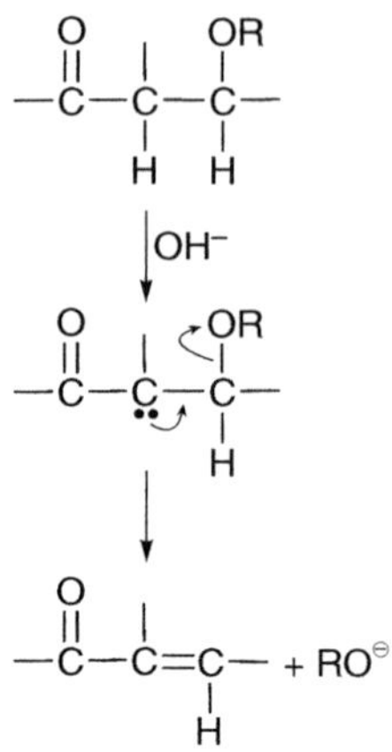

Figure 4.7 General scheme of a beta-elimination process occurring upon removal of the acidic proton from the carbon atom alpha (that is, adjacent) to the carbonyl group.

of glycosyl units and, therefore, the polymer chain is cleaved at this point. Different types of chain cleavage via beta-elimination are given in Figs. 4.8–4.12.

Depolymerization via elimination may be important in the use of poly(uronic acids) such as pectin (Chapter 15). Aldehydic groups may subsequently be oxidized to carboxylate groups.

As already mentioned, hydrocolloid producers often deliberately depolymerize their products to make a variety of lower viscosity products (Chapter 5). In making cellulose derivatives (Chapter 8), the polymer is steeped in a strongly alkaline solution and "aged" in a manner that promotes oxidation of hydroxyl groups by dissolved oxygen from air. Subsequent alkali-induced base-catalyzed eliminations cleave the chain. Acids, enzymes, and the beta-elimination process are used to produce a variety of depolymerized starch products (Chapter 7).

Physical modifications

Deliberate physical treatments for modification of properties and functionalities of polysaccharides are practiced with starches (Chapter 7) and cellulose (Chapter 8). However, physical treatments are mentioned here because the same effects are experienced by other polysaccharides, and various physical treatments are being investigated as nonthermal food processing techniques. Therefore, a product developer needs to understand what might happen to the polysaccharide components during such treatments.

Some thermal depolymerization may occur at higher temperatures, especially if shear is also applied. Such depolymerization is most likely to occur with larger (higher-molecular weight) molecules (as discussed below). At extremely high temperatures (for food processing), thermal degradation (a chemical process) will occur.

Ultrasonic treatments, which may be used for food processing, can result in depolymerization of dissolved polysaccharides. The depolymerization (viscosity reduction)

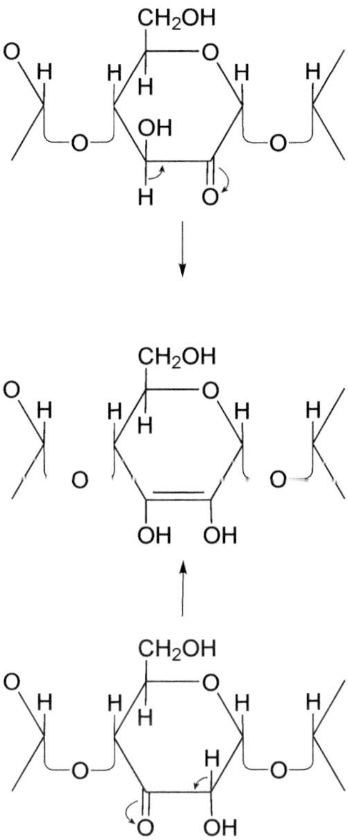

Figure 4.8 Base-catalyzed interconversions of C2 and C3 carbonyl (ketone) groups in a (1 → 4)-linked α-glucan.

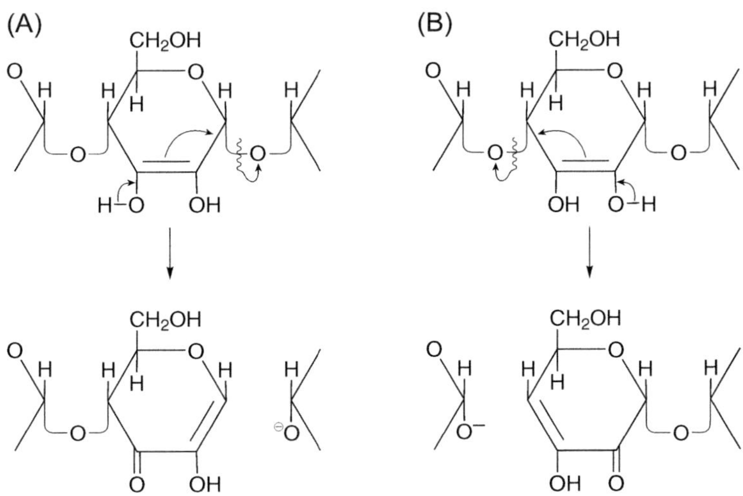

Figure 4.9 Two mechanisms of chain cleavage via beta-elimination after oxidation of the (1 → 4)-α-glucan (Fig. 4.8) at C2 or C3.

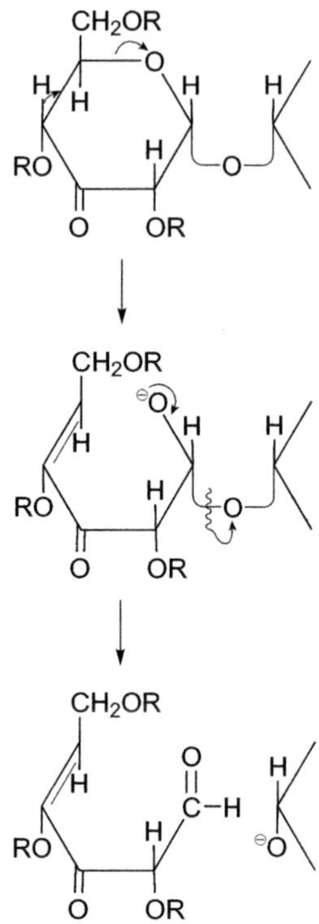

Figure 4.10 An alternative mechanism for chain cleavage via beta-elimination after oxidation at C3.

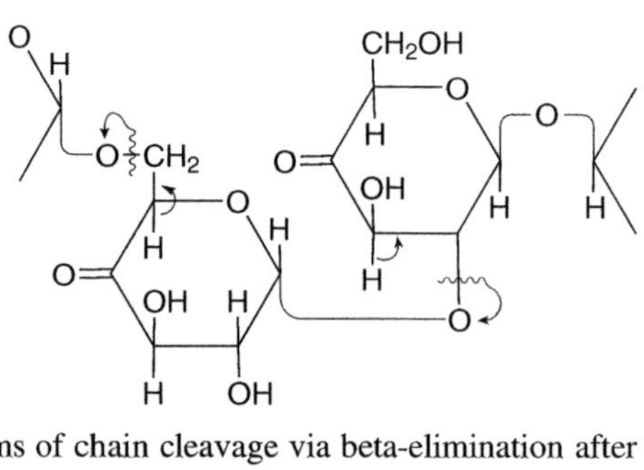

Figure 4.11 Mechanisms of chain cleavage via beta-elimination after oxidation at C4 in a glucan containing (1→6) and/or (1→2) linkages.

that occurs as a result of exposure to ultrasound has been attributed to both mechanical effects and to generation of OH radicals. Evidence that OH radicals are involved comes from the fact that the presence of radical scavengers reduces the rate of depolymerization. Evidence that mechanical force is involved comes from the fact that polysaccharide molecules are also depolymerized by shear generated by other energy

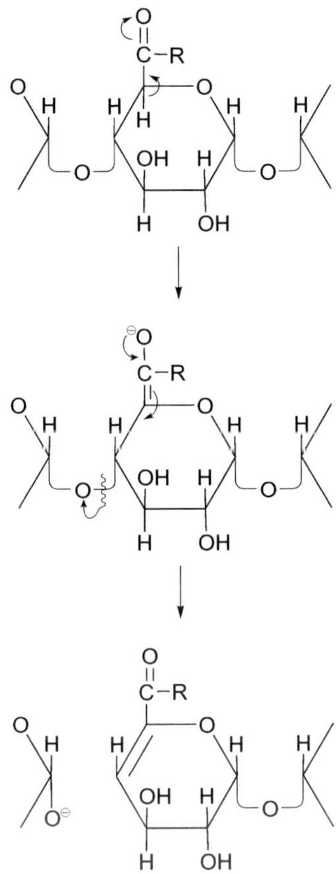

Figure 4.12 Chain cleavage in a (1→4)-linked glucan via beta-elimination after oxidation of C6. C6 may be either an aldehydic group or an ester group (see Fig. 14.1), both of which are electron withdrawing, making the proton on C5 acidic, but not a carboxylate group, which is not electron withdrawing.

sources. Whatever the source of the mechanical force, only the largest polymer molecules are effected and they tend to be cleaved near their main chain midpoints where the stress on the molecules is greatest. Ultimately, smaller polysaccharide molecules with a narrower range of sizes are produced.

Mechanical shear during food processing is also imparted via use of homogenizers, especially high-pressure homogenizers; so polysaccharides undergo some depolymerization when homogenizers are used in product preparation, for example, to create emulsions. The loss of thickening power that accompanies depolymerizations effected by any means can be compensated for by using more of the starch or hydrocolloid in the formulation.

Gamma ray and ultraviolet irradiation and cold plasma treatment, while physical treatments, generate free radicals that result in chemical modifications of polysaccharides.

Multiple modifications

More than one type of modification may be used to produce a single food polysaccharide ingredient. For example, the preparation of a modified food starch may begin with waxy maize starch (a natural genetic modification of corn/maize [Chapter 6]), which may be esterified, etherified, subjected to acid-catalyzed hydrolysis, and/or heat treated. The possibilities are outlined in Fig. 4.13, although no genetically engineered polysaccharide is currently used in any food product.

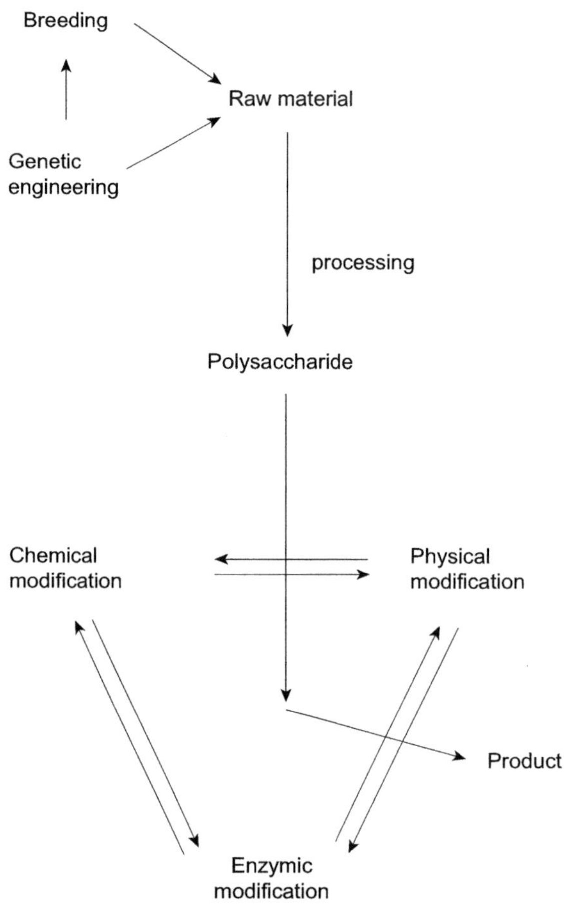

Figure 4.13 Summary of the ways to modify the structures and/or properties of hydrocolloids and starches. Classic breeding and genetic engineering are types of biological modifications.

Additional Resources

BeMiller, J.N., 2001. Classification, structure, and chemistry of polysaccharides in foods. In: Cho, S.S., Dreher, M.L. (Eds.), Handbook of Dietary Fiber. Marcel Dekker, New York, pp. 603–611.

BeMiller, J.N., 2001. Polysaccharides: occurrence and significance in nature. In: Fraser-Reid, B.O., Tatsuta, K., Thiem, J. (Eds.), Glycoscience: Chemistry and Chemical Biology, Vol. III. Springer-Verlag, Heidelberg, pp. 1865–1881.

BeMiller, J.N., 2016. Hydrocolloids/food gums: analytical aspects. In: Eliasson, A.-C. (Ed.), Carbohydrates in Food. CRC Press, Boca Raton, pp. 257–283.

Cui, S., 2005. Structural analysis of polysaccharides, Chap. 3. In: Cui, S.W. (Ed.), Food Carbohydrates: Chemistry, Physical Properties, and Applications. CRC Press, Boca Raton.

Dumitriu, S. (Ed.), 2005. Polysaccharides, second ed. CRC Press, Boca Raton.

Imeson, A. (Ed.), 2009. Food Stablisers, Thickeners and Gelling Agents. Wiley-Blackwell, Oxford.

Steinbüchel, A., Rhee, S.K. (Eds.), 2005. Polysaccharides and polyamides in the food industry. Properties, Production, and Patents. Polysaccharides, Vol. 1. Wiley-VCH Weinheim.

Vandamme, E.J., DeBaets, S., Steinbüchel, A. (Eds.), 2002. Biopolymers. Polysaccharides I and II, Vols. 5 and 6. Wiley-VCH, Weinheim.

Polysaccharides: Properties

5

Chapter Outline

Carbohydrate Chemistry for Food Scientists. https://doi.org/10.1016/B978-0-12-812069-9.00005-4

Key information and skills that can be obtained from study of this chapter will enable you to

1. Describe and discuss the relationship of polysaccharide structure to solubility, solution stability, and viscosity.
2. Give the molecular explanation for the differences in solution viscosity of linear versus branch-on-branch polysaccharides of the same molecular weight (DP) at the same concentration.
3. Discuss the factors involved in, and a general procedure for, making a polysaccharide solution/sol.
4. Define or identify

body	Bingham plastic
bulk	Newtonian flow
glass transition temperature (Tg)	pseudoplastic flow, pseudoplasticity
sol	short flow
"fish eyes"	long flow
viscosity	thixotropic flow, thixotropy
hydrodynamic volume	hysteresis
rheology	overlap concentration (c*)
stress	gel (two definitions)
strain	junction zone
viscoelastic behavior	syneresis
elastic modulus (G′)	hydrogel
storage modulus (G′)	xerogel
viscous modulus (G″)	emulsifier
loss modulus (G″)	emulsion stabilizer
loss tangent (tan δ)	amphoteric
yield stress (yield value)	steric stabilization

5. Describe and give a molecular explanation for the effects of the following on polysaccharide solution viscosity:

chemical structure	pH
molecular shape	ionic strength*
temperature	concentration of solutes*
concentration	

*Explains why it increases solution viscosity of some polysaccharides and decreases solution viscosity of others.

6. Describe the relationship of DP to viscosity of a linear polysaccharide.
7. Describe and compare the two general kinds of shear-thinning rheology.
8. Describe, with examples of applications, how use can be made of pseudoplastic rheology in processing and formulating food products.
9. Discuss the molecular bases for pseudoplasticity and thixotropy.
10. Discuss the effects of concentration and DP on the pseudoplasticity of solutions of a specific polysaccharide.
11. Explain the relationship of the degree of pseudoplasticity to mouthfeel.
12. Describe the molecular basis for a yield value (yield stress).
13. Discuss how polysaccharides that produce solutions with yield values can be applied (using examples).
14. Identify applications of yield value.
15. Describe the concept of c*
16. Describe the molecular bases for gel formation.
17. List the five types of junction zones related to hydrocolloid use in foods.
18. Describe the molecular basis for syneresis and a means to control it (with an explanation).
19. Describe the molecular mechanism of gel textural properties.
20. When given the structure of a polysaccharide, predict its properties. Such properties should include (a) solubility, (b) viscosity, (c) the effect of pH on viscosity and solubility, and/or (d) the effect of added salts on viscosity.

Introduction

The principle biopolymers in the great majority of foods are polysaccharides. (Meat- and dairy-based foods are exceptions.) As you learned in the previous chapter, when water-soluble, nonstarch polysaccharides are used in the formulation of food products, they are referred to as hydrocolloids or food gums. Hydrocolloids and starches are widely available, have a relatively low cost per imparted functionality, and impart a wide range of functionalities. Hydrocolloids also have physiological effects beneficial to human health (Chapter 17). The properties of polysaccharides in food processing and in determining the quality of food products are so overwhelming and so important that a considerable portion of this book is devoted to the subject.

The physical properties of polysaccharides and other polymers are in large part determined by the shapes of their molecules. Their shapes in turn are a function of their chemical structures (the natures of their monomer units, the sequential arrangements of the different monomer units within the polymer chain(s), the connections between the monomer units, any branching) and their environment. The nature of that aqueous environment (amount of moisture present; pH; temperature; presence and concentration of salts; presence and concentration of other solutes, such as sugar) impacts the conformations of the proteins and polysaccharides present. Processing treatments, especially heating and application of mechanical energy, often change the shapes and intermolecular associations of both proteins and polysaccharides. After heating in an aqueous environment, the polymers present are usually in an amorphous, metastable[1] state. Water is the primary plasticizer in food systems and the primary plasticizer for polysaccharides. The amount of water present and the temperature of the system determines polymer mobility, phases present, and product characteristics such as organoleptic qualities (texture) and stability.

Polysaccharide solutions are usually compositionally, spatially, and dynamically complex. They are compositionally complex because polysaccharide structures are both polymolecular and/or polydisperse. They are spatially complex because most polysaccharide molecules can adopt a large number of shapes and because interactions with water and other molecules in their environment influence their conformations/shapes. As a result, for polysaccharides, statistical methods must be used to describe the distributions of conformations. (In contrast, the folding of other biopolymer molecules [protein, DNA, and RNA molecules] is more specific so that their distributions of conformations are much narrower.) The dynamics of polysaccharide molecules (motions of the polymer chain, its segments, and its monomer units) are complex because the same molecular chain may pass in and out of solution and crystalline and amorphous solid regions. In other words, some segments may be in crystalline regions of different sizes, others may be in regions of amorphous solids of different sizes (and perhaps types), and yet others may be in solution. Segments in these regions will have different degrees of mobility, with their motions being a function of temperature and restrictions of motion due to the chemical structure of the molecule and associations with other molecules, including water and other molecules of the same or a different polysaccharide. The motions of polysaccharide chain segments in semisolid food systems cover a wide range of frequencies and are major contributors to the mechanical responses and thermodynamics of the system.

From this brief introduction, it is obvious that much of the chemistry of polysaccharides in foods is physical chemistry. In this chapter, the physicochemical properties of polysaccharides will be presented in a descriptive way.

Even when naturally occurring and not added as ingredients, food polysaccharides are very often determinants of processing conditions and product attributes, including shelf life. Hydrocolloids may be added as ingredients to control or minimize a defect (that is, to improve the appearance, body,[2] rheological properties,[3] bulk,[4] or stability of a food product), to develop new products, or to allow use of new processes. New products might include those that are more convenient, more pleasurable, more nutritious/healthful, or more fun to eat. An example of a new process might be extrusion, during which it is desirable to reduce the energy requirement via the lubricating effect of an added hydrocolloid.

[1] A metastable state is a state in a dynamic system that is not the system's state of least energy.

[2] **Body** refers to the quality of a food or beverage; it is the sum of rheological[3] properties that contribute to mouthfeel.

[3] Rheological properties are those related to types of flow of liquids and deformation of semisolids, such as gels.

[4] **Bulk** refers to the solids content that contributes to the texture and palatability of a food product.

In summary, the shapes of polysaccharide molecules in foods are complex because the polysaccharide molecules themselves are polymolecular and/or polydisperse, because polysaccharide molecules in solutions and gels are in a constant state of dynamic flux, and because segments of molecules may pass in and out of different physical states. The nature of the aqueous environment surrounding polysaccharide molecules also affects the conformations of individual molecules, and the amount of water present and the nature of aqueous phase (especially the temperature) determine the mobility of polymer chain segments. The shapes and mobilities of the polysaccharide molecules determine properties and functionalities of the product.

Water sorption by polysaccharides

Polysaccharides control the structure and mobility of liquid water. Much (but not all) of the usefulness of polysaccharides in processed foods arises from their ability to influence the properties of aqueous systems, which in turn is a function of their interactions with water. Polysaccharides may attract and hold water (act as humectants,[5] reduce evaporation) or thicken or gel aqueous systems. In addition, water is the best plasticizer for polysaccharides, so the amount of water present controls polymer mobility and phase transitions, such as glass transitions (section on Glass transitions of polysaccharides in this chapter). Thus, water and polysaccharides together determine and/or control many of the functional properties of foods such as crispness, other textural characteristics, and shelf life (for example, water migration from the fillings of filled bakery products). Because of their high DP and because they are usually used at concentrations of less than 2%, polysaccharides generally have little effect on the water activity (a_W) of a food product, in contrast to nonpolymeric carbohydrate molecules, which often have a large effect on a_W because they are used at higher concentrations and have lower DPs so that there are larger numbers of molecules present in solution.

Most polysaccharides are composed of glycosyl units that, on average, possess three hydroxyl groups. Thus, polysaccharide molecules are polyols, each hydroxyl group of which can form hydrogen bonds with one or more water molecules. Also, the ring oxygen atom and the glycosidic oxygen atom connecting one sugar ring to another can hydrogen bond with water. Polysaccharides have a strong affinity for water and readily hydrate with available water because each glycosyl unit in the chain has the capacity to hold water molecules. Under conditions of normal humidity, polysaccharides generally equilibrate to contain 8%–12% moisture. In high moisture or fluid water systems, particles of a powdered polysaccharide take up water, swell, and often dissolve (section on Polysaccharide dissolution in this chapter). During drying from

[5] A humectant is a substance that absorbs or retains moisture.

highly moist conditions, polysaccharide molecules lose water relatively slowly, allowing hydrated chains time to replace hydrogen bonding with water to hydrogen bonding with hydroxyl groups of other polysaccharide molecules, creating extensive intermolecular hydrogen bonding that creates hard/gritty particles. Such intermolecular associations are difficult to break, even when the particles are placed in excess water and the mixture is heated. Consequently, drying from a hydrated state is avoided in water-soluble polysaccharide production and handling.

Polysaccharides affect the mobility and structuring of water to a thickness of several molecules around each polymer chain. This water of hydration has been referred to as plasticizing water. While motions of these so-called "bound" water molecules that solvate carbohydrate molecules are retarded, they are still able to exchange freely and rapidly with bulk water molecules, so they are not immobilized. The water of hydration makes up only a small part of the total water in gels and fresh tissue foods. Water in excess of that involved in hydration is entrapped in capillaries and cavities of various sizes in the tissue or food product. Water of hydration that is naturally hydrogen-bonded to and, thus, solvates polysaccharide molecules does not freeze. On freezing of the free/bulk water, soluble components concentrate in the unfrozen water surrounding polysaccharide chains.

Glass transitions of polysaccharides

For amorphous or partially amorphous—partially crystalline polymers, what is known as the **glass transition (Tg)** is an important property. Most often, a partially crystalline polysaccharide is primarily in a glassy, amorphous state, with the glassy amorphous (noncrystalline) domains connecting smaller crystalline domains. (Cellulose [Chapter 8] and chitin [below; Chapter 17] are exceptions because, in them, the crystalline domains are larger than the amorphous domains.) Only small vibrational motions are possible in the glassy amorphous regions. Raising the temperature of a sufficiently plasticized glassy region will bring about a change from the glass to a supercooled viscous liquid (a rubber) (that is, it will effect a softening of the amorphous region) because of the increased motion of the polymer chains. This change in the nature of the system takes place over a temperature range rather than a specific temperature. (The Tg is often reported as a single temperature rather than a temperature range, in which case T_g will refer to either the onset temperature or the midpoint temperature, most often the latter.) Further heating will change the rubbery material to a fluid. Plasticizers decrease the T_g, so a material can go from a glass to a rubber either by increasing the temperature or by increasing the plasticizer content. Therefore, water, being the best plasticizer for carbohydrates, strongly influences the temperature at which T_g occurs. Compounds with hydroxyl groups such as glycerol, sorbitol, and propylene glycol may also be plasticizers, but these same substances often exhibit an antiplasticization effect when moisture is present because they are less effective plasticizers than water and compete with the polysaccharide molecules for hydrogen bonding with water molecules.

The behaviors of food products during processing, storage, and consumption are a function of both their physical state(s) and their physicochemical properties. The T_g is often the most important physicochemical property in this regard, and in addition to the water content, the kind and amount of carbohydrate present in the food is most often the determinant of T_g. When dry and low-moisture foods are heated, so that their temperatures pass through the Tg range (which is a function of the moisture content of the food), virtually all mechanical properties will change. An example of the relation of T_g to product properties is the difference between the crispness of crackers or cookies in low- and high-humidity conditions. When subjected to high humidity, the product may pick up enough moisture to lower the T_g below room temperature so that, at room temperature, the product will be in a rubbery state, rather than a brittle glass (that is, it will lose its crispness). Cooling a substance that is at a temperature above its T_g (a melt) or cooling a hot concentrated solution at a rate that is more rapid than the rate of crystallization of the key molecules will result in the formation of amorphous glassy (rather than crystalline) materials. (Hard candies are glasses formed by cooling of a hot sugar solution.) Crystallization is usually a rather slow process because molecular mobility is required for molecules to orient themselves in crystal lattices; so if the system reaches a highly viscous state in which molecular mobility is inhibited before extensive formation of crystal nuclei and crystal growth occurs, it will be "frozen" in that amorphous (glassy) state. This is another way of saying that most food products are in a nonequilibrium state and are prevented from attaining equilibrium by the greatly reduced mobility of their molecules as a result of the extremely high viscosities of their predominant glassy regions.

The list of processing operations and product characteristics in which T_g plays the major role is lengthy. A few additional examples follow: drying and handling powders (caking, stickiness, and flow of such things as hydrocolloid powders); crystallization (growth of ice and/or sugar [lactose or sucrose] crystals in ice creams, other frozen products, and confections, which would produce sandiness/grittiness); sugar bloom in chocolate; gelatinization of starch (Chapter 6); producing crisp snacks such as corn curls and breakfast cereals in which rapid cooling occurs following extrusion; and changes in other attributes of quality such as hygroscopicity,[6] changes during aging, crispness, and hardness.

Polysaccharide solubility

Most, if not all, polysaccharides (except those with very bushlike, branch-on-branch structures) and certainly most, if not all, homoglycans with an essentially linear structure exist in some sort of helical shape. Many polysaccharides are totally amorphous, but segments of molecules of certain linear homoglycans that have uniform, flat, ribbon-like structures (like cellulose [Chapter 8]) and certain single-, and double-, or

[6] Hygroscopicity is the sorption of water from the atmosphere that results in a change in physical characteristics.

triple-helical structures of polysaccharide molecules with regular, repeating unit structures (like κ-carrageenan [Chapter 13] and gellan [Chapter 12]) associate in an ordered manner that allows hydrogen bonding and formation of crystallites.

Cellulose (a linear homoglycan consisting of a chain of β-D-glucopyranosyl units), mannan (a linear homoglycan of β-D-mannopyranosyl units), and chitin (a linear homoglycan of 2-acetamido-2-deoxy-β-D-glucopyranosyl units) are present in the organisms that make them as highly ordered, highly insoluble substances. They contain crystallites intermingled with amorphous regions. In the crystallites, the linear, homogeneous chains are packed in microcrystalline domains of varying lengths, then pass into less-ordered amorphous domains, then again enter into and become part of another group of crystallized chain segments, and so on. Crystalline regions of this type have no clear surfaces (as opposed to single crystals). It is these crystallites of linear chains that give the strength and insolubility to the trunks and branches of trees and shrubs, crab and lobster shells, and insect exoskeletons. They also make the same structures quite resistant to breakdown (as compared to other biomaterials) because their crystalline regions, which predominate over the amorphous regions, are nearly inaccessible to penetration by enzymes. These highly ordered polysaccharides with orientation and crystallinity are the exception, rather than the rule. Most polysaccharides are not sufficiently crystalline to make them insoluble in water but readily hydrate and dissolve in water, even though they may contain segments that have ordered conformations that facilitate cooperative interactions between them.

The crystalline domains of some polysaccharides will melt if sufficient plasticizer is present. (Water is always the most effective plasticizer for polysaccharides.) Good

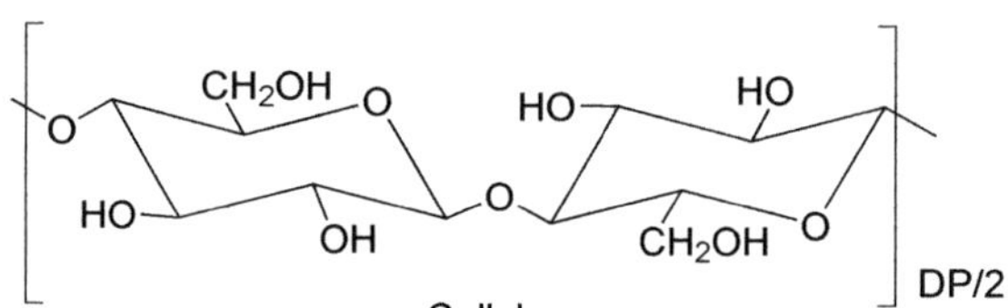

Cellulose

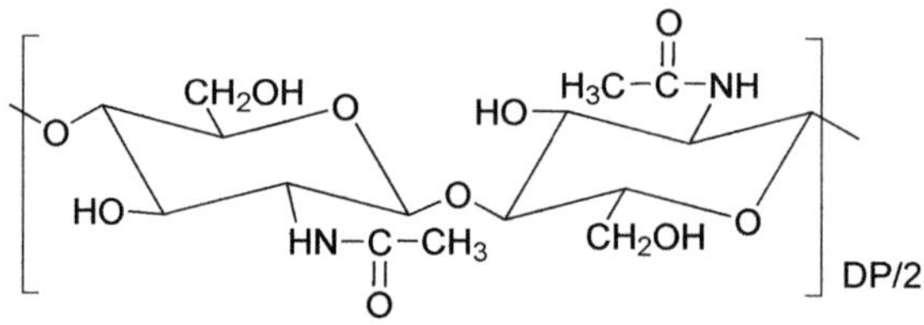

Chitin

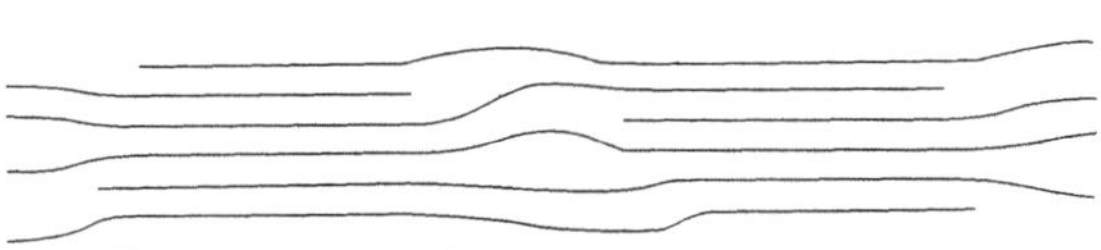

Representation of a partially crystalline, partially amorphous segment of chitin and/or cellulose

examples of crystallite melting are the gelatinization of starch (Chapter 6) and the melting of gels (section on Gels in this chapter). Highly crystalline polysaccharides, such as cellulose and chitin, will decompose before melting, even if water molecules are present.

Linear diheteroglycans containing nonuniform blocks of glycosyl units and branched glycans cannot form crystalline regions because chain segments cannot be packed together over lengths necessary to form sufficient numbers of intermolecular hydrogen bonds to hold them together. Such chains have some degree of solubility. The greater the degree of chain irregularity of a polysaccharide, the more soluble it is.

Polysaccharide dissolution

Most polysaccharide preparations do not form true solutions like small molecules do. Rather they form molecular dispersions. The term hydrocolloid indicates that, in many cases, each hydrated molecule is large enough to be a colloidal particle. (By definition, a colloidal particle has at least one dimension in the 1–500 nm range, and most hydrated hydrocolloid molecules meet this definition.) Because polysaccharide solutions are not true solutions, they are sometimes called **sols.** (In this book, what is formed when polysaccharides are well dispersed in water are referred to as polysaccharide [or hydrocolloid or starch] solutions, rather than polysaccharide dispersions or sols.)

A crystalline substance such as sugar dissolves by dissolution of molecules from the surfaces of the crystals so that the crystals become smaller and smaller and eventually disappear. In contrast, hydrocolloids "dissolve" by particle swelling, followed by some degree of swollen particle disruption, until ideally a dispersion of individual hydrated molecules is formed (Fig. 5.1). The rate of particle hydration can be controlled. The

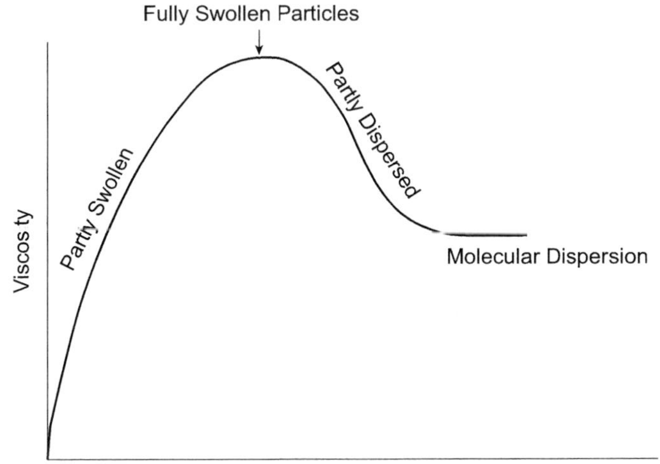

Figure 5.1 Idealized curve of the development of viscosity with time as polysaccharide particles swell and disintegrate in a stirred (sheared) aqueous system.

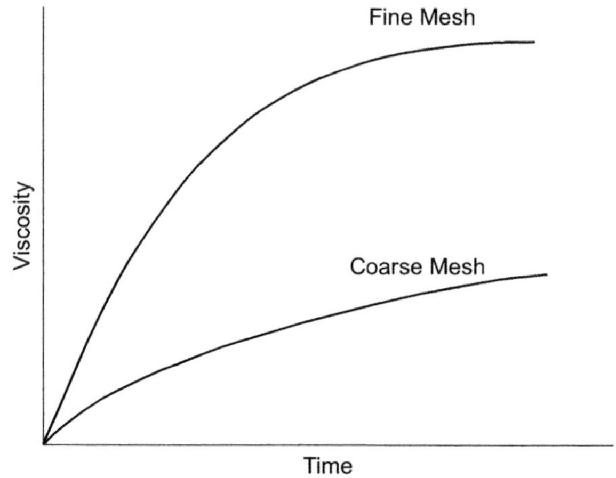

Figure 5.2 The influence of particle size on the development of viscosity with time/degree of hydration.

degree and rate of dissolution of a hydrocolloid is influenced by its nature (chemical nature, viscosity grade, particle size (Fig. 5.2), any surface treatment given particles, the type and amount of counter ions present [if an anionic hydrocolloid]), the nature of the aqueous system (temperature, pH, presence of other solutes), and the means of dispersion.

To prepare good molecular dispersions ("solutions") of hydrocolloids, the polysaccharide particles must be well dispersed in the aqueous medium before substantial hydration and swelling of the particles begins. It is almost always best to dissolve the hydrocolloid in water first and then add other ingredients. A fine mesh powder is likely to lead to clumping and has a greater requirement for good dispersion before swelling. In other words, it is more difficult to make a smooth solution from fine hydrocolloid particles than coarser particles, so for dispersion of a dry mix containing a hydrocolloid in the home kitchen (where special equipment is not available), it is best to use coarse particles.

Solubilities of polysaccharides are decreased by the presence of substances that compete with them for water molecules. In general, ionic hydrocolloid molecules are more soluble, dissolve faster, and hold more molecules of hydration than do neutral hydrocolloid molecules of the same size and shape, but the solubility of ionic hydrocolloids is more depressed by salts than is the solubility of neutral hydrocolloids. It is unlikely that full viscosity will be obtained if a hydrocolloid, especially an ionic one, is added to a solution of sugar, salt, or another solute (Fig. 5.3); so the general recommendation is that, whenever possible, a hydrocolloid should be dispersed and dissolved in water before substantial amounts of salts, sugars, and other strongly hydrophilic components (that will compete with the hydrocolloid for water molecules and will restrict hydration of the hydrocolloid molecules) inhibit formation of a molecular dispersion and prevent attainment of the hydrocolloid's full functionality.

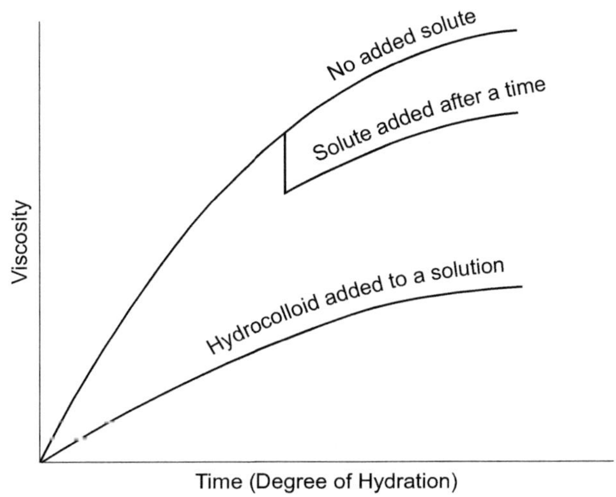

Figure 5.3 The effect of solutes added before or after the polysaccharide/hydrocolloid is added to water on the development of solution viscosity as a function of time (compare Fig. 5.2).

Methods for dissolving polysaccharides

Generally, to realize their functionality, water-soluble polysaccharides must be thoroughly hydrated, but dissolution in an aqueous system is often a problem encountered when using hydrocolloids. Commercial hydrocolloids are powders of different particle sizes and some (such as carrageenans [Chapter 13] and pectins [Chapter 15]) are often sold as mixtures of the polysaccharide and finely ground sucrose to make solution preparation easier and to standardize them with respect to viscosity or gel strength.

For homogeneous dispersion, polysaccharides must be added to water under controlled conditions because all hydrocolloids will form lumps if added to water that is not thoroughly and rapidly agitated and if the powder is not added slowly to ensure rapid mixing. In most cases, if a hydrocolloid powder is added to only slightly stirred water, some particles may dissolve, but the surfaces of many particles and agglomerates of individual particles will hydrate quickly, producing a sticky, gelatinous coating. This coating slows the rate of water diffusion through the surface layer, leaving a dry interior. The agglomerates are called **"fish eyes"** and are always undesirable. Fish eyes (agglomerates) form under insufficient agitation when partially hydrated particles collide and adhere to each other via their sticky, gel-coated surfaces before complete molecular hydration and dissolution occurs.

To prevent this problem and effect good dissolution of a hydrocolloid, one of several different procedures may be employed. The most important is use of an efficient mixer. A hydrocolloid powder may be sifted slowly into the vortex of rapidly stirred water, where high shear forces ensure that particles become dispersed before significant hydration occurs. The shear forces also tear away partially hydrated molecules and prevent the formation of a gelatinous outer layer. So high-shear mixing equipment that provides rapid dispersion and mixing of particles in water is essential to making a uniform hydrocolloid solution. A high-speed propeller mixer can achieve these conditions if the hydrocolloid powder is added slowly to vigorously stirred water. Otherwise, various types of commercial equipment are designed for this purpose. One such is a powder funnel (called an eductor) that allows a fine stream of powder to fall into a jet of turbulent water, giving very rapid mixing.

Alternatively, the hydrocolloid may be mixed thoroughly with a portion of another dry ingredient (such as sucrose). This method enhances the probability that hydrocolloid particles are separated from each other before significant hydration occurs. Generally, the ratio of the other ingredient to the hydrocolloid is at least 5:1 w/w. (The method, while successful because it ensures that the hydrocolloid particles remain separated before they hydrate, is contrary to the admonition to always dissolve the hydrocolloid first to achieve maximum functionality.) In addition to sugar and other solid ingredients, liquids such as glycerol, propylene glycol, vegetable oil, or a glucose syrup (Chapter 7) may be used. In this procedure, the hydrocolloid powder is thoroughly dispersed in the liquid using a liquid to hydrocolloid ratio of about 7:1 w/w; then the slurry is added to rapidly stirred water. Specially prepared slowly hydrating forms of gums with treated particle surfaces and agglomerated gums are also employed for easier dispersibility.

Even under the best conditions, absolute molecular separation of hydrocolloid molecules is not readily attained; and some continued hydration of the molecular agglomerates with consequent development of higher viscosity may occur over a period of several minutes to several hours. Raising the temperature after the dispersion is made usually aids in dissolution. For some hydrocolloids, raising the temperature above room temperature is essential to complete dissolution; examples of necessary elevated temperatures are ι-carrageenan (40–45°C [105–115°F]), κ-carrageenan (55–60°C [130–140°F]) (Chapter 13), and locust bean gum (85–95°C [185–205°F]) (Chapter 9). (See, however, methylcelluloses and hydroxypropylmethylcelluloses in Chapter 8.) High-shear conditions and, in some cases, elevated temperatures are necessary to achieve a molecular dispersion because the forces (largely hydrogen bonds) that bind together chain segments must be overcome; but the temperature should not be raised until the hydrocolloid particles are dispersed (rather than the powder added to hot water) because in the latter case the particles would hydrate before being completely dispersed and form fish eyes.

Hydrocolloids such as λ-carrageenan (Chapter 13), guar gum (Chapter 9), sodium alginate (Chapter 14), some pectins (Chapter 15), and xanthan (Chapter 11) will dissolve in water at temperatures of 30°C (85°F) or less (given sufficient time). Starches (other than special types called pregelatinized starch [Chapter 7]) must be cooked for dissolution.

Properties of polysaccharide solutions

Polysaccharides (hydrocolloids and starches) are used primarily to thicken and/or gel aqueous solutions and otherwise to modify and/or control the flow properties and textures of liquid food and beverage products and the deformation properties of semisolid foods. They are generally used in food products at concentrations of 0.1%–2.0%, indicating their great ability to produce viscosity and to form gels.

Viscosity is the tendency of a fluid to resist flow (section on Rheology of polysaccharide solutions in this chapter). The viscosity of a polymer solution is a function of both the sizes and the basic structures (linear or bushlike, for example) of the polymer molecules and the conformations they adopt in the solvent system. In foods and beverages, the solvent system is an aqueous solution of other solutes. As already pointed out, most polysaccharide molecules can adopt a very large number of conformations/shapes, and the conformations attainable are a function of the chemical structure of the polysaccharide (namely, the natures of the monosaccharide units [the specific sugar, including whether it is charged, and the ring type], the natures of the glycosidic linkages (Fig. 5.4) and the environment surrounding the polysaccharide. The nature of the glycosidic linkages plays the largest role because the greater the freedom of motion around each glycosidic bond, the greater the number of conformations available to the segment containing that linkage and the more the polysaccharide will have the characteristics of a completely flexible chain. Greater chain flexibilities lead to disordered (random coil) states; however, most polysaccharides form somewhat stiff coils, the specific nature of which is a function of its chemical structure. An anionic polysaccharide with otherwise flexible glycosidic linkages will assume a more extended conformation because any chain folding would result in bringing like negative charges closer to each other, and if they get too close to each other, they repel each other.

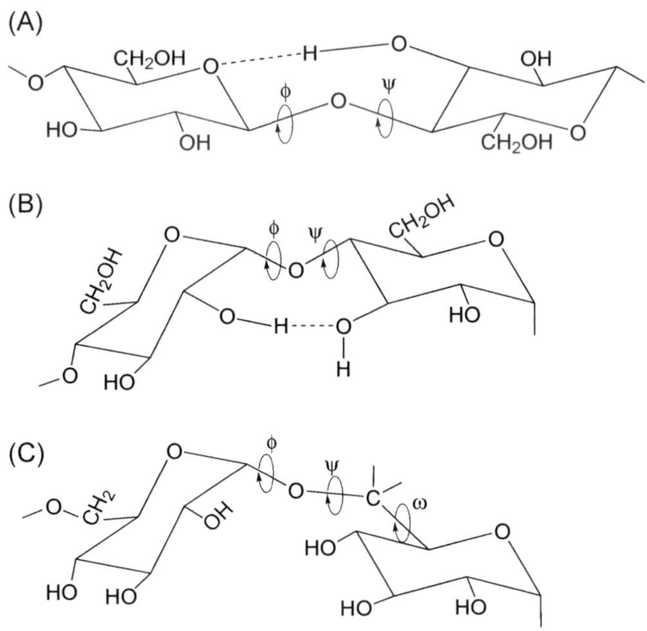

Figure 5.4 Rotation around the two bonds of the glycosidic linkage (Φ, ψ) of cellulose (A), amylose (B), and a (1 → 6)-linked α-glucan (dextran, C) and the C5–C6 bond (ω) of the latter and the hydrogen bonding stabilizing the conformations of cellulose and amylose. The structures (A, B, C) are ordered in terms of deceasing flexibility, with structure C being much more flexible than A or B because of the extra hinge point in the molecule.

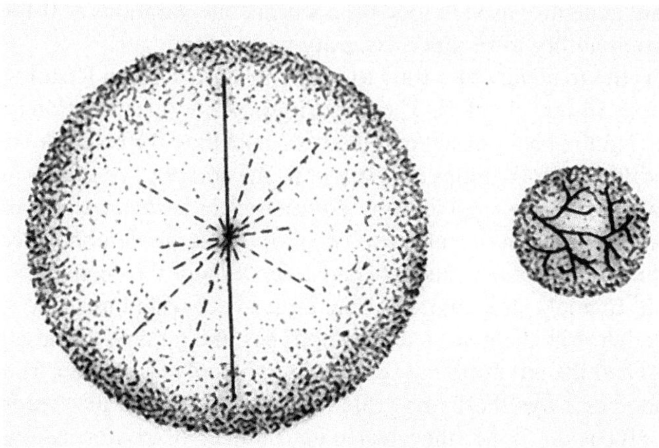

Figure 5.5 Relative volumes occupied by a linear polysaccharide and a highly branched polysaccharide of the same molecular weight (assuming that the linear molecule has the characteristics of a rigid rod, although it probably has the shape of a random coil to some degree). *Figure reproduced with permission from* Whistler R.L., Introduction to industrial gums, In: Whistler R.L. and BeMiller, J.N. (Eds.) Industrial Gums, third ed., 1993, p. 9.

The dynamic motion of linear polysaccharide molecules in solution results in their sweeping out a large space called the **hydrodynamic volume.**[7] When molecules collide with one another, energy is consumed and friction is created; the result is viscosity. The larger the hydrodynamic volume occupied by polysaccharide molecules in solution, the greater is the viscosity they generate (at equal concentrations). The volume swept out by the dynamic motions of a polysaccharide molecule in solution is a function of both its size and shape. A linear polysaccharide will usually occupy a greater theoretical volume than a polysaccharide with a branch-on-branch structure of even greater DP, depending on the folding of the linear polysaccharide (Fig. 5.5). But even considering two polysaccharides that behave as linear polymers (Table 5.1), their shapes and their average DPs must be considered in evaluating their ability to thicken an aqueous system because a highly flexible molecule in a random-coil shape will occupy a smaller domain than will a stiffer, more extended molecule of equal DP. Thus, polysaccharides capable of forming helices (single, double, or triple) will produce more viscous solutions at lower concentrations than will those that cannot form such ordered, hydrogen-bonded, stiffened structures.

[7] The **hydrodynamic volume** refers to a model in which the actual polysaccharide molecule is replaced by the volume of a sphere whose radius corresponds to the root mean square radius of gyration of the polymer chain in solution. The concept of occupied space is related to both viscosity and separation of polysaccharide molecules by size-exclusion chromatography (Chapter 4). Hydrodynamic volume is related to the DP of the polymer, its nature (linear vs. bushlike; neutral vs. anionic, etc.), polymer–solvent interactions, and the temperature of the solution.

Table 5.1 Classification of some food polysaccharides according to their solution properties

Polysaccharides that behave as linear polymers	
Agarose	Curdlan
Alginates	Dextran
Amylose	Galactomannans
Arabinoxylans	Gellans
Beta glucans	Pectins
Carrageenans	Pullulan
Cellulose derivatives	Xanthan
Chitosan	
Polysaccharides that behave as bushlike polymers	
Gum arabic	
Psyllium seed gum	
Polysaccharide with intermediate properties	
Amylopectin (waxy maize starch)	

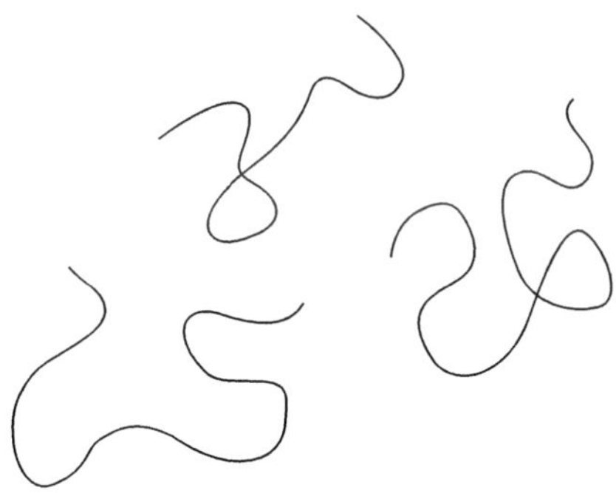

Random coils

Random coils

Linear polysaccharides produce highly viscous solutions, even at low concentrations. A 1% aqueous solution of a polysaccharide can easily have a viscosity of 10,000 mPa·s (centipoise, cP) (section on Rheology of polysaccharide solutions in this chapter). Viscosity is a function of both the DP and the extension and rigidity (that is, the shape) of the solvated polymer chain. Rigidity is influenced by whether the glycosidic bonds are (1 → 3), (1 → 4), or (1 → 6), for example, and their anomeric configuration (that is, whether the linkages are α or β). Chain extension is also dependent on the extent of chain hydration and any charge on the monosaccharide units, which may be a function of the amount and type of derivatization. Linear polysaccharide chains bearing a charge (in hydrocolloids used in foods, always a negative charge is imparted by ionized carboxylate or sulfate half-ester groups) assume a more extended configuration due to repulsion of the like charges,[8] increasing the end-to-end chain length and, thus, increasing the volume swept out by the polymer. Therefore, these polymers tend to produce solutions of high viscosity, but the viscosities of solutions of anionic polysaccharides with carboxylate groups varies as the pH approaches the pk_a values of the carboxyl groups, which is generally in the pH 3.3–3.8 range (section on Effect of pH in this chapter). (Those with sulfate half-ester groups are anionic at all pH values.)

As already mentioned, the degree of chain extension and/or coiling of polysaccharide molecules is influenced in large part by the locations and configurations of their glycosidic bonds. Although a large number of conformations around a particular glycosidic bond may be possible, certain specific conformations within a range of conformations are preferred for thermodynamic reasons. The range of preferred conformations may be narrow or broad. Cellulose (Chapter 8), because of its equatorial–equatorial glycosidic bonds, has its (1 → 4)-linked β-D-glucopyranosyl units positioned so that its adjacent rings can form hydrogen bonds between the ring oxygen atom of one glucosyl unit and the hydrogen atom of the C3 hydroxyl group of the preceding ring (Fig. 5.4A). These intramolecular hydrogen bonds hinder free rotation of the rings about their connecting glycosidic bonds and result in a stiffening of the chain. The flat, ribbon-like character of the entire molecule allows adjacent cellulose chains to fit closely together into ordered crystalline domains. In comparison, a chain of (1 → 4)-linked α-D-glucopyranosyl units (as in the amylose component of starch [Chapter 6; Fig. 5.4B and 5.6]) does not form a ribbon-like structure. Rather the axial–equatorial linkages cause the molecules to coil (Fig. 6.5).

A solvated linear molecule without strong intra- and intermolecular hydrogen bonding naturally forms a random structure of, on average, a football-like shape (that is, it adopts a random coil shape). A coiled configuration is easily adopted by (1 → 6)-linked polysaccharides because the extra carbon atom (methylene group) between sugar rings moves them too far from each other to form hydrogen bonds between successive units that could restrict ring rotation. The C–C bond also provides

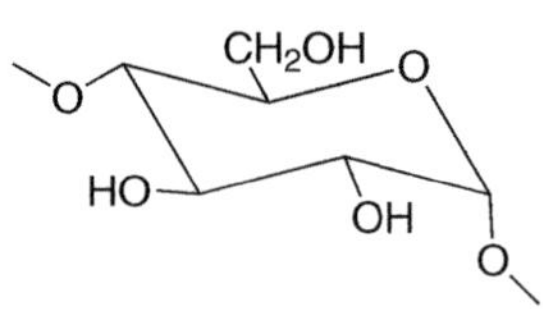

Figure 5.6 An α-D-glucopyranosyl repeating unit of amylose showing the axial glycosidic bond at the anomeric center (C1) and the equatorial bond to the oxygen atom at C4 (O4).

[8] As already mentioned, any folding of the molecule brings the charges closer together, and if they are close enough to each other, they repel each other.

an extra degree of flexibility (Fig. 5.4C). Derivatization of chain units (either naturally or industrially) and chain branching, because each introduces protrusions that hinder association of chain segments, increases polysaccharide solution stability. Formation of single-, double-, or triple-helical structures stiffens polymer chains (that is, makes them less flexible and more extended).

The nature of the aqueous system in which the polysaccharide is hydrated or dissolved also affects its hydrodynamic volume and solution viscosity. Linear polysaccharide molecules in a good solvent (usually pure water) will tend to be extended to maximize polymer—water contacts. The same polysaccharide molecules in a poorer solvent (usually an aqueous system containing other dissolved substances) will have a greater number of polymer—polymer contacts. In this case, the molecules may either coil back on themselves, creating intramolecular hydrogen bonds (decreasing polymer-solvent contacts and viscosity), or aggregate via intermolecular bonding (again decreasing polymer—solvent contacts, but increasing viscosity). It is for these same reasons that polysaccharide particles generally do not fully hydrate when placed in solutions of sugars, salt, and other water-soluble substances (Fig. 5.3).

A highly branched polysaccharide of the same DP as a linear polysaccharide will sweep out far less space (Fig. 5.5). Thus, highly branched molecules will collide less frequently and will produce much less viscosity than will linear molecules of the same DP (at the same concentration). This also implies that a highly branched polysaccharide must be significantly larger than a linear polysaccharide to produce the same viscosity at the same concentration.

Most polysaccharides that are useful in formulating food products have properties that are intermediate between those of a random coil and those of a single-, double-, or triple-helical structure (that is, they have the properties of a segmented, wormlike coil which are intermediate between those of a random coil and those of a rodlike structure). A very general categorization of some food polysaccharides as to whether they behave mostly as linear polymers (even though they may have short branches [Table 4.1]) or as branch-on-branch polymers is given in Table 5.1. However, whether linear molecules of a given polysaccharide behave more as extended polymer chains, as coiled polymer chains, as rigid rods, as flexible random coils, or as intermediate structures often cannot be stated in absolute terms because the nature of the solvent system has a strong effect on their conformation. For example, an anionic polysaccharide may behave more as a coiled polymer without intermolecular associations at pH 7, where it contains charged carboxylate groups (CO_2^-), but behave as a self-associating polymer (section on Molecular Associations of Polysaccharides) that forms rigid supramolecular structures (also known as supermolecular structures), when the pH is lowered to a value near the pK_a of the carboxyl groups (pH 3.3—3.8), where about half of the ionic carboxylate groups are changed into uncharged carboxyl ($-CO_2H$) groups (Chapters 14 and 15). Also, higher ionic strengths increase the flexibility of an anionic polysaccharide, as monovalent cations shield the anionic charges, decrease electrostatic repulsion, and allow the chains to be more flexible. Smaller (low-DP) molecules are more likely to form intermolecular associations of low flexibility. Specific conformations of different hydrocolloid molecules are presented with descriptions of their structures in subsequent chapters.

Summary of structure property relationships

Polysaccharide conformations are functions of both the structure of the polysaccharide and the nature of the solvent system (pH, ionic strength, type of cations, temperature). Ionic polysaccharides are more sensitive to electrolytes (salts) and pH than are neutral polysaccharides. The conformation of a polysaccharide molecule is dynamic, and therefore, the result of any measurement will be that of an average structure under the given conditions at that point in time. Linear polysaccharides with "bumps" along the chain, in the form of either mono- or oligosaccharide side groups or noncarbohydrate ether or ester groups, while technically branched, behave as linear polymers (Table 5.1).

Property–use relationships

The ability of polysaccharides to produce high viscosity at low concentrations is a major property of polysaccharides that gives them valuable and widespread use. The relation of chain length to viscosity development was explained earlier. A single cleavage of a glycosidic bond at the center of the chain by hydrolysis, oxidation, or mechanical action (shear) will produce two polymer molecules of one-half the original DP and much lower viscosity-producing potential. If both the starting molecule and each of the two molecules produced by cleavage of the starting molecule in the middle were rigid rods, each half of the starting molecule will sweep out a sphere of, but 1/8th the volume of, the original molecule (because the radius of the volume it sweeps out upon gyration is one-half that of the original molecule), resulting in a viscosity of only 1/4th the original value.[9] As a consequence, care must be taken to prevent chain cleavage during preparation or use of polysaccharides if their ability to produce high viscosity at low concentration is to be maintained. Most hydrocolloids are available in a wide range of viscosity grades produced by deliberate depolymerization (later).

Both high- and low-molecular weight carbohydrates (with the exception of starches) are generally effective in protecting food products stored at freezer temperatures (typically −18°C [0°F]) from destructive changes in texture and structure with various degrees of effectiveness. The improvement in product quality and storage stability is a result of controlling both the amount (particularly in the case of low-molecular weight carbohydrates) and the structural state (particularly in the case of polysaccharides) of the freeze-concentrated, amorphous matrix surrounding ice crystals.

Using starch as an example of a polysaccharide, when a starch gel is frozen, a two-phase system of crystalline water (ice) and a glass consisting of 76% starch molecules and 24% nonfreezable water is formed. (A sucrose glass will contain 83% sucrose, and a sorbitol glass will contain 88% sorbitol.) As in the case of solutions of low-molecular

[9] $r = 1/2$. $r^3 = 1/8$. Therefore, each half would contribute 1/8th the original viscosity, and together they would contribute 1/4th the original viscosity.

weight carbohydrates, the nonfreezable water is in a highly concentrated carbohydrate solution in which the mobility of the water molecules is restricted by the extremely high viscosity. However, while most polysaccharides provide some degree of cryostabilization (that is, protection against structural damage caused by freezing) by producing this matrix which limits molecular mobility, there is evidence that some provide cryrostabilization by restricting ice crystal growth through adsorption onto ice crystal surfaces. Some other polysaccharides may even be ice nucleators.

Polysaccharides do not significantly increase the osmotic pressure or depress the freezing point of water because they are large, high-molecular weight molecules used at low concentrations, and osmotic pressure and freezing point depression are colligative properties.[10]

Molecular associations of polysaccharides

Many linear glycans, when dissolved in water by heating or by aid of solubilizing agents, such as a base (which is then neutralized), form unstable molecular dispersions that precipitate or gel (section on Gels in this chapter). This occurs as segments of the long molecules collide and form intermolecular hydrogen bonds over the distance of a few units. Initial short alignments may then extend in a zipper-like fashion to strengthen intermolecular associations. (Polysaccharide molecules associate with one another whenever possible because they generally prefer polysaccharide—polysaccharide contacts over polysaccharide—water contacts.) Segments of other chains colliding with this organized nucleus bind to it, increasing the size of the ordered, crystalline phase. Linear molecules continue to bind, producing particles that may reach a size where gravitational forces effect precipitation. For example, starch amylose, when dissolved in water with the aid of heat and then cooled to below 65°C (150°F), undergoes molecular aggregation and precipitates, a process called retrogradation (Chapter 6). During cooling of bread and other baked products, amylose molecules associate to produce firming. Over a longer storage time, the branches of amylopectin associate, resulting in staling (Chapter 6).

Formation of associative regions between polysaccharide chains is a crystallization process. In general, all linear, neutral homoglycans have an inherent tendency to associate and partially crystallize. However, if linear glycans are derivatized (for example, cellulose ethers [Chapter 8]) or occur naturally derivatized (for example, guar gum [Chapter 9], with single-unit glycosyl branches along a backbone chain, and xanthan [Chapter 11], with trisaccharide units on a cellulose chain), segments are prevented from association and stable solutions result.

Stable solutions are also formed if the linear chains contain charged groups. In this case, repulsion of like charges prevents segments from approaching each other. (As already mentioned, charge repulsion also results in extension of chains, which increases the end-to-end chain length, the volume swept out by the polysaccharide,

[10] Colligative properties are properties that are a function of the number of molecules in solution, irrespective of their nature.

and the viscosity it produces.) Such highly viscous, stable solutions are seen with sodium alginate (Chapter 14), in which each glycosyl unit is a uronic acid unit (that is, has a carboxylate group), in xanthan (Chapter 11), where one out of five glycosyl units is a uronic acid unit, and in CMC (Chapter 8), which has a protruding group that contains an anionic carboxylate group, thus preventing chain association in two ways. But, if the pH of an alginate solution is lowered to 3, where ionization of carboxylic acid groups is somewhat repressed (that is, more than half of the negatively charged $-COO^-$ groups are converted to neutral $-COOH$ groups) because the pk_a values of the two constituent monomers are 3.38 and 3.65, the resulting less ionic molecules can associate and precipitate or form a gel as expected for linear, neutral glycans (section on Effect of pH in this chapter).

Carrageenans are mixtures of linear chains that have a negative charge due to numerous ionized sulfate half-ester groups along the chain (Chapter 13). These molecules do not precipitate at low pH because the sulfate group remains ionized at all practical pH values.

As would be expected, multivalent cations such as calcium ions can interact with anionic carboxylate groups on two different chains and bind them together. More will be discussed about this phenomenon in the section on gels in this chapter, but cross-linking with di- and multivalent cations can also influence viscosity.

Summary of molecular associations

If given the opportunity, polysaccharide molecules will associate with one another. Identical segments of polysaccharide molecules usually bond together quite well. Oriented linear molecules will form ordered two- and three-dimensional structures (often in the form of crystallites), but they generally must be neutral (either naturally or by neutralization or shielding of the charge) to do this. Non-uniformity along the polymer chains decreases intermolecular interactions.

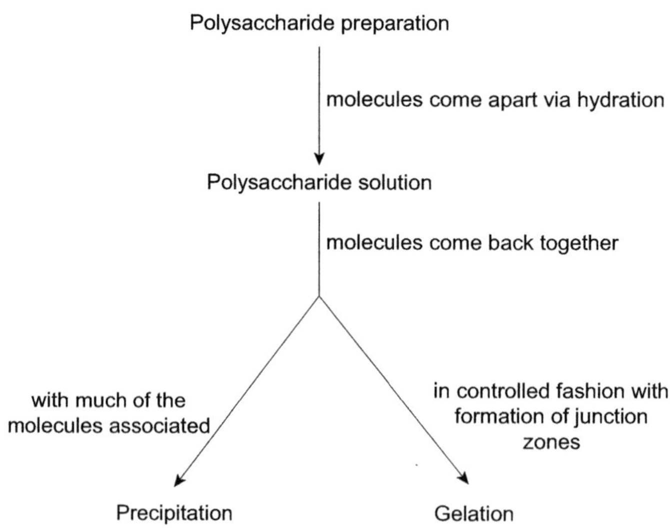

The viscosities of solutions of polysaccharides are functions of not only the shapes of the molecules in solution (which can be influenced by any anionic charges on the molecules) but also on any molecular associations, which in turn may be a function of the shapes of the molecules, the presence and concentrations of ions and other solutes, and in the case of hydrocolloids containing carboxylate groups, solution pH. Of course, the concentration and DP distribution of the polysaccharide and the temperature of the solution also determine its viscosity.

Characteristics of polysaccharide solutions

To repeat, polysaccharides modify the structure and mobility of liquid water, and the ability of polysaccharides to thicken and/or gel aqueous systems and otherwise to modify and/or control their rheological (below) properties is the basis of most of their use in food systems (Table 5.2). They also have other useful properties, as will be discussed in context with specific polysaccharides in Chapters 6–16.

Table 5.2 General classification of some major commercial polysaccharides according to their ability to modify the properties of aqueous systems

Hydrocolloids that form gels	Algins
	Carrageenans
	Curdlan
	Gellans
	Gum arabic
	Hydroxypropylmethylcelluloses
	Methylcelluloses
	Pectins
	Starches, including modified starches
	Locust bean gum + xanthan
Hydrocolloids used as thickeners and stabilizers[a]	Hydroxypropyl alginates
	Carboxymethylcelluloses (certain types)
	Hydroxypropylmethylcelluloses
	Modified starches
	Xanthan
Hydrocolloids used primarily as thickeners	Carboxymethylcelluloses (certain types)
	Guar gum

[a]Because their solutions exhibit yield stress values and/or Newtonian behavior at low shear rates.

Rheology is the science of flow of liquids and deformation of gels and other semisolids. Rheological properties are the mechanical properties of materials that flow (liquids) and deform (solids) and are expressed in terms of the effects of stress, strain, and time. **Stress** is the intensity of the force acting on a body and is expressed in units of force per unit area. **Strain** is the change in size or shape of a body in response to the applied force; it is a nondimensional parameter representing movement that is expressed either as a ratio or as the percentage change in relation to the original size or shape. Of course, time is a component of the rate of strain or stress, the behavior of a material under constant stress or strain, and the rate of return to the original state when the stress is removed.

The rheological properties of foods are related to both their processibility and their mouthfeel. Most foods are semisolids that contain some characteristics of solids and some characteristics of liquids. They are described as being **viscoelastic** because they behave as if they were both a viscous fluid and an elastic solid. When a force is applied to an ideal elastic solid, the shape changes, with the amount of deformation being proportional to the applied force. When the force is removed, the material returns to its original shape. (Think of a rubber band.) In other words, in an ideal elastic material, deformation occurs the moment stress is applied, is directly proportional to the stress, and disappears almost instantly and completely when the stress is removed. In an ideal viscous material, deformation occurs the moment stress is applied, but is proportional to the rate of strain, and is not recovered when the stress is removed. In other words, when a force is applied to a liquid, it flows, with the rate of flow being proportional to the applied force. Polysaccharide solutions and gels have both solid-like and liquid-like properties (that is, they are viscoelastic).

A common way of characterizing a liquid is by measurement of its viscosity. **Viscosity** is the internal friction of a fluid (that is, as already mentioned, it is the fluid's tendency to resist an applied force [that is, to resist flow]). It is calculated by dividing the applied shear stress by the measured shear rate (that is, the rate of change of strain). Shear stress is the applied force per unit area produced by such actions as pouring, mixing, pumping, chewing, or swallowing. When stress is applied, a shearing strain (movement) results. Shear rate is a value expressing how fast the liquid flows.

$$\text{Apparent viscosity } (\eta) = \frac{\text{unit force}}{\text{unit flow}} = \frac{\text{shear stress}}{\text{shear rate}} = \frac{\text{applied force}}{\text{flow speed}}$$
$$= \frac{\text{stress}\left(\text{dynes}/\text{cm}^2\right)}{\text{unit flow}}$$

The basic SI unit of solution viscosity is millipascals (mPa·s) × seconds mPa·s). (1 mPa·s = 1 cP = 1 cp = the viscosity of water at room temperature.) Dissolved substances (especially polymers) increase viscosity. There are several ways to present viscosity graphically. One is to plot shear stress versus shear rate (Fig. 5.7).

Flow properties and viscosities of liquids can be determined using one or more of a variety of instruments called viscometers and rheometers. Viscosities and rheological properties of polysaccharide dispersions are most often measured with rotational viscometers that measure torque (the resistance to a spindle or cylinder rotating at a given speed in a fluid). Shear rates (spindle speeds) can be changed, so one can obtain both

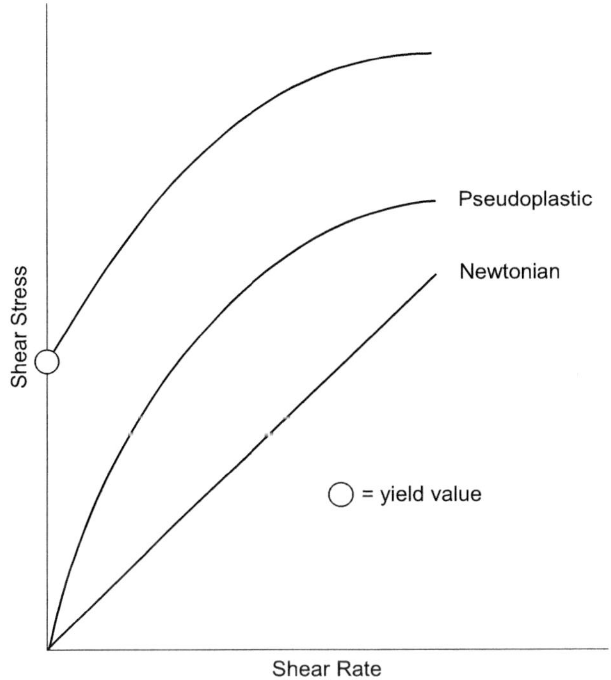

Figure 5.7 Idealized flow curves of three different rheological systems: Newtonian flow (seldom exhibited by polymers), pseudoplastic flow, and pseudoplastic flow with a yield stress value.

readings at a given shear rate and shear stress versus shear rate plots. (The latter feature is a must for polysaccharide solutions [see below]). (Remember that aqueous solutions of starches and hydrocolloids often contain aggregates of hydrated molecules in addition to individual hydrated molecules and, therefore, are not complete molecular dispersions [true solutions].)

Instruments that create rapid oscillatory shear provide information that characterizes the viscoelastic response of a system subjected to dynamic (sinusoidal) shear deformations in terms of the basic rheological parameter, which is the shear modulus (G = shear stress/strain), which is a measure of how much force is required to produce a given amount of deformation. A computer collects the stress and strain data from response measurements and calculates the **elastic modulus (G′)** and the **viscous modulus (G″).** G′ is the elastic component of a viscoelastic response of a material subjected to a dynamic (sinusoidal) shear deformation and is a quantitative measure of the elasticity of the sample (that is, it is a measure of the proportion of the deformation energy that is reversibly stored in the system). Because G′ is related to the energy stored in the material to which a stress is applied, it is also known as the **storage modulus**. G″ is the viscous component obtained in the same way and a quantitative measure of the amount of energy introduced to the system that is not stored in the system. G″ is also a measure of the viscosity of the system. Because G″ is related to the energy that is dissipated as heat when a stress is applied to the system, it is also known as the **loss modulus**. Relative values of the two indicate the degree to which a system is liquid-like and/or solid-like. When a sol gels, G′ increases dramatically and G″ decreases, so that G′ > G″. The computer also calculates the ratio of energy lost to energy stored (G″/G′ = tan δ, called the **loss tangent** or the loss factor) and other parameters. The loss tangent directly reflects the ratio of the viscous properties to the elastic properties of the system. When tan δ is less than 1, the system is predominately elastic (that is, solid-like); when it is greater than 1, the system is predominately viscous (that is, liquid-like).

Cooked starch dispersions are generally called pastes. The viscosity of starch pastes is most often determined with a Rapid Viscoamylograph or a similar instrument that measures viscosity as starch suspensions are cooked to form pastes, while the pastes are held hot, and while the pastes are then cooled. Their use is described in Chapter 6.

Some hydrocolloid solutions appear to have **yield stress** (a **yield value**), which is the force that must be applied before the solution begins to flow (see Fig. 5.7). Systems that exhibit yield values are known as **Bingham plastics.** A Bingham plastic is a material that behaves like a solid until enough force is put on it that it begins to flow and behaves like a liquid. This is an important property in suspension and emulsion stabilization (section on Hydrocolloids as stabilizers in this chapter). To stabilize a suspension, the resistance to movement (flow) must be greater than the downward force (gravity). To stabilize an oil-in-water emulsion, the resistance to flow must prevent the oil droplets from rising to the surface owing to their lower density. Yield values of hydrocolloid dispersions occur when they contain molecular associations that rupture when sufficient force is applied. An example of a Bingham plastic is a spoonable salad dressing, which is an emulsion thickened with a modified waxy maize starch (Chapter 6). Such a dressing behaves like a soft solid at rest but, like mustard or ketchup in a squeeze bottle, can be made to flow when sufficient force is applied.

Types of flow

In **Newtonian flow**, the rate of shear is directly proportional to the applied force. In Newtonian fluids, viscosity is independent of both shear rate and time (Fig. 5.8). Solutions of low-molecular weight carbohydrates and other small molecules that form true solutions are Newtonian. Few polysaccharide solutions are Newtonian. Solutions of gum arabic and low-viscosity grades of other hydrocolloids at low shear rates can be essentially Newtonian.

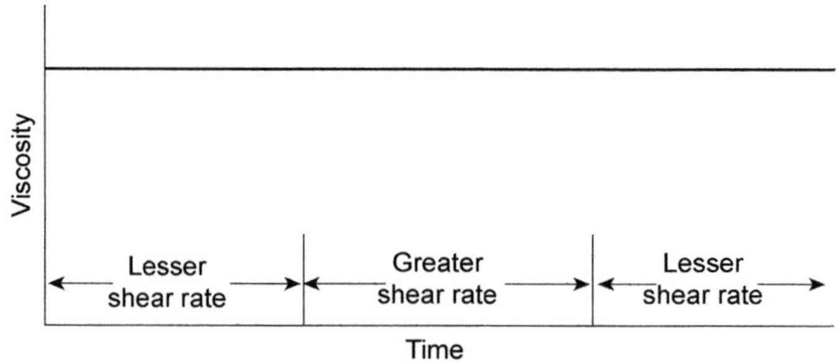

Figure 5.8 Idealized representation of Newtonian solution rheology: viscosity as a function of shear rate and time.

Most solutions of polymers exhibit non-Newtonian flow. The viscosity of non-Newtonian liquids varies with the rate of shear, but not in a directly proportional manner. The flow behavior of non-Newtonian dispersions is determined by the properties of the particles present. The particles in a hydrocolloid solution are hydrated molecules and aggregates of molecules that vary in shape, size, flexibility, degree of hydration, and the presence and magnitude of charges, all of which influence flow behavior. There are two general kinds of non-Newtonian flow: pseudoplastic and thixotropic, both of which exhibit shear thinning.

Pseudoplastic flow

Pseudoplastic liquids are shear-thinning liquids. In **pseudoplastic flow**, an increase in shear stress results in more rapid flow (that is, the greater the applied force, the less viscous the fluid becomes). The rate of flow changes instantaneously as the shear rate is changed (that is, the change in rate of flow is independent of time) (Fig. 5.9). Because shear rate and shear stress do not vary in a linear manner, and because viscosity is the shear stress divided by the shear rate, viscosity measurements must be made at rates of shear that are encountered during processing or consumption, and the shear rate must be specified when reporting a viscosity value if comparisons are to be made between viscosities of solutions of different hydrocolloids.

Pseudoplastic flow is a characteristic of linear polymers. In general, the stiffer the polymer molecules, the more pseudoplastic will be their solutions. Solutions of polymers that impart pseudoplastic rheology may have a significant yield value. With this combination of properties, there is no flow until a force is applied, and the at-rest viscosity returns immediately on cessation of shear (Fig. 5.9). As already mentioned, this behavior is a requirement for suspensions and emulsions that are to be poured and/or pumped (section on Hydrocolloids as stabilizers in this chapter). Such shear-thinning dispersions, solutions, suspensions, and emulsions require significantly less energy for stirring, pumping, and mixing. Shear thinning is nonlinear, having a log–log relationship (Fig. 5.10). Pseudoplastic hydrocolloid solutions only exhibit their pseudoplasticity over a range of shear rates.

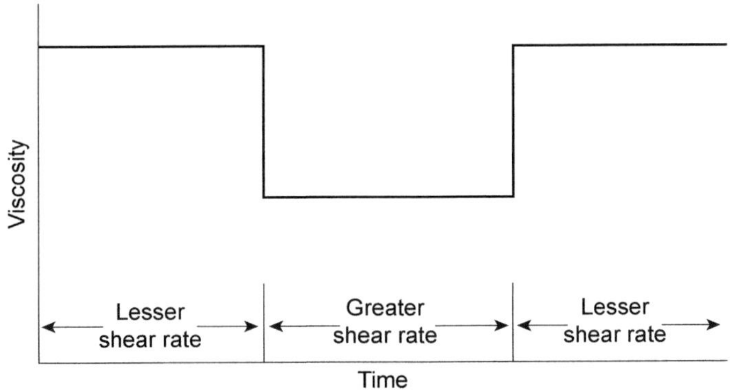

Figure 5.9 Idealized representation of pseudoplastic solution rheology: viscosity as a function of shear rate and time.

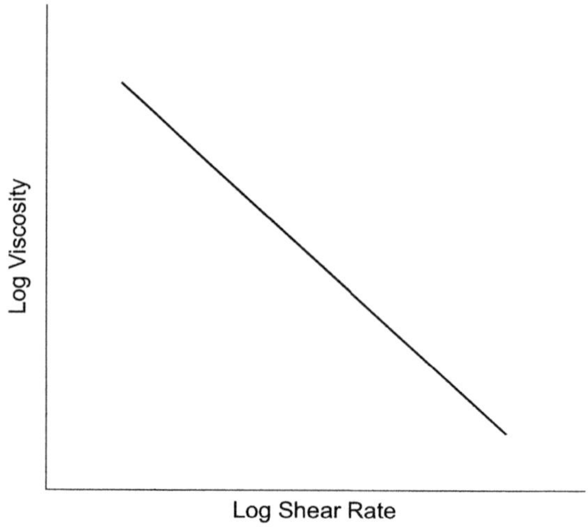

Figure 5.10 Idealized pseudoplastic solution rheology: the logarithm of the viscosity as a function of the logarithm of the shear rate.

The degree of pseudoplasticity of a hydrocolloid solution is also dependent on the concentration of the hydrocolloid, its salt form (if it is anionic), and its DP. Thus, a hydrocolloid dispersion may have Newtonian flow at low concentrations and pseudoplastic flow after the "break point" at the concentration used is reached (Fig. 5.11). In general, high-molecular weight hydrocolloids are more pseudoplastic and are, therefore, more affected by shear (Fig. 5.11).

The sensory properties of liquid and semisolid foods are, in large part, determined by their rheological properties. In food technology, a slimy material is one that is thick, coats the mouth, and is difficult to swallow. Sliminess is inversely related to pseudoplasticity (that is, the more pseudoplastic is a system, the less it is perceived as being slimy). Viscous polysaccharide solutions that are less pseudoplastic are said to give long flow[11] and are perceived as being slimy. Viscous solutions that are more pseudoplastic are described as having short flow[11] and are perceived as being nonslimy (Fig. 5.12). To be perceived as being nonslimy, the system must thin sufficiently at the low shear rates produced by chewing and swallowing that it will clear the mouth. The degree of pseudoplasticity of some hydrocolloid solutions can be modified by controlling the concentration, viscosity grade, and/or salt form of the hydrocolloid or the solution pH, which determines the degree of ionization of the polysaccharide if it contains uronic acid units. A slight degree of sliminess is at times desirable, in that the

[11] **Short flow** and **long flow** refer to draining behaviors from a pipet or funnel. As the forming drop gets larger, making it heavier, the force of gravity on it becomes greater. This causes the liquid's flow rate to increase as the mass of the drop increases. In a pseudoplastic fluid, the increased flow rate causes the viscosity and the diameter of the stream to decrease. Finally, the stream breaks. The result is that the fluid exits the pipet in "short" drops. Fluids without shear-thinning behavior do not exhibit the increase flow rate and the stream exits the pipet in "long" strings.

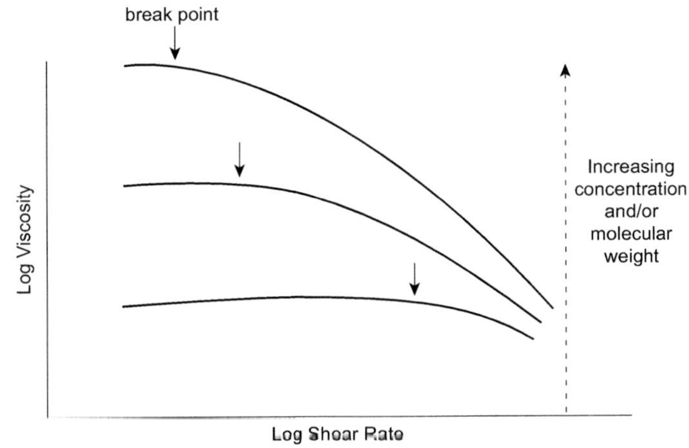

Figure 5.11 Idealized pseudoplastic solution rheology: the logarithm of the viscosity as a function of the logarithm of the shear rate as influenced by polymer concentration and/or molecular weight.

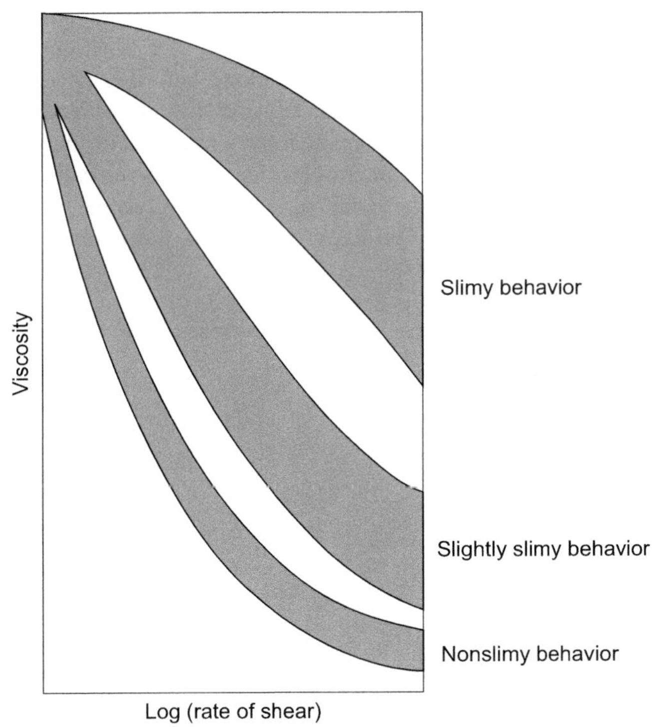

Figure 5.12 The general relationship of the degree pseudoplasticity to the perception of sliminess.
Adapted from A.S. Szczesniak and E. Farkas (1962); Journal of Food Science 27, 381–385.

coating left in the mouth mimics that left by a fatty food, providing a low- or no-fat food with the perception of richness or creaminess.

Processing conditions that are influenced by flow properties include such characteristics as energy required for mixing and pumping, rate of flow through pipes, behavior during pasteurization and on scraped surface heat exchangers, and in any other operation in which an applied force causes flow.

Thixotropic flow

Thixotropic flow is another type of non-Newtonian, shear-thinning flow. The viscosity of thixotropic solutions (like that of pseudoplastic solutions) decreases as the rate of shear increases, but it changes in a time-dependent manner rather than instantaneously. Also, like pseudoplastic fluids, thixotropic fluids regain their original viscosity after cessation of shear, but again only after a measurable time interval (which can range from fractions of a second to hours), rather than instantaneously (Fig. 5.13). This behavior is due to a gel → sol → gel transition. A thixotropic solution at rest is a weak, often pourable, gel. In other words, a thixotropic solution has properties of a gel (at rest) and properties of a fluid when a shear force is applied. Alginates (Chapter 14).are examples of hydrocolloids whose solutions can exhibit thixotropic behavior.

Solutions of certain cellulose derivatives (especially certain types of CMCs) may exhibit thixotropic behavior. The degree of thixotropy is a function of the degree of substitution (DS), the uniformity of substitution, and the DP of the polymer (Chapter 8). Cellulose molecules associate easily and strongly with each other, so non-uniformly derivatized cellulose molecules that have stretches of unmodified β-D-glucopyranosyl units can form limited intermolecular associations called junction zones (section on Gels in this chapter). These interactions are responsible for the weak, easily broken gel structure. Energy may be required to break the junction zones and start

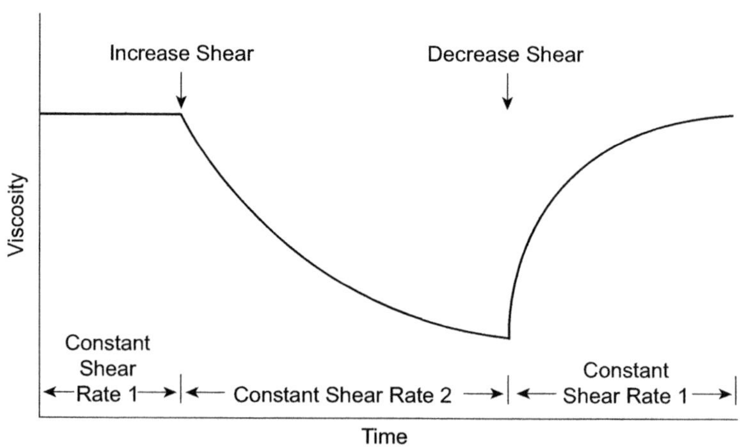

Figure 5.13 Idealized representation of thixotropic rheology showing the time dependence of the change in viscosity with a change in shear rate. Compare with pseudoplastic rheology (Fig. 5.9).

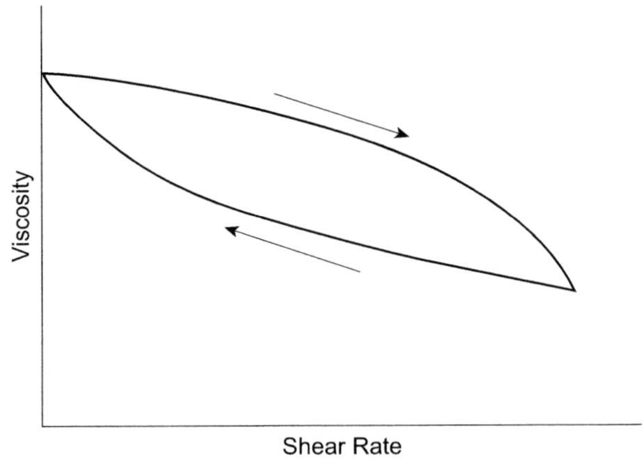

Figure 5.14 One type of graphical presentation of the hysteresis loop of thixotropic solutions: viscosity versus shear rate.

flow. Hysteresis[12] loops are characteristic of thixotropic flow (Fig. 5.14). Because the pulp in fruit juices provides a slight thixotropic characteristic, a dry mix for a fruit drink should mimic this behavior.

Viscosity grades of hydrocolloids

From what has already been presented, it is rather easy to see that the volume (space) occupied by polysaccharide molecules in solution is a function of the number of molecules (that is, the concentration of the polysaccharide) and the volume occupied by each molecule, which is a function of its degree of extension or coiling, which in turn is a function of its structure and the nature of the solvent. However, polysaccharides thicken aqueous systems differently at different concentrations of the polysaccharide. At very dilute concentrations, the viscosity of a polysaccharide solution is a function of the sum of the effects of individual hydrated molecules with the corresponding restructuring of water molecules at the polysaccharide–water interface. At concentrations more typical of food use (0.1%–2%), the probability for polymer molecules to collide and/or for the domains swept out by their constant, dynamic motion to overlap is increased. Such collisions and overlap of hydrodynamic volumes result in entanglements and dramatic increases in the volumes occupied by the entangled hydrated molecules, internal friction, and viscosity (Fig. 5.15). The concentration at which interpenetration of polymer chains (overlap of hydrodynamic volumes) occurs and the slope of the viscosity versus polymer concentration curve begins to increase rapidly is known as the **overlap concentration (c*)**.

Most hydrocolloids are available in a wide range of viscosity grades (Fig. 5.16), which are provided by their manufacturers for different specific applications. As an

[12] **Hysteresis** is the dependence of the state of a system on its previous history, generally (as in this case) in the form of a lagging of a physical effect behind its cause. Hysteresis loops (Fig. 5.14) are formed in the graphs produced by a rheometer as the shear rate is increased, then decreased.

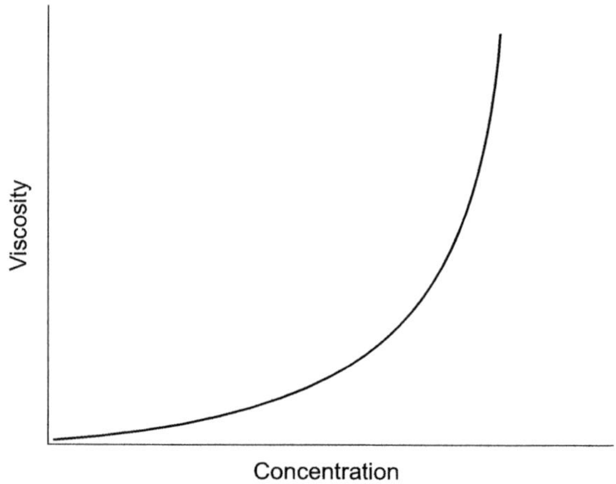

Figure 5.15 Idealized curve of the relation of viscosity to concentration of a polysaccharide in water. At "higher" concentrations, which are still generally low (less than 5%), random coils of polymer begin to overlap one another to form an entangled network, giving a much greater concentration dependence to the viscosity.

example, cellulose derivative products (Chapter 8) may have solution viscosities at 2% concentration that vary from less than 5 to more than 100,000 mPa·s, indicating products with a wide range of average DPs produced by depolymerization of the parent-derivatized polysaccharide.

It is almost as important to select the proper viscosity grade of a hydrocolloid as it is to select the most efficacious hydrocolloid for a particular application. If viscosity is the attribute desired, a high-viscosity hydrocolloid at a low solids concentration is used. If binding, stabilization, or coating, for example, is the goal, a low-viscosity hydrocolloid at a high solids concentration is used. Several viscosity grades of a given hydrocolloid should be tried for each application to find the one that works best.

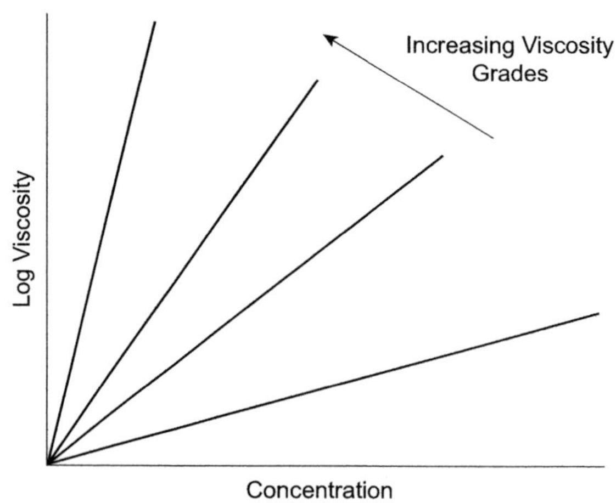

Figure 5.16 Idealized curves of the logarithm of the viscosity as a function of hydrocolloid concentration for four viscosity grades of a single hydrocolloid.

The same process should be used in all comparisons because viscosity is a function of the degree of molecular dispersion (that is, the degree of hydration of the hydrocolloid particles and molecules, which in turn is determined by the conditions of dissolution/dispersion). The manufacturer's data or the data of others can be used for comparisons of the various viscosity grades of a single product, but comparisons of different hydrocolloids in an application should be done by the user.

The viscosity of hydrocolloid solutions generally increases approximately logarithmically with concentration (that is, doubling the concentration will often give about a ten-fold increase in viscosity).

Effects of temperature

There are four ways that temperature affects polysaccharide solutions: (1) effect of preparation temperature on the hydration rate of hydrocolloid particles, (2) effect of preparation temperature (and shear) on completeness of dissolution and viscosity of the sol, (3) effect of changing temperature on the viscosity and rheological characteristics of the sol, and (4) effect of holding temperature on stability of the hydrocolloid. All will be discussed as the properties of the various hydrocolloids are discussed.

Most hydrocolloid solutions decrease in viscosity as the temperature is increased (Fig. 5.17). (Xanthan is an exception [Chapter 11].) Often, therefore, hydrocolloid solutions are made by first dispersing the hydrocolloid at a low temperature where its rate of hydration is slow, then elevating the temperature to a point where solubility is high and viscosity is low. After thorough dispersion/dissolution of the hydrocolloid, the solution is cooled for thickening. Prolonged heating, such as during retorting, may cause some degradation of the polysaccharide, the degree of which is controlled by the temperature, pH, and its inherent stability (Fig. 5.18). When viscosity loss occurs, more hydrocolloid may be added initially so that the final viscosity

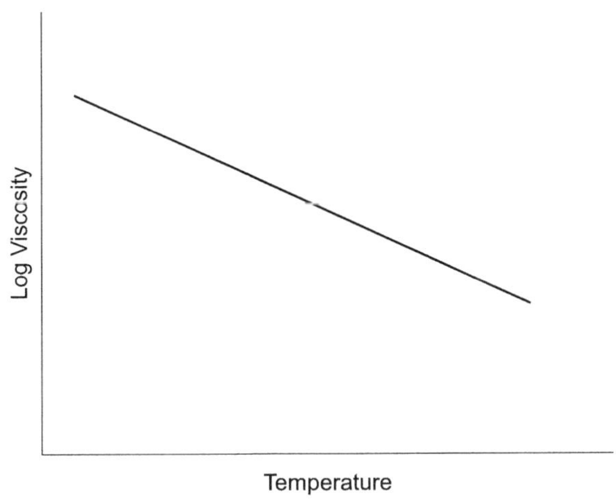

Figure 5.17 Idealized relation of the logarithm of the viscosity to solution temperature. For most hydrocolloids, this relationship is reversible. The slope of the log viscosity versus temperature line is different for each polysaccharide.

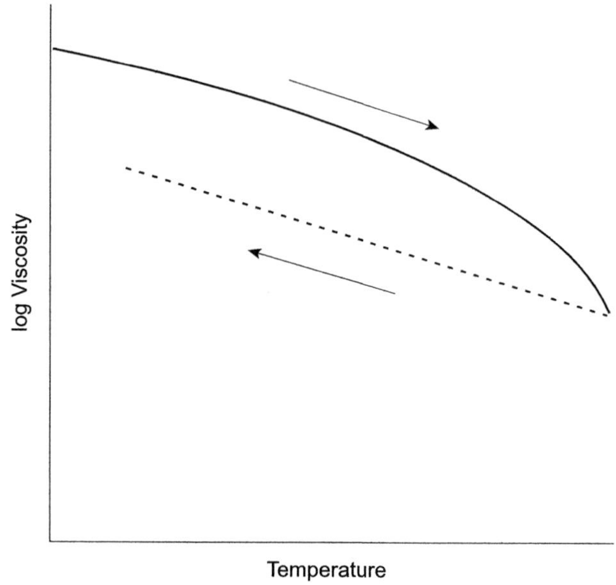

Figure 5.18 Idealized relation of the logarithm of the viscosity to solution temperature for a hydrocolloid that is inherently unstable to heat at the solution pH (compare Fig. 5.17). Note that viscosity increases as the solution is cooled (*dotted line*), but the original viscosity is not recovered because of polymer degradation during the heating cycle.

(after degradation) is the desired viscosity. As already pointed out, with iota- and κ-type carrageenans (Chapter 13) and locust bean gum (Chapter 9), some heating is necessary to obtain full viscosity (Fig. 5.19), so viscosity may increase during processing of food formulations containing them. Solutions of methyl- and

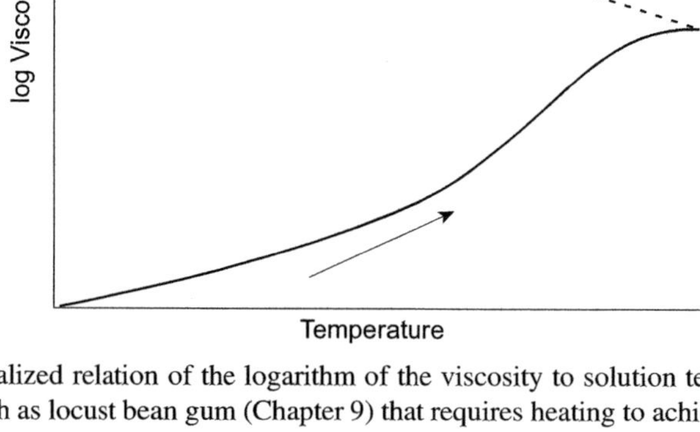

Figure 5.19 Idealized relation of the logarithm of the viscosity to solution temperature for a hydrocolloid such as locust bean gum (Chapter 9) that requires heating to achieve full hydration and dispersion.

hydroxypropylmethylcelluloses are unique in that they first thin when heated and then gel at particular temperatures (that is, they exhibit thermal gelation) (Chapter 8).

Increasing the solution temperature decreases the hydrodynamic volumes occupied by polysaccharide molecules. Temperatures that are high enough to disrupt the intramolecular hydrogen bonds that stabilize single-, double-, and triple-helical structures or other associations of a particular polysaccharide will result in a transition from the ordered structure to a disordered (random coil) conformation. This generally is a reversible transition (that is, upon cooling of the solution, the polysaccharide molecules are usually able to regain their ordered structure). Increased temperatures may also disrupt the hydrogen bonding of less-ordered aggregates of molecules.

Effects of pH

The viscosity of solutions of ionic hydrocolloids[13] containing carboxylate groups is affected by pH. Usually, the viscosity of a solution of a poly(uronic acid) increases markedly when its pH is lowered to values below pH 3.0–2.5 (Fig. 5.20). Some highly carboxylated anionic hydrocolloids become insoluble at low pH; solutions of some others gel at low pH. (As already pointed out, as the pH is lowered to values near or below the pK_a values of the carboxylate groups present, some anionic $-COO^-$ groups are converted into neutral $-COOH$ groups, thus removing the hindrance to

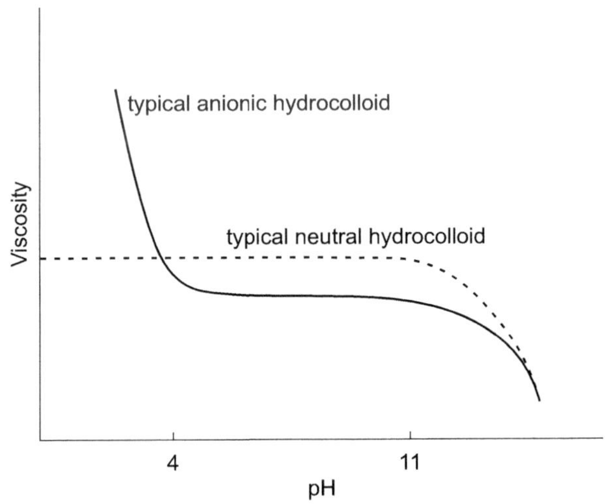

Figure 5.20 Idealized relation for the effect of solution pH on the viscosity of solutions of neutral hydrocolloids and poly(uronic acids). Alkaline pH values (above pH 8) are rarely, if ever, encountered in foods.

[13] All hydrocolloids are either neutral or anionic. Hence, ionic hydrocolloids are anionic polymers, most often, but not in every case, containing carboxyl ($-COOH/-COO^-$) groups.

chain associations.) (The same hydrocolloid may either precipitate or form a gel depending on its concentration and how the pH is adjusted.) The pH of a hydrocolloid solution should be adjusted to an acidic pH only after the hydrocolloid is dissolved because it will not become fully hydrated in the low-pH system in which it is less soluble. Highly carboxylated hydrocolloids are generally much less soluble in acidic systems than they are in neutral or alkaline systems.

Effects of solutes

Like solubility of hydrocolloids, the viscosity of hydrocolloid dispersions is affected by the presence of substances that compete with the polysaccharide for water molecules. Salts and other solutes decrease polysaccharide hydration. With some hydrocolloid solutions, salts/solutes increase viscosity or effect gelation (section on Gels in this chapter) as a result of increased intermolecular interactions. With other hydrocolloid solutions, salts/solutes decrease viscosity as a result of increased intramolecular interactions (that is, as a result of more molecular coiling that decreases their hydrodynamic volume). Addition of polyols and sugars often increases viscosity.

Interactions with other polysaccharides or proteins

Viscosities of hydrocolloid solutions are also influenced by interactions with other polymers (section on Molecular associations of polysaccharides in this chapter). Some synergistic relationships in which the viscosity or gel strength of the combination is greater than that predicted by summing the individual properties are given in Table 5.3. Interaction of a polysaccharide molecule with a molecule of a different polysaccharide molecule or with a nonpolysaccharide polymer, such as a protein molecule, produces the same effect as intermolecular interactions between like polysaccharide molecules.

Table 5.3 Examples of synergisms

Viscosity enhancing combinations	Carboxymethylcelluloses + guar gum
	Carboxymethylcelluloses + casein
	Carboxymethylcelluloses + soy protein
	κ-type carrageenans + kappa-casein
	Xanthan + agarose
	Xanthan + κ-type carrageenans
	Xanthan + guar gum
Gelling combinations	κ-type carrageenans + locust bean gum
	Xanthan + locust bean gum

Summary of characteristics of polysaccharide solutions

General relationships of polysaccharide structures to the viscosities and other rheological properties of their solutions are the following:

- The viscosity of a solution of a polysaccharide is a function of both the size of the molecules and their shapes in solution because solution viscosity is a function of the volume swept out by the motions of the molecules.
- At equal concentrations, polysaccharide molecules that occupy (sweep out) a larger solution volume will generate more viscosity than those that occupy a lesser volume.
  - Linear molecules occupy greater solution volumes than do bushlike molecules of the same DP.
  - Anionic, linear polysaccharide molecules tend to adopt more extended conformations due to mutual repulsion of the charges and, therefore, generally produce higher viscosities than do neutral polysaccharides of the same DP.
  - Neutralization of the charges on anionic polysaccharide molecules may either decrease solution viscosity by allowing the molecules to fold back on themselves, thus reducing their overall dimensions, or increase solution viscosity by allowing molecular aggregation (depending on the specific structure of the polysaccharide).
- Aggregation of polysaccharide molecules creates particles of increased size that occupy larger solution volumes, resulting in greater viscosities. This effect is produced by intermolecular associations of molecules of a particular polysaccharide and molecules of another polysaccharide or a protein.
- A polysaccharide in a good solvent system (in which it is well hydrated) is highly extended to maximize polymer—solvent contacts and, therefore, produces high viscosity. Conversely, a polysaccharide in a poor solvent system will either coil back on itself to decrease the number of polymer—solvent contacts or will never become fully hydrated and will never form a complete molecular dispersion. Both effects result in lower viscosities.
- The rheology of polysaccharide solutions is generally shear-thinning (usually pseudoplastic, less often thixotropic) rheology.
  - The more linear are the molecules and the more extended and stiff are the linear molecules, the more pseudoplastic is the solution.
- Mouthfeel is related to the degree of solution pseudoplasticity.
- Viscosity increases with concentration in a nonlinear manner.
- The viscosities of most solutions of polysaccharides decrease with increasing temperatures.

Characteristics of polysaccharide gels

Although much is known about gels and how they are formed, there are so many different kinds of gels, that a gel is not easy to define. Most of us, however, have a mental picture of gels being deformable, soft solids. One definition of a **gel** is that it is a viscoelastic semisolid that deforms under stress and requires a finite time to respond to the applied stress. (As pointed out in the discussion of liquid behavior, viscoelastic materials have both liquid-like and elastic behaviors, so they do not fully recover to their original state when the applied stress is removed.)

Structurally a gel is a three-dimensional network of connected molecules or particles entrapping a large volume of a continuous liquid phase, which in the case of food products is an aqueous solution of low-DP solutes and dissolved portions of polymer chains (Fig. 5.21 and 5.22). The three-dimensional network, which is continuous throughout the entire volume of the gel, is formed by cross-links between polymer molecules and/or particles. The network entraps the liquid phase and gives the gel some degree of both rigidity and elasticity. Interactions forming the cross-links may be polymer molecule to polymer molecule, particle to particle, or polymer molecule to particle interactions. In most food products, the gel network consists of polysaccharide and/or protein molecules (or bundles of these polymer molecules). The bonding may be noncovalent or covalent (almost exclusively noncovalent in polysaccharide food gels). So a **gel** can also be defined as being comprised of a continuous, three-

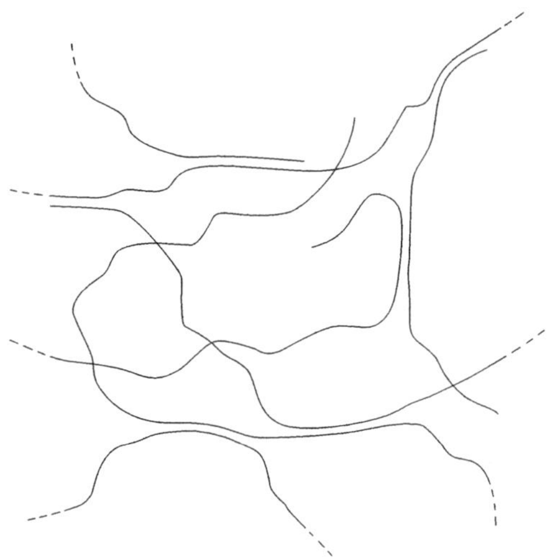

Figure 5.21 Diagrammatic representation of the three-dimensional network structure found in many polysaccharide gels. Parallel side-by-side chains indicate the ordered, crystalline structure of a junction zone. The holes formed by the junction zones contain an aqueous solution of dissolved segments of polymer chains and other solutes.

Figure 5.22 Diagrammatic representation of the three-dimensional network structure found in gels when double helix formation makes the junction zones. The double helices may then be packed together in crystalline structures.

dimensional network of polymer molecules, bundles of polymer molecules, or particles (connected for the most part by noncovalent bonds entrapping a large volume of a continuous liquid phase.

The region where polymer molecules (or supramolecular bundles of polymer molecules) are joined by noncovalent interactions is called a **junction zone**. In a junction zone, glycosyl units of two or more polysaccharide chains are associated with each other in an ordered way (generally in a microcrystalline structure). Junction zones need not be large (that is, the number of ordered units that may be required to form the junction zone may be only a small percentage of the total number of units). In addition to being small, junction zones are dynamic (that is, transient to some degree), especially with respect to size. However, at any instant, there is enough of sufficient size to give the gel network whatever strength and permanence it has under the given conditions (temperature, for example). Segments of polysaccharide molecules not involved in junction zone structures remain completely solvated by water molecules. These soluble segments of polymer chains between junction zones can be thought of as springs that give some freedom of motion to the somewhat solid structure. Thus, a gel can be thought of as a system of polysaccharide molecules in which segments of the molecules are dissolved and other segments are associated with each other in nondissolved, microcrystalline structures.

There may be particles within a gel network that may be swollen starch granules or fragments of starch granules (Chapter 6), emulsion droplets, molecular aggregates (such as fibrils or other supramolecular structures), or microcrystals. Particle sizes may range from submicroscopic to macroscopic.

Gels have some of the characteristics of solids and some of the characteristics of liquid (that is, they are viscoelastic materials). The properties of a gel are determined by the nature of the network structure (that is, by the natures of the substances or materials composing the framework structure and the interactions between them), the nature of the aqueous phase, and interactions between the two phases. When polysaccharide molecules or supramolecular bundles of polysaccharide molecules in solution interact over portions of their lengths (forming junction zones and a three-dimensional network), the fluid solution is transformed into a solid, sponge-like structure that retains its shape to some degree and entraps a large volume of the solution (that is, the solution is transformed into a gel). The three-dimensional network structure provides resistance to applied stress, causing it to behave like an elastic solid. The continuous liquid phase, in which molecules are completely mobile, makes a gel less stiff than an ordinary solid, causing it also to behave somewhat like a viscous liquid. Thus, a gel is a viscoelastic semisolid because the response of a gel to stress is partly that which is characteristic of an elastic solid and partly that which is characteristic of a viscous liquid. Hence, gels exhibit some of the behaviors of elastic solids and some of the behaviors of polymer solutions. However, by definition, for a gel, $G' > G''$ (usually $G' >> G''$), that is, a gel has more of the characteristics of an elastic solid than those of a viscous liquid.

Gel formation

The process of gelation is the process of forming junction zones. Gels are usually produced by direct intermolecular collisions and binding of short segments of otherwise soluble polysaccharide chains. The formed junction will have a stability that is primarily a function of its length (that is, the number of weak intermolecular bonds [usually hydrogen bonds] that develop). Enlargement of the junction zone is effected by movement of the chains that allows adjacent segments of the molecules to align. Thus, the junction zone may grow in a zippering fashion. (Reduced movement that would pull segments away from the junction zone is effected by lowering the temperature.) Further binding of segments is aided by restricting the number of water molecules available for solvation of the polysaccharide molecules. Reduced hydration can be effected by the addition of salts, sugars, or other substances that become highly hydrated and thus limit the number of water molecules available to solvate the polysaccharide molecules. As junction zones develop, the result is formation of a three-dimensional network throughout the system that fills the entire container in which the gel is formed. The degree of overlap of chains in the junction zone determines the strength of the gel, with increasing overlap increasing intermolecular binding and strength.

An increase in junction zone lengths results in the three-dimensional network shrinking. In some gels, this results in some liquid being squeezed out of the gel and accumulating on the gel surface, a process called **syneresis**. Syneresis is usually accelerated by freeze-thaw cycles because ice crystal growth in the aqueous phase forces polymer chains together and promotes associations. Addition of a nongelling hydrocolloid that binds water is often an effective way to reduce syneresis.

Gels are produced by normally soluble polysaccharides if junction zone formation is restricted to rather short chain segments because a gel contains both polymer chain or polymer bundle segments that are in solution and short segments that are in an associated, crystalline, solid phase. If the associations are long (large), insolubility (that is, precipitation rather than gel formation) results; so for gel formation to occur, there must be discontinuities in the polysaccharide chain structure that limit the size of the associations.

Intermolecular forces involved in junction zone formation in systems containing polysaccharides may be hydrogen bonds, van der Waals attractions (hydrophobic interactions), or ionic cross-bridges involving a di-, tri-, or polyvalent cation. Gels may also be formed by simple chain entanglements, examples being starch amylopectin (Chapter 6) and gum arabic (Chapter 16).

The nature of the aqueous phase affects the nature of the junction zone. Hydrogen bonds, hydrophobic interactions, and ionic interactions involved in junction zones depend on the unique nature of water molecules, which are both hydrogen bond donors and hydrogen bond acceptors. Solutes modify the properties of water. Sugar, for example, may increase the size and number of junction zones and raise the setting and melting temperatures of a gel by making water a poorer solvent and increasing polymer-to-polymer intermolecular interactions (Chapter 15).

Some polysaccharides can form what are described as weak gels. Weak gels do not have the characteristics one usually associates with a gel because they do not support their own weight, but measurements reveal that $G' > G''$ and that these moduli are only weakly dependent on frequency, as is the case with gels. In weak gels, the polymer molecules are believed to be both entangled and bound to one another through weak (probably short) associations. As a result, they exhibit greater shear-thinning behavior than does a system of only entangled polymer molecules. There is little, if any, distinction between a thixotropic solution and a weak gel. An example is chocolate milk thickened with a carrageenan (Chapter 13), which is a weak gel at rest (keeping the cocoa particles in suspension), but which turns into a sol in a time-dependent manner with application of the forces involved with pouring, drinking, and swallowing.

Uniformly derivatized CMC products (Chapter 8) form stable solutions because the anionic carboxymethyl groups along the cellulose chain block the intermolecular associations necessary for junction zone formation. However, nonuniformly derivatized types of CMC, in which sections of the cellulose chain are unsubstituted, produce weak gels because the bare (or sometimes called naked) sections of chains hydrogen bond together, forming short junction zones (Chapter 8).[14] (The junction zones are short because the DS must be high enough to make the molecule soluble in water.) Such behavior is desired for toothpaste; tubes can be filled with a liquid which will subsequently gel, in this case after several hours, but then return to a fluid state when a force is put on it by squeezing the tube.

Sometimes mixed junction zones are formed from association of naked sections of different polysaccharide molecules. Such synergistic interaction either increases viscosity through an increase in total molecular size or, if the increase in the number of junction zones is sufficient, results in gel formation (Chapters 9, 11, and 13).

Gel structures can also be produced by chemically cross-linking polymer chains. This may occur with certain pectins (Chapter 15) and arabinoxylans esterified with ferulic acid (Chapter 17). Poly(uronic acid) molecules can produce gels with divalent cations, such as calcium ions, or polyvalent cations, such as protein molecules, which bind together individual chains (Chapters 14 and 15).

Summary of gel formation

Gelation of polysaccharide solutions may be caused by cooling a hot solution, by additives (generally acids or cations), or in the case of some methyl- and hydroxypropylmethylcelluloses (Chapter 8), by heating a cool solution. Junction zones found in food products can be any one of several different types:

1. Intermolecular interactions between linear segments of molecules of the same polysaccharide. Examples are methylcelluloses and hydroxypropylmethylcelluloses (Chapter 8) and high-methoxyl pectins[15] (Chapter 15) (via hydrophobic interactions/van der Waal's forces), and starches (Chapter 6) and alginic acid[16] (Chapter 14) (via hydrogen bonding).

[14] The thixotropy of solutions of high-DP, nonuniformly substituted CMC (Chapter 8) is due to the same mechanism.

[15] See Chapter 15 for definitions.

[16] Lowering the pH of a solution of a poly(uronic acid) so that the charge is removed from the carboxylate groups ($-COO^- \rightarrow -COOH$) converts the anionic polysaccharide to a neutral polysaccharide and allows hydrogen bonding between chain segments. Either gelation or precipitation can occur depending on the structures of the molecules, the conditions under which the acidification is done, and the nature of the overall system.

2. Intermolecular interactions between linear segments of polyanionic polysaccharide molecules effected by cations. Examples are alginates (Chapter 14) and low-methoxyl pectins[15] (Chapter 15) (via single-chain interactions) and carrageenans (Chapter 13) and gellans (Chapter 12) (via interactions of double-helical segments).
3. Intermolecular interactions between linear chain segments of two different polysaccharide molecules. Examples are locust bean gum plus a κ-type carrageenan (Chapter 13) and locust bean gum plus xanthan (Chapter 11).
4. Intermolecular interactions between anionic polysaccharide molecules and protein molecules. An example is a κ-type carrageenan plus κ-casein (Chapter 13), with the protein molecules serving as polyvalent cations.
5. Chain entanglements, especially in the case of branch-on-branch polymers. Examples are gum arabic (Chapter 16) and starch amylopectin (Chapter 6).

As a general rule, regular, linear segments of polysaccharide molecules are in the form of a helix. Intermolecular interactions result either in associations of single helices or in the formation of double[17] (or, in the case of curdlan [Chapter 12], triple) helices. The relatively stiff, linear, double-helical segments may then interact (pack together) to form a super junction zone and a three-dimensional gel network.

To form a gel from polymer molecules, the molecules must first be in solution, then associate in junction zone regions to form the three-dimensional gel network structure. To convert a polysaccharide sol to a gel (that is, to form one of the several types of junction zones), one may have to add a solute that competes with the polysaccharide molecules for water, add a di- or polyvalent cation, add another polymer that associates with the polysaccharide, change the temperature, change the pH, or evaporate some of the water, the latter being especially true of gels formed by molecular entanglements.

Junction zones between molecules must be of limited size because, if molecules interact over a major portion of their length, precipitation/insolubility results. Therefore, regular, linear chain segments must be interrupted by irregularities so that interactions take place over only limited segments of a molecule and a limited junction zone is formed. The end result is molecules that are partially in solution and partially in an aggregated, undissolved state. In general, the larger the associated segments, the more difficult it is to separate molecules by application of shearing forces and/or thermal energy. Also, in general, if junction zones grow after formation of the gel, the network becomes more compact, the structure contracts, and syneresis occurs.

Food gels (junction zones) can be formed in several ways (Table 5.4). The most common is cooling of a hot solution to a temperature below its gelling temperature. (Hysteresis is often observed [that is, the melting temperature of a gel is usually higher than its gelling temperature].)

Gellike or salvelike materials can also be formed by high concentrations of particles (tomato paste is an example).

[17] A double helix consists of two chains coiled around a common axis. It is the structure obtained if one takes two rods made out of something flexible like rubber, places them side-by-side, holds one end steady, and twists the other end to make helices out of each of the two rods. DNA, for example, is a double helix of antiparallel chains.

Table 5.4 Examples of formation of polysaccharide food gels

Gelling conditions	Hydrocolloids
With acids	Sodium alginates High-methoxyl (HM) pectins[a] + sugar
With cations	Sodium alginates + Ca^{2+} Carrageenans + protein (above pI) ι-type carrageenans + Ca^{2+} κ-type carrageenans + K^{+} Gellan Low-Methoxyl pectins[a] + Ca^{2+}
Upon Cooling of Hot Solutions	Agar Calcium alginates[b] (low conversion) κ-type carrageenans + locust bean gum Gellan[c] Deaceylated konjac glucomannan HM pectins[a] + acid + sugar Calcium pectinates Starches Xanthan + κ-type carrageenans Xanthan + locust bean gum Xanthan + agarose
Upon heating solutions	Curdlan (irreversible) Hydroxypropylmethylcelluloses (thermoreversible) Methylcelluloses (thermoreversible)
With addition of solutes	HM pectins + acid

[a]See Chapter 15 for definitions.
[b]Under certain specific conditions. May not be thermoreversible.
[c]A K^{+} gellan gel does not melt when heated. A Na^{+} gellan gel melts when heated.

Gel textures

Many polysaccharide gels contain 1% or less of polymer (that is, they may contain 99% or more of water). Such gels can be quite strong, even though they contain only small concentrations of polymer. Examples of polysaccharide gels are aspics, confections (for example, gum drops and other "gummy" confections), dessert gels, jams, jellies, meat analog pet foods, structured fruit pieces, and structured onion rings. To understand why and how polysaccharides form strong gels at such low concentrations, the food gum gellan (Chapter 12) can used as an example. (Calculations for other polysaccharides will give essentially the same results.) Consider 200 mL (a small jelly jar) of a gel made by cooling a hot 1% solution of gellan. (The gel formed will be quite rigid, like an agar gel.) X-ray fiber diffraction reveals that gellan molecules form parallel double helices[17] and that the double helices pack together in an antiparallel manner in crystalline arrays (Fig. 5.23). (Polysaccharides with regular repeating-unit structures, as does gellan, have a natural tendency to form helical conformations, which may evolve into double or triple helices.) (It can be seen in Fig. 5.23 that, in

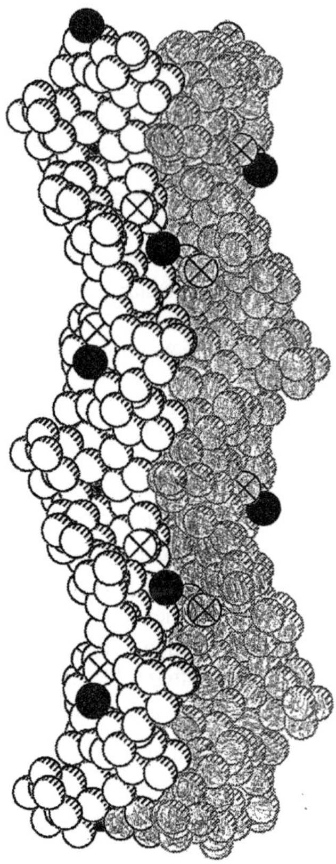

Figure 5.23 The native gellan double helix. Potassium ions forming salt bridges holding the chains together are shown as black circles. Circles with crosses are the oxygen atoms of carboxylate groups.
Courtesy of R. Chandrasekaran and A. Radha.

the case of gellan, the packing of the two chains in double helices is so perfect and so tight that the two chains can hardly be distinguished from one another.) Chemical structural analysis, X-ray fiber diffraction analysis, and molecular modeling shows that the polysaccharide contains a tetrasaccharide repeating unit with the distance from the start of one repeating unit to the start of another being 18.8 Å. Therefore, there are eight sugar units in each 18.8 Å of the double helix. From the DP of the tetrasaccharide repeating unit, it is easy to calculate the number of moles of repeating units in 2 g (the amount in 200 mL of a 1% solution) of gellan; it is about 3 mmol. Using Avogadro's number and the 18.8 Å length of two repeat units in the calculation discloses that the 200 mL of solution contains about 1.7×10^9 km (1 billion miles) of double helix. That is about 5000 times the distance from you to the moon. Even if the entities forming the gel structures are bundles of five or six polymer chains and the number has to be divided by five or six, it is still a very long length. Thus, there is enough length of polymer in the 200 mL to form a substantial gel network. In this case, the junction zones are probably formed between fibrils formed by the packing together of double-helical structures held in place by cations as depicted in Fig. 5.23.

Because solutes that compete for water increase interactions between hydrocolloid molecules, they may affect the texture of the gel. κ-Type carrageenans (Chapter 13) and gellans (Chapter 12) form gels with potassium ions. Certain CMCs, gellans, ι-type carrageenans, alginates, and LM pectins[15] form gels with di- and multivalent cations (most often Ca(II) ions) (Chapters 8, 13, 14, 15). As would be expected, the nature and concentration of the cation required for gel formation also affects gel texture.

A soft, thixotropic gel forms when a κ-type carrageenan (an anionic hydrocolloid) interacts with the κ-casein of milk which takes place even though the pH of milk is above the isoelectric pH value (pI) for casein and casein has a net negative charge (Chapter 13). Other anionic gums (for example, CMC) will interact with proteins when the polysaccharide is in the anionic form (pH value above its pK_a) and the protein is in a cationic form (pH value below its pI).

Two general kinds of rheological measurements can be made on gels, each giving different information. One type of measurement is that related to small deformations; the other type of measurement is related to large deformations. Gentle pressure on the surface of a gelatin gel (a gel made from a protein rather than a polysaccharide) or a jar of grape jelly (a gel made with a pectin [Chapter 15]), both of which are elastic semi-solids, will deform the structure. Release of the slight pressure results in the gel returning to its original shape. If more pressure is applied to a gelatin gel, it will fracture because it is brittle. In contrast, the grape jelly can be spread if sufficient force is applied to it. Both types of behavior can be characterized with a single instrument, namely a texture analyzer. Parameters that can be determined and used to describe a gel are the following: (1) shear modulus (also known as the modulus of elasticity or Young's modulus), which is a measure of the firmness of a gel (usually correlating with a sensory perception of firmness); (2) hardness, which is correlated with what is commonly called gel strength; (3) brittleness (also known as fracturability); (4) elasticity (also called springiness), which correlates with how rubbery a gel feels in the mouth; (5) cohesiveness, which is a measure of how tough and difficult to break up in the mouth a gel is; (6) adhesiveness, which is a measure of how sticky, tacky, or gooey a gel is; (7) chewiness, which is a combination of hardness, cohesiveness, and elasticity; and (8) gumminess, which is the energy required to disintegrate a semi-solid food product to a state ready for swallowing and is a combination of hardness and adhesiveness. The most important of these parameters to food product formulators are usually hardness, firmness, and brittleness.

It is obvious that the characteristics described above are related to such things as eating quality, use properties (for example, cuttability and spreadability), and handling properties (for example, shape retention) under different conditions and time scales.

All starches and hydrocolloids mask flavor to some degree. As a general rule of thumb, gel-forming hydrocolloids have less of an effect than do thickening gums (Table 5.2). The variable characteristics of gels as presented in this section are summarized in Table 5.5.

Summary of gel textures

The texture of a gel may be brittle, elastic, plastic, rubbery, or tough. Its strength may be firm or soft. It may be cuttable, spoonable, or spreadable. Most stand-alone gels are

Table 5.5 Variable characteristics of food gels

1. Means of gelation **a.** Chemical (acids, inorganic cations, proteins, other gums, other solutes) **b.** Thermal (cooling, heating)
2. Reversibility **a.** Reversible (via heating, cooling, or shear[a]) **b.** Irreversible
3. Texture—brittle, elastic, plastic, rubbery, tough
4. Gel strength—rigid, firm, soft, mushy, spreadable, pourable
5. Degree of syneresis
6. Clarity
7. Freeze-thaw stability

[a]Shear reversibility is not common. Only a few types of gels exhibit shear reversibility.

not shear reversible, but thixotropic fluids can be considered to be pourable, reversible gels because, even in this case, G' is greater than G''. Achieving the desired properties is a matter of formulating a proper gelling system.

Hydrogels

Removal of water from a polysaccharide solution or gel forms a **hydrogel (xerogel)**. Hydrogels may contain as much as 30% moisture while appearing to be dry. When placed in water, hydrogels absorb many times their weight of water and form a gel. Upon absorbing this water, hydrogels may swell up to several hundred times their dry volume without disintegrating. Polysaccharide films are two-dimensional hydrogels. Polysaccharide films require a plasticizer (most often glycerol) to impart flexibility.

Hydrocolloids as stabilizers

Most food emulsions require both an emulsifier and an emulsion stabilizer. An **emulsifier** is a single substance or a mixture of substances that is capable of promoting emulsion formation and short-term stabilization via interfacial activity. An **emulsion stabilizer** is a single substance or a mixture of substances capable of conferring long-term stabilization to an emulsion. Without emulsification and emulsion stabilization, fat and oil droplets (which have lower densities than water) will float to the surface due to the Archimedes force.

Only a few carbohydrates are **amphoteric** (that is, only a few contain both hydrophilic and hydrophobic portions and thus have surface activity [surfactant/emulsifying properties]). They are sorbitan esters (Chapter 2), certain sucrose fatty acid esters (Chapter 3), and the polysaccharides gum arabic (Chapter 16), octenylsuccinylated starches (Chapter 7), and methyl- and hydroxypropylmethylcelluloses (Chapter 8).

However, many polysaccharides have the ability to stabilize emulsions—in many cases simply by thickening the aqueous phase of an oil-in-water emulsion. However, high viscosity will only retard movement and, therefore, coalescence of oil droplets (emulsion breakdown), not prevent it. A hydrocolloid like xanthan (Chapter 11) that produces solutions with a high yield value that must be overcome before the solution can flow around droplets (that is, before the droplets can move) is very effective in emulsion stabilization.

Certain polysaccharides will coat droplet surfaces with a thick protective layer that inhibits coalescence. They do this either by displacing the primary emulsifier or by absorbing to the hydrophilic ends of the emulsifying, surface-active molecules that protrude from the oil droplets. Hydrocolloids that are effective in forming such layers are often anionic; such coated droplets repel one another because their coating gives them like charges and because of other effects. Emulsions in which the fat or oil droplets are coated with a polysaccharide are said to be **sterically stabilized**. Thick film formation and good **steric stabilization** is most often achieved by a high-molecular weight hydrocolloid and is most effectively achieved by polymers that have groups with a strong affinity for the surface of the oil droplet. These groups anchor the polymer chains to the surface of the droplets. An example is the octenylsuccinate ester of starch (Chapter 7).

Gum arabic and octenylsuccinylated starch products are unique in having both emulsifying and emulsion-stabilizing properties. Well-dispersed, colloidal cellulose microcrystals (Chapter 8) collect at oil—water interfaces and can add to emulsion stabilization.

Other food products may contain heavy particles (for example, spice or relish particles in salad dressings) which tend to settle because they have greater densities than water. Stabilization of particle suspensions involves the same general principles as stabilization of emulsions (that is, particles suspended in solutions with a sufficient yield stress value will not settle out if the force of gravity on the particles is not sufficient to overcome the yield value and allow the solution to flow around the particles). Likewise, coating particles with a polysaccharide that gives them a negative charge will keep them separated. As with emulsions, simple thickening of the medium in which particles are suspended will delay, but not prevent, settling.

Hydrocolloids can be very effective in these applications. As examples, 0.02% of a carrageenan is used to prevent settling of cocoa particles in chocolate milk and fat separation in low-fat UHT cream (Chapter 13); alginates (Chapter 14) and LM pectins[15] (Chapter 15) are used to prevent settling of the pulp in fruit juices; and 0.5% of a mixture of propylene glycol alginate (Chapter 14) and xanthan (Chapter 11) is used to prevent both oil separation and settling of spice and/or relish particles in pourable salad dressings.

Choosing a hydrocolloid or starch product as a thickening, gelling, or stabilizing agent

Some questions that should be asked when choosing a thickening or gelling agent from the available and approved hydrocolloids and starch products for development of a particular product include the following:

- **I.** Will the product be canned, bottled, frozen, or dry?
- **II.** What other ingredients will be used (water, milk, salts, sugars, acids, fats, emulsifiers, proteins, another hydrocolloid(s), a starch product, flavors)?
- **III.** If a hydrocolloid, what equipment is available for its dispersion?

IV. What stability to processing and storage conditions will be required (time, temperature, amount of shear, and pH at each processing and storage step)?

V. What is the necessary stability (shelf life)? Under what shipment and storage temperatures? Is freeze-thaw stability required?

VI. Is a hydrocolloid or starch product being considered to create or modify/improve texture? If so,

- **A.** Is thickening the desired result? If so,
  1. Will the product be water-based or milk-based?
     a. Will the product be a high- or low-acid product?
  2. Will the product be an instant (dry mix) product or a cooked product?
     a. If a dry mix, will it be designed to be added to water (hot or cold) or milk (hot or cold)?
  3. What is the target viscosity and rheology?
  4. Must the product be clear or is a cloudy product desired?
- **B.** Is gel formation the desired result? If so,
  1. When do you want the system to gel?
  2. Under what conditions do you want the system to gel?
  3. How fast do you want the system to gel?
  4. What gel texture profile is desired?
  5. What other characteristics, such as freeze-thaw stability, clarity, or lack of syneresis, are needed?
  6. Will significant amounts of protein or other ingredients be present? At what pH?
  7. Must the gel be thermal or shear reversible?

VII. Is a hydrocolloid being considered to stabilize a system? If so,

- **A.** What is being stabilized?
  1. An emulsion?
  2. A suspension?
  3. A foam?
  4. A protein (against low pH or thermal denaturation)?
  5. Crystals (ice or sugar) against growth in size?
- **B.** What will be the conditions of thermal processing?

VIII. Is a hydrocolloid or starch product being considered for other reasons?

- **A.** For example, as a binder in processed meat products and/or to reduce purge, improve surface appearance, improve sliceability, increase tenderness, increase freeze-thaw stability, increase brine uptake/product yield, reduce cooking loss, reduce tumbling time, and/or for its lubricating effect on the mix?
- **B.** As a fat mimetic (Chapter 17)?

IX. What other factors need to be considered?

- **A.** Legal restrictions (standards of identity)
- **B.** Cost (should be based on cost per amount required in the formulation, not on cost per unit weight of the ingredient)
- **C.** Kosher status or other ethnic requirements
- **D.** Availability of quality food-grade product[18]

[18] Most hydrocolloids have nonfood as well as food applications. In fact, for many, the quantities used in nonfood applications far exceeds those used to make food products. It is important, therefore, that a food-grade hydrocolloid or starch product be used. (The same starch products are seldom used in both food and nonfood applications. The exception is in pharmaceutical applications.)

E. Nutritional aspects (digestibility, prebiotic)
F. Effect on color, clarity, aroma, and taste
G. Microbiological load and stability

It is clear from the above questions that a single food gum or starch product could not satisfy the variety of requirements that might need to be met, so selection of a polysaccharide product or a mixture of polysaccharide products to meet the demands of processing conditions, final texture, compatibility with other ingredients, storage conditions, etc., requires knowledge of the relationships of structures to physical and functional properties. In choosing the best hydrocolloid or starch product for a particular application, consideration of rheological characteristics of solutions or gels often dominates the criteria. In selecting a hydrocolloid or starch product for its ability to produce viscosity or a gel, the following factors must be considered:

1. Available viscosity grades (Highest viscosity grades are often not the best gel formers.)
2. Amount needed of the proper viscosity grade to give a specified viscosity or gel strength
3. Cost per functionality (cost of the amount needed to impart the desired functional characteristics)
4. pH of the system (effect on hydration rates and stability)
5. Temperatures during processing and times at those temperatures (effects on hydration rates and stability)
6. Times at each storage temperature (effects on stability).
7. Interactions with other ingredients and competition with other dissolved substances for water
8. Desired texture
9. Ease of dispersion with available equipment (particle form and size, salt form if ionic.)
10. Color
11. Desired sweetness; other flavors and aromas present
12. Microbiological load

In a few products, hydrocolloids enhance flavor and sweetness through their coating effect, which holds flavor in the mouth longer. However, most often, the perceived intensity of flavor and sweetness is suppressed as the viscosity increases, presumably due to reduced rates of transport of fresh flavor or sweetener to taste buds. As with the perception of sliminess (Fig. 5.12), the more shear-thinning is the solution rheology, the less is the flavor/taste suppression.

As already discussed, gelation of polysaccharide solutions may be caused by additives (generally acids or cations), by cooling a hot solution, or in the case of some methyl- and hydroxypropylmethylcelluloses (Chapter 8), by heating a cool solution. Most gels are thermally reversible, and gelling and melting temperatures are often important gel characteristics. With regard to gel textures, polysaccharide gels may be brittle, elastic, plastic, rubbery, or tough. Polysaccharide gel strengths may be firm or soft. They may be cuttable, spoonable, or spreadable. Most are not shear reversible, but thixotropic fluids can be considered to be pourable, reversible gels. So achieving the desired properties is a matter of formulating a proper gelling system.

Hydrocolloids and starch products do not provide a single functionality, but rather a combination of functionalities. As has been pointed out, one provided functionality is usually, but not always, thickening or gel formation, and there are a variety of kinds of thickened and gelled systems. When characteristics in addition to, or other than, specific flow and gel properties and gel formation conditions are required, the unique properties of specific gums most be considered. Other properties of polysaccharides are responsible for their ability to provide

Body[2]
Bulk[4]
Flavor release control
Release from a mold
Water migration control
Water retention

and to function as

Adhesives
Binders
Clarifying agents
Cloud agents
Coating materials
Crystal formation modifiers
Dietary fiber
Emulsifiers
Emulsion stabilizers
Encapsulating agents
Fat sparers
Film formers
Flocculating agents
Foam stabilizers
Protein stabilizers
Suspension stabilizers
Swelling agents
Syneresis inhibitors
Texturing agents
Whipping agents

Each food starch or hydrocolloid generally has one or more outstanding properties related to a functional characteristic that is the basis for its choice. Properties and functionalities of specific food starches and hydrocolloids will be presented and discussed in subsequent Chapters 6–16. A much abbreviated summary of structural features and key properties and functionalities of the hydrocolloids and modified food starches covered in Chapters 6–16 is given in Table 5.6 and Chapter 20. Even the abbreviated summary in the table reveals that (1) most hydrocolloids have a unique property that makes it the hydrocolloid of choice in specific applications, (2) there are overlapping properties in some cases that allow one hydrocolloid to be substituted for another (in which case, the choice is usually made on the basis of economics), and (3) combinations of hydrocolloids or a hydrocolloid and a starch product are often used with the combination resulting in (a) synergistic interaction that produces a functionality neither product provides alone, (b) a modification of a functionality produced by a

Table 5.6 Significant characteristics of hydrocolloids and modified food starches[a]

	Significant and/or unique properties	Significant structural features	Other characteristics
Agar (13)[b]	Forms firm gels. Gels require high temperatures for remelting. Gels are compatible with high solute concentrations.	Contains two polysaccharides. Essentially a galactan with 3,6-An-L-Gal*p* units. Slightly anionic.	Soluble only at high temperatures.
Alginates (14)	Solutions gel on addition of Ca^{2+}.	Blocks of L-guluronopyranosyl units, D-mannuronopyranosyl units, and mixed L-Gul*p*A and D-Man*p*A units. Anionic.	
Propylene glycol alginates (PGA)	Mild surface activity. Do not gel or precipitate in acidic systems. Nongelling with Ca^{2+}.	Partial propylene glycol esters of alginates. Anionic.	
Carrageenans (13)			
κ-type Carrageenans	Thicken cold milk. Form soft, thixotropic gels that tend to synerese. Gel most strongly with K^+. Synergistic gelation with locust bean gum (LBG).	Sulfated galactan with 3,6-AnGal*p* units. Anionic.	Synergism with starch. K^+ and Ca^{2+} salts are insoluble. Frequently a component of blends with other carrageenans.
ι-type Carrageenans	Insoluble in cold milk. Gel most strongly with Ca^{2+}. Gels are soft, resilient, have good freeze-thaw stability, and do not synerese. Synergistic gelation with LBG.	Sulfated galactan with 3,6-AnGal*p* units. Anionic.	Ca^{2+} salts give thixotropic dispersions. Frequently a component of blends with other carrageenans.

Continued

Table 5.6 Significant characteristics of hydrocolloids and modified food starches[a]—cont'd

	Significant and/or unique properties	Significant structural features	Other characteristics
λ-type Carrageenans	Nongelling	Sulfated galactan. Anionic.	K^+ and Ca^{2+} salts are soluble. Frequently a component of blends with other carrageenans.
Carboxymethylcelluloses (CMC) (8)	Form clear, stable solutions	Linear. Anionic.	Solutions can be either pseudoplastic or thixotropic. Wide range of viscosity types. Stabilize proteins.
Curdlan (12)	Irreversible thermogelation	β-Glucan. Neutral.	
Gellans (12)	Form gels with a range of textures. Good suspension stabilizers. Solutions of native gellan are thixotropic and effective emulsion and suspension stabilizers. High-acyl (native) types form soft, elastic, nonbrittle gels. Low-acyl types form firm, brittle, nonelastic gels.	Linear. Anionic. Native gellan is acylated. Low-acyl types and blends are available.	Gel with any cation. Blends give a range of gel textures.
Guar gum (9)	Relatively low-cost thickener. Salt tolerant.	Galactomannan structure. Neutral.	Frequently used in combination with CMC, LBG, or κ-carrageenan.

Gum arabic (16)	Emulsifier and emulsion stabilizer for flavor oils	Amphiphilic. Highly branched. Anionic.	Forms low viscosity solutions at relatively high concentration. Good for spray drying. Will form gels. Compatible with high concentrations of sugar.
Hydroxypropylcelluloses (8)	Form nontacky, heat-sealable packaging films. Good coating material.	Hydroxypropyl ether of cellulose. Linear. Neutral.	Wide range of viscosity types.
Inulin (10)	Solutions with concentrations as high as 50% can be made. Solutions of concentrations >25% will gel when cooled. Gels have a creamy, fatlike texture.	Linear. Neutral.	Very susceptible to hydrolysis. A prebiotic.
Locust bean gum (LBG) (9)	Synergistic interaction with κ-carrageenan and xanthan results in gel formation	Galactomannan with unsubstituted regions of mannan backbone. Neutral.	Requires elevated temperature for dissolution. Frequently used in combination with xanthan, κ-carrageenan, or guar gum.
Methylcelluloses and Hydroxypropylmethylcelluloses (HPMC) (8)	Reversible thermogelation. Soluble in cold water; insoluble in hot water.	Linear with hydrophobic substituents. Neutral.	Emulsion and foam formation and stabilization. Can provide fatlike mouthfeel. Can reduce adsorption of fat in products being fried. Wide range of viscosity types.

Continued

Table 5.6 Significant characteristics of hydrocolloids and modified food starches[a]—cont'd

	Significant and/or unique properties	Significant structural features	Other characteristics
Pectins (15)			
Amidated pectins	Gelation with very low concentrations of Ca^{2+}	Low-methoxyl (LM) pectin with some carboxamide groups. Anionic.	
High-methoxyl pectins	Form spreadable fruit gels	Polygalacturonate with >50% methyl ester content. Anionic.	Requires high soluble solids conc. and pH ~3 for gelation
Low-methoxyl-pectins	Gelation with Ca^{2+}, i.e., without sugar and acid	Polygalacturonate with <50% methyl ester content. Anionic.	
Xanthan (11)	Forms highly pseudoplastic solutions. Solution viscosity does not change with temperature. Good emulsion and suspension stabilizer. Stable in highly acidic systems; pH has no effect on solution viscosity.	Stiff, linear molecules. Anionic.	Salt tolerant. Synergistic thickening with guar gum. Synergistic gelation with LBG. Imparts freeze-thaw stability to products.
Modified starches (7)			
Cross-linked	Increased pasting temperatures. Increased resistance of pastes to shear. Increased stability of pastes to low-pH conditions. Increased paste viscosities.	Contain phosphate or adipate diester groups or a glyceryl diether group.	Cross-linking frequently practiced along with stabilization and thinning.

Stabilized	Reduced pasting temperatures. Reduced tendency to retrograde. Improved freeze-thaw stability. Improved clarity of pastes.	Contain hydroxypropyl ether, acetate ester, or monoester phosphate groups.	Easier to redisperse when pregelatinized. Frequently practiced with along with cross-linking and thinning.
Cross-linked and stabilized	Reduced pasting temperatures. Produce pastes with increased viscosities.		Other attributes of cross-linking and stabilization
Octenylsuccinylated	Surface activity. Emulsifier and emulsion stabilizer.	Contain ocetenylsuccinate ester groups. Amphiphilic.	Some products produce low-viscosity solutions and can be used for spray-drying. Can provide fatlike mouthfeel. Higher-DS products do not hydrate and can be used for dusting.
Oxidized	Whiter. Reduced pasting temperatures and paste viscosity. Produce softer, clearer gels.	Hypochlorite oxidized. Contain carbonyl and carboxyl groups.	Usually have reduced average molecular weights (DP).
Thinned	Reduced pasting temperatures and paste viscosity. May produce firmer gels.	Average DP reduced slightly	Frequently practiced along with cross-linking and stabilization.

[a]Only commercial hydrocolloids and modified food starch products are included.
[b]Numbers in parentheses are chapter numbers.

polysaccharide product or a combination of products by a second or third ingredient, or (c) achievement of the functionalities imparted by each product separately. Structure–function relationships are described in each subsequent chapter. Also, described are a few examples of the use of several different food hydrocolloids and modified starch products in the preparation of a single food product.

Additional resources

General

Doublier, J.-l, Garnier, C., Cuvelier, G., 2016. Gums and hydrocolloids: functional aspects. In: Eliasson, A.-C. (Ed.), Carbohydrates in Food. CRC Press, Boca Raton, pp. 285–332.

Hoefler, A.C., 2004. Hydrocolloids. Eagen Press, St. Paul.

Wang, Q., Cui, S., 2005. Understanding the physical properties of food polysaccharides. In: Cui, S.W. (Ed.), Food Carbohydrates: Chemistry, Physical Properties, and Applications. CRC Press, Boca Raton, pp. 168–224.

Williams, P.A., Phillips, G.O., 2009. Introduction to food hydrocolloids. In: Phillips, G.O., Williams, P.A. (Eds.), Handbook of Hydrocolloids. CRC Press, Boca Raton, pp. 1–22.

Water Activity and Glass Transition

Levine, H., Slade, L., 2004. Influence of hydrocolloids in low-moisture foods - a food polymer science approach. In: Williams, P.A., Phillips, G.O. (Eds.), Gums and Stabilisers for the Food Industry 12, Special Publication No. 294. The Royal Society of Chemistry, Cambridge, pp. 425–436.

Slade, L., Levine, H., 2002. Progress in food processing and storage, based on amorphous product technology. In: Levine, H. (Ed.), Amorphous Food and Pharmaceutical Systems, Special Publication No. 281. The Royal Society of Chemistry, Cambridge, pp. 139–144.

Slade, L., Levine, H., 2003. Food polymer science approach to studies on freshness and shelf life of foods. In: Cadwallder, K.R., Weeman, H. (Eds.), Freshness and Shelf Life of Foods, American Chemical Society Symposium Series No. 836. American Chemical Society, Washington, pp. 214–222.

Rheology

Lapasin, R., Pricl, S., 1995. *Rheology of Industrial Polysaccharides*. Chapman & Hall, New York.

Nishinari, K., 2006. Polysaccharide rheology and in-mouth perception. Chap. 16. In: Stephen, A.M. (Ed.), Food Polysaccharides and Their Applications. Marcel Dekker, New York.

Schenz, T.W., 1997. Using rheology of weak gels to improve fluid foods. Food Technology 51, 83–85.

Steffe, J.F., 1996. *Rheological Methods in Food Process Engineering*, second ed. Freeman Press, East Lansing.

Gels

Clark, A.H., 2000. Biopolymer gelation - the structure-property relationship. In: Gums and Stabilisers for the Food Industry - 10, Special Publication No. 251. The Royal Society of Chemistry, London, pp. 91–110.

Lee, C.M., 2002. Role of hydrodynamically active biopolymeric ingredients in texture modification and physical stabilization of gel-based composite foods. Journal of Food Science 67, 902–908.

Polysaccharides as Stabilizers and Other Applications

Dickinson, E., 1993. Emulsion stability. In: Nishinari, K., Doi, E. (Eds.), Food Hydrocolloids. Plenum Press, New York, pp. 387–398.

Dickinson, E., 1995. Emulsion stabilization by polysaccharides and protein-polysaccharide complexes. Food Science and Technology 67, 501–515.

Dickinson, E., 2003. Hydrocolloids at interfaces and their influence on the properties of dispersed systems. Food Hydrocolloids 17, 25–39.

Pegg, A.M., 2012. The application of natural hydrocolloids to foods and beverages. In: Baines, D., Seal, R. (Eds.), Natural Food Additives, Ingredients and Flavourings. Woodhead Publishing, Oxford, pp. 175–196.

Williams, P.A., Phillips, G.O., 2003. The use of hydrocolloids to improve food texture. In: McKenna, B.M. (Ed.), Texture in Food, Semi-solid Foods, vol. 1. Woodhead Publishing, Cambridge, pp. 251–274.

Starches: Molecular and Granular Structures and Properties

6

Chapter Outline

Key information and skills that can be obtained from the study of this chapter will enable you to do:

1. Define or identify:

Amylose	Gelatinization
High-amylose corn starches	Gelatinization temperature (T_{gel})
Amylomaize starch	Gelatinization temperature range
Apparent amylose content	Birefringence end-point temperature (BEPT)
Amylopectin	Melting endotherm
A chains	Pasting
B chains	Paste
C chain	Granule ghost
Waxy starches	Breakdown
Phytoglycogen	Setback
Hilum	Retrogradation

Carbohydrate Chemistry for Food Scientists. https://doi.org/10.1016/B978-0-12-812069-9.00006-6

2. Describe in general terms the (a) chemical and (b) molecular structures of amylose.
3. Describe in general terms the structure of amylopectin.
4. Write the Haworth structure for the following and similar segments of amylopectin:

$$\begin{array}{c} -(\mathrm{Glc})_2-\mathrm{Glc} \\ \downarrow \\ -\mathrm{Glc}-\mathrm{Glc}-(\mathrm{Glc})_2- \end{array}$$

5. Describe, using a labeled diagram, what happens when a starch slurry (excess water) is heated with some shear.
6. Describe the process of amylose and amylopectin retrogradation based on polymer crystallization theory.
7. Explain why amylose and amylopectin form complexes of different colors with iodine.
8. Describe the effects of polar lipids on starch pasting and starch pastes.

Introduction

Starch's unique chemical and physical characteristics and nutritional attributes set it apart from all other carbohydrates. Starch, products derived from starch, and sucrose constitute most of the digestible carbohydrate in the adult human diet (the lactose of milk being the only other carbohydrate digestible by humans). Starch is the predominant food reserve substance in plants, and the majority of starch consumed by humans occurs naturally within cereal grains (much of that being consumed after the kernels have been reduced to flours) and vegetables. Starch is the most abundant nutrient in the diet of humans—supplying 70%–80% of the calories consumed by humans worldwide. Of the edible dry matter in the world's food supply, 70% is in the form of cereals (corn/maize, 22%; wheat, 21%; rice, 16%; other cereals, 11%), 7% is in the form of roots and tubers (primarily potatoes, sweet potatoes, and cassava), 6% is in the form of sugar, and 5% is in the form of fruits and vegetables. So only about 12% is in the form of commodities that are not primarily composed of carbohydrates (animal products, 8%; legumes and oilseeds, 4%).

You will learn that starch is also unique in other ways. It is a substance that strongly interacts with water and other very polar compounds but can form complexes with hydrophobic compounds. It is usually thought of as a mixture of water-soluble polysaccharides, which it is, but can also behave like a thermoplastic material (that is, melts under the proper conditions).

Commercial isolated starches are obtained from seeds, particularly from corn, waxy corn (waxy maize), high-amylose corn (amylomaize), wheat, and rice of various amylose contents, and from tubers or roots, particularly potato and tapioca (cassava/manioc/yuca). The amount of starch used in the preparation of food products greatly exceeds the amount of all food hydrocolloids combined, and this does not include the starch in flours used to make bread and other bakery products, that naturally occur in grains used to make breakfast cereals, or that consumed in vegetables, such as potatoes.

Starch is converted into sugars during the ripening of fruits. As an example, the starch content of a mature, unripe (green) banana is about 20% of its fresh weight, but in a ripe (yellow) banana, it is about 2%. During the ripening period, the sucrose concentration increases from about 2.2% to about 8.5% of the fresh weight, the D-fructose concentration increases from about 0.5% to 5.3%, and the D-glucose concentration increases from about 0.4 to about 4.3%. Also, potato tubers, like many other plant tissues, undergo a phenomenon known as low temperature sweetening (that is, when they are stored at low temperatures, starch is converted into glucose, fructose, and sucrose). The ratios of the three sugars vary with both the potato cultivar and the storage temperature, greater amounts being produced at lower temperatures. The glucose is often (but not always) present in the greatest amount.

Starches and modified starches (Chapter 7) have an enormous number of food uses. In different formulations, they may provide adhesion, binding, clouding, dusting, film-formation, foam strengthening, gel formation, glazing, moisture retention, stabilizing, texturizing, and/or thickening. Certain modified starches prepared with particular specific physical characteristics can even play a role in lowering the fat content of prepared foods by providing a sensory perception similar to that of fat.

Starch granules

Starch is unique among carbohydrates because it occurs naturally as discrete particles, called granules (Figs. 6.1–6.4). Appearances of granules obtained from different plant sources are usually so different that the crop source of most commercial starches can be determined with an ordinary light microscope (that is, corn/maize, potato, rice, tapioca/cassava, wheat, etc.).

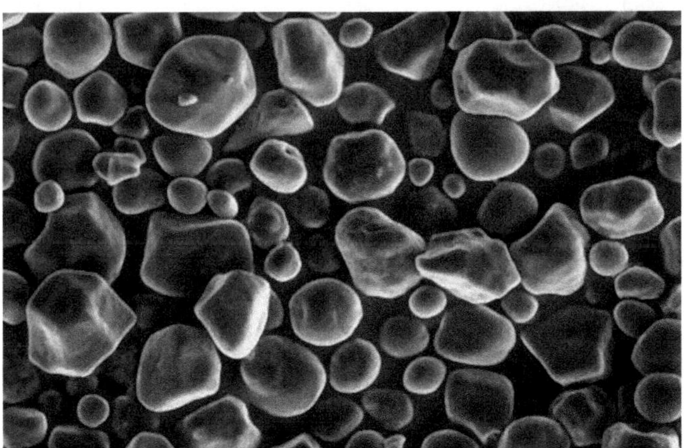

Figure 6.1 Corn/maize starch granules viewed with a scanning electron microscope. (Normal and waxy maize starch granules [below] look the same.) (Reprinted with permission from Fitt L.E. and Snyder E.M., Photomicrographs of starches, In: Whistler R.L., BeMiller J.N. and E.E. Paschall, (Eds), *Starch: Chemistry and Technology,* 1984, Academic Press, New York, 675-689.)

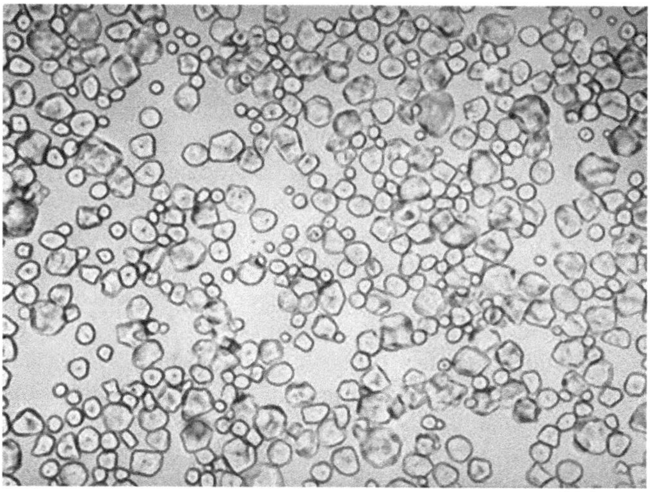

Figure 6.2 Corn/maize starch granules viewed with a light microscope. (Normal and waxy maize starch granules look the same.) (Reprinted with permission from Fitt L.E. and Snyder E.M., Photomicrographs of starches, In: Whistler R.L., BeMiller J.N. and E.E. Paschall, (Eds), *Starch: Chemistry and Technology*, 1984, Academic Press, New York, 675-689.)

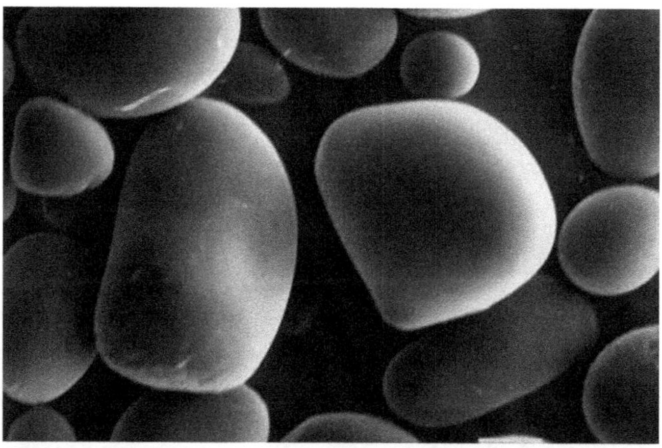

Figure 6.3 Potato starch granules viewed with a scanning electron microscope. (Reprinted with permission from Fitt L.E. and Snyder E.M., Photomicrographs of starches, In: Whistler R.L., BeMiller J.N. and E.E. Paschall, (Eds), *Starch: Chemistry and Technology*, 1984, Academic Press, New York, 675-689.)

Starch granules are relatively dense and insoluble but hydrate in room temperature water. They can be dispersed in water, producing low-viscosity slurries that can be easily mixed and pumped, even at concentrations of 35% or greater. The thickening power of starch is realized only when a slurry of granules is heated. A 5% slurry of most unmodified starch granules heated to about 80 °C (175°F) with stirring produces

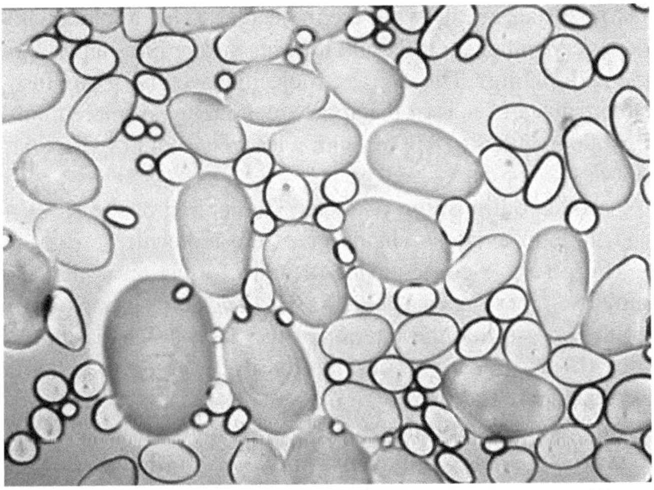

Figure 6.4 Potato starch granules viewed with a light microscope. (Note the differences in size for potato and corn starch granules in Figs. 6.1–6.4.) (Reprinted with permission from Fitt L.E. and Snyder E.M., Photomicrographs of starches, In: Whistler R.L., BeMiller J.N. and E.E. Paschall, (Eds), *Starch: Chemistry and Technology*, 1984, Academic Press, New York, 675-689.)

a high viscosity paste (Sections Granule Structure and Starch Gelatinization, Pasting, Pastes, and Gels).

A second uniqueness is that most starch granules are composed of a mixture of two polymers: an essentially linear polysaccharide called amylose and a highly branched polysaccharide called amylopectin.

Polysaccharide components

Amylose

Amylose is essentially a linear, polymeric molecule of (1 → 4)-linked α-D-glucopyranosyl units (Fig. 6.5) (Chapter 4). However, some amylose molecules have a few α-D-(1 → 6) branches (0.3%–0.5% of the total linkages). The branches are generally either very long or quite short and either near the reducing end of the molecules or

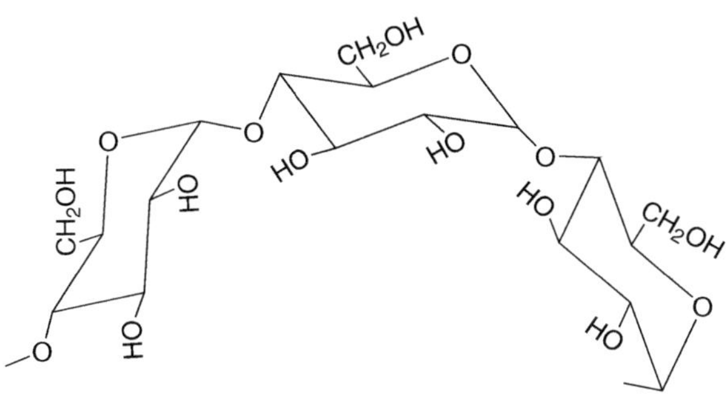

Figure 6.5 A trisaccharide segment of an unbranched portion of amylose and amylopectin molecules.

separated by large distances, allowing the molecules to act as essentially linear molecules, associating/crystallizing easily and forming strong films and fibers and (Section Retrogradation and Staling). The average molecular weight of amylose molecules of different commercial starches is in the approximate range of more than 10^5 to at least 3×10^6 (DP > 600 to at least 18,000), with wide ranges of polydispersity (Chapter 4) in all preparations.

The axial → equatorial position (Chapter 1) coupling of the (1 → 4)-linked α-D-glucopyranosyl units in amylose chains gives the molecules a right-handed helical (spiral) shape (Fig. 6.5). A consequence of helix formation is that films and fibers made from amylose are somewhat elastic. The hydroxyl groups are positioned on the exterior of the coil so that the interior of the helix is lined with hydrogen atoms and is lipophilic (hydrophobic). About one-fourth of the polysaccharide content (by weight) of most starches is amylose.

Two common commercial corn starches, called **high-amylose corn starches** and **amylomaize starches**, have **apparent amylose contents**[1] of 50%–60% and 70%–80%, respectively. These starches contain an amylose-like fraction (that is, one that analyzes as amylose, but which is different than the "true" amylose) that replaces approximately one-half and two-thirds, respectively, of the amylopectin molecules. They also contain some "true" amylose (20%–25%), but it is of lower molecular weight (DP 250–300) than the amylose of normal corn starch.

Amylopectin

Amylopectin, is a very large, highly branched glucan. An amylopectin molecule consists of one chain, called the **C chain,** which carries the one reducing end group and numerous branches (termed B chains) to which A chains or other B chains are attached. **A chains** are those chains that are connected to another chain via a (1 → 6) linkage and are unbranched. **B chains** are chains that are connected to the C chain or another B chain via a (1 → 6) linkage and have one or more A or other B chains attached to them via (1 → 6) linkages (that is, they are branched) Each of the chains of amylopectin molecules has the same right-handed α-helix conformation found in amylose molecules, but the C and B chains of amylopectin molecules have other linear chains of α-D-glucopyranosyl units joined by (1 → 4) linkages attached to them at the O6 position of glucopyranosyl units (that is, in (1→6) linkages) as branches. Branch–point linkages constitute 4%–5% of the total linkages. The evidence is strong that the branches of amylopectin molecules are clustered. The classic structure for amylopectin molecules is given in Fig. 6.6. In it pairs of nonreducing chain ends in clusters are entwined around each other in parallel double helices,[2] which

[1] Apparent amylose content refers to the amylose content that is measured as a complex with iodine (Section Complexes). However, some of the relatively long linear chains that are part of amylopectin or other rather highly branched molecules may analyze as amylase by this method without being true amylose.

[2] A double helix consists of two chains coiled around a common axis. It is the structure obtained if one takes two rods made out of something flexible such as rubber, places them side-by-side, holds one end steady, and twists the other end to make helices out of each of the two rods. DNA, for example, is a double helix of antiparallel chains.

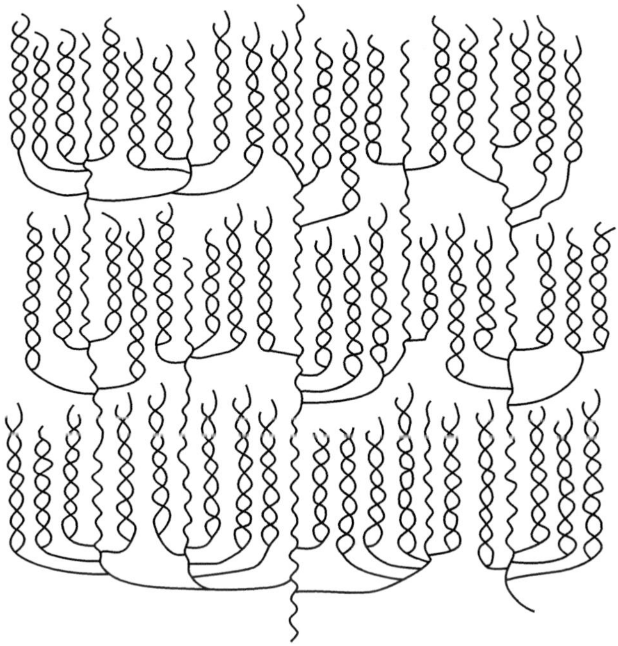

Figure 6.6 A classic diagrammatic representation of portions of amylopectin molecules showing how clusters could be packed in crystalline arrays in a starch granule.[3] Individual chains are helical; pairs of chains are double helical. The reducing end is at the bottom. (Redrawn from A. Imberty, A. Buléon, V. Tran, and S. Pérez, *Starch/Stärke* 43: 375–384 [1991].) The figure actually shows two molecules. The one on the right has single clusters on a vertical molecule, The one on the left has more than one cluster coming off horizontally from the same point on the vertical molecule.

are packed together to form the crystallites in starch granules. An alternative and quite probable general structure of amylopectin molecules (Fig. 6.7) maintains the cluster arrangement of the branches, but has them arranged differently along a C chain. This model is congruent with known features of amylopectin molecules, such as evidence that they have supermolecular helical structures.

Average structures, average molecular weights, molecular weight ranges, and, perhaps, shapes of amylopectin molecules vary with the botanical source. However, most amylopectin molecules have a trimodal distribution of A and B chain lengths. For potato starch, the average degrees of polymerization (DP) of these three fractions are 13–15, 23–32, and 34–45, the smallest of which consists of the outermost chains

[3] It is not actually known whether there is one cluster per unit of the C chain as in the right-side structure or several clusters wide as in the left-side structure, or even if this general structure is correct (as described in the text). It is known, however, that amylopectin molecules are in the shape of a flat ellipsoid with the length (long axis) being 30–40 times greater than the width (short axis), so it is most likely that a structure of a type shown in the left diagram and the right side of the structures in the right diagram is closer to the actual structure.

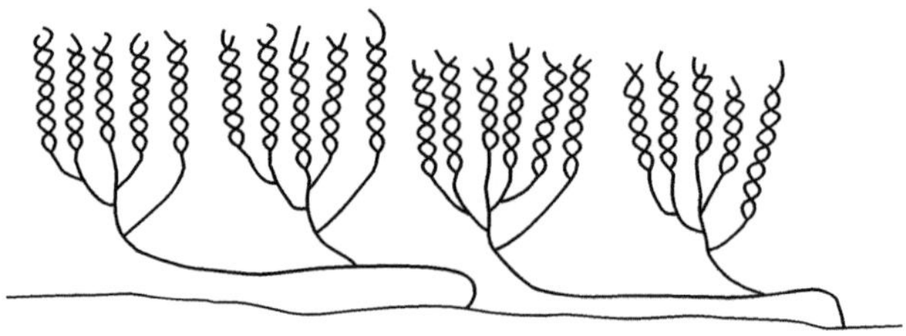

Figure 6.7 A newer diagrammatic representation of a portion of an amylopectin molecule as proposed by E. Bertoft.
Adapted from E. Bertoft (2013) *Cereal Chemistry* 90: 294–311.

(that is the A chains). The other two fractions are generally referred to as short B chains and long B chains. Average molecular weights of amylopectin molecules of commercial starches are often $>10^7$ (DP about 10^5) and have been reported to be as large as 2×10^9 (DP about 10^7), making them among the largest molecules in nature (with perhaps only lignin being larger).

Amylopectin is present in all known starches, constituting about three-fourths (by weight) of most normal starches. However, some starches (commonly called **waxy starches**) consist entirely of amylopectin. Waxy maize (waxy corn) was the first grain recognized as containing only amylopectin in its starch. It was so termed because, when the kernel is cut, the new surface appears vitreous or waxy, but wax is not present in the kernel. Since the discovery of waxy maize starch, other all-amylopectin starches have also been called waxy. However, the trend is to call the all-amylopectin potato starch just that (a more descriptive term). Because the enzyme that synthesizes amylose is on three gene loci in wheat and barley, some cultivars of these grains can be termed, "partial waxys" (that is, they can have amylose contents between those of normal starch [22% for wheat starch] and a complete waxy type [0% amylose]). Amylose contents of normal rice starches vary considerably. Apparent amylose contents[2] of rice starch can vary from 0% (waxy rice starch) to at least 40% (that is, from 60% or less to 100% amylopectin), depending on the cultivar and the growing conditions.

Fine structures of the amylopectin molecules, especially the ratios of the longer chains to the shorter chains, appear to be a determinant of the properties of the starch and foods that contain it. For example, hard-cooking rice has more long chains (DP 92–98) as compared with soft-cooking rice, which has more short chains (DP < 25).

Potato amylopectin is unique in having phosphate ester groups attached to one in about every 200–300 α-D-glucopyranosyl units (that is, about 0.33%–0.50% of the units are phosphorylated). The phosphate ester groups are located near branch points and most often (60%–70%) at an O6 position; the remainder are on O3. About 88% are found on B chains. Because of its phosphate ester content, potato starch has a negative charge. The resulting slight coulombic repulsion likely contributes to the rapid swelling of potato starch granules in warm water and to the good clarity of potato starch pastes. (In cereal starches, less than 0.01% of the α-D-glucopyranosyl units are phosphorylated.)

Phytoglycogen, which also is a highly branched glucan containing α-D-glucopyranosyl units joined by (1 → 4) and (1 → 6) linkages, has a higher degree of branching than does amylopectin (about 10% of all linkages, about twice that of amylopectin). It is water soluble and is not present in granular form. It is present in sweet corn in amounts of up to 25%.

Granule structure

Starch molecules are synthesized in a plant cell organelle called an amyloplast that eventually becomes filled with a starch granule or a cluster of granules. As starch molecules form in an amyloplast via the action of synthesizing enzymes, they combine with one another to form a compact ordered semicrystalline mass. The region of ordered molecules continues to grow in a radial direction from the growth center, called the **hilum**.[4] Completed granules contain molecules arranged radially,[5] and both polycrystalline[6] and noncrystalline (amorphous) regions. The clustered branches of amylopectin occur as packed double helices (perhaps as in Fig. 6.7). These double-helical structures form the many small crystalline regions in dense layers of starch granules that alternate with less dense (less crystalline) regions, generally called amorphous layers. (The layers are something like the layers in an onion except that they cannot be separated from each other.) Amylose molecules reside in amorphous regions and can diffuse from partially water-swollen granules more readily than do amylopectin molecules.

Amylopectin molecules, which form the principal structure of the granule, are arranged with their reducing ends toward the center of the granule. This radial,[5] ordered arrangement of amylopectin molecules in a granule makes granules anisotropic.[7] Evidence that starch granules are anisotropic and, therefore, birefringent[8] is given by the polarization cross (cross of isocline) seen when granules are examined using plane polarized light (Fig. 6.8). The center of the cross is at the hilum. Cereal starches produce an X-ray diffraction pattern (called an A-type pattern) that is indicative of parallel, double helices separated by interstitial water. In potato starch granules (which produces a B-type X-ray diffraction pattern), a column of water molecules replaces one of the double helices (Fig. 6.9).

[4] The hilum is the original growing point of a granule. It is usually a cavity in isolated granules and usually, but not always, near the center of granules.

[5] Like spokes in a wheel in a three-dimensional sense.

[6] Polycrystalline in this case refers to the presence of a large number of small crystallites.

[7] The word anisotropic means not the same (from an- meaning not and isotropic meaning the same). Anisotropic substances show different properties when measurements are made in different directions. In the case of starch granules, that would be in the direction of the orientation of the amylopectin molecules (radially) versus across the orientation of the molecules. When a light beam is passed through a granule, the light traveling in the direction of orientation of the amylopectin molecules will travel at a different velocity than the light traveling across the orientation of the molecules. This phenomenon is known as birefringence.[8]

[8] In the context of starch, birefringence is the splitting of parallel waves of plane polarized light into two waves perpendicular to each other by passing through a granule (an optically anisotropic[7] medium), resulting in a dark cross on a light background as seen in Fig. 6.9.

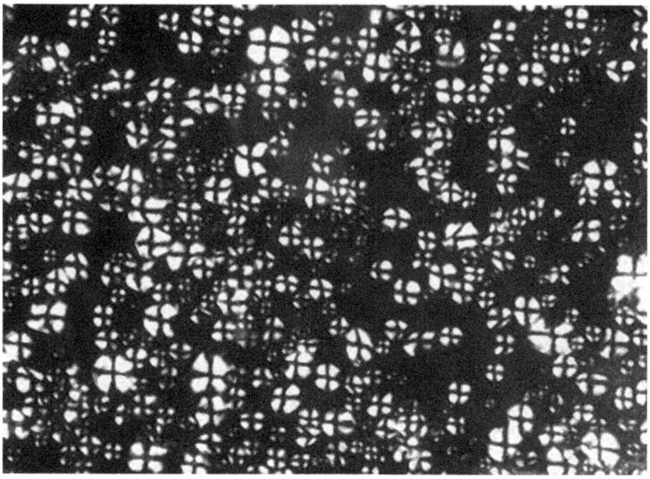

Figure 6.8 The same field of corn starch granules shown in Fig. 6.2 viewed through crossed polarizers using a polarizing microscope. The polarization cross is an indication of the ordered nature of native starch granules. (Reprinted with permission from Fitt L.E. and Snyder E.M., Photomicrographs of starches, In: Whistler R.L., BeMiller J.N. and E.E. Paschall, (Eds), *Starch: Chemistry and Technology*, 1984, Academic Press, New York, 675-689.)

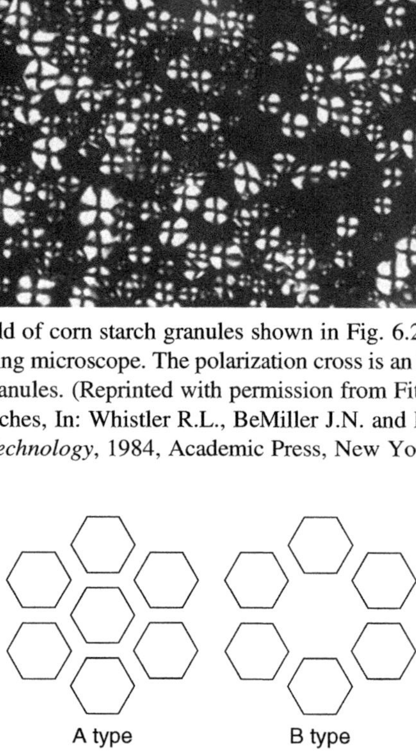

Figure 6.9 A diagrammatic representation of the arrangement of six parallel double helices in starches that give a type A pattern and starches that give a type B pattern. Water molecules replace the center double helix in the B-type starches.

Other components of granules

The granules of all starches contain nonstarch components, including ash, lipid, and protein (Table 6.1). The phosphorus content of potato starch (0.06%−0.1%)[9] is due to the presence of phosphate monoester groups on the starch polysaccharide molecules (primarily, if not exclusively, on amylopectin molecules). Calcium, magnesium, sodium, and potassium ions are present as cations associated with the phosphate groups, increasing the ash content. The amylose and amylopectin molecules of cereal starches either do not have phosphate ester groups or have very much smaller amounts.

[9] The phosphorus content varies with cultivar and growing conditions.

Table 6.1 General properties of granules and pastes of major food starches

	Normal Corn Starch	Waxy Maize Starch	High-Amylose Corn Starch	Potato Starch	Tapioca Starch	Wheat Starch
Granule size (μm)	2–30	2–30	2–24	5–100	4–35	0.5–45
% Amylose (approximate)	28	0	50–70	21	17	28
Gelatinization/pasting temperature range (°C)[a]	75–80	65–70	66–170[b]	58–65	52–65	52–85
Relative viscosity	Medium	Medium–high	Very low[b]	Very high	High	Medium-low
Paste rheology (body)	Short	Long	Short	Very long	Long	Short
Paste clarity	Opaque	Slightly cloudy	Opaque	Clear	Clear	Cloudy
Tendency to gel/retrograde	High	Very low	Very high	Medium–low	Low	High
Gel consistency	Firm	Nongelling	Very firm	Salvelike	Soft	Soft
Lipid (% ds)[c]	0.8	0.2	–	<0.1	<0.1	0.9
Protein (% ds)[c]	0.35	0.25	0.5	0.1	0.1	0.4
Starch-bound phosphorus (% ds)[c]	0	0	0	~0.08	0	0
X-ray diffraction pattern type	A	A	B	B	A	A

[a]From the initial temperature of gelatinization to complete cookout.
[b]Under ordinary cooking conditions, where the slurry is heated to 95–100°C, high-amylose corn starch produces essentially no viscosity. Gelatinization of high-amylose corn/maize starch with about 50% apparent amylose content may begin as low as 66°C and end as high as 166°C, depending on moisture content. Gelatinization of high-amylose corn/maize starch with about 70% apparent amylose content may begin as low as 66°C and end as high as 171°C, depending on moisture content.
[c]ds = dry solids.

Only cereal starches contain significant endogenous lipids in the granules (Table 6.1). These internal lipids are primarily free fatty acids (FFA) and lysophospholipids (LPL), largely lysophosphatidyl choline (89% of the LPL in normal corn starch), with the ratio of FFA to LPL varying from one cereal starch to another. The lipids in normal corn starch are 51%–62% FFA and 24%–46% LPL, whereas the lipids of wheat starch are 86%–94% LPL and only 2%–6% FFA. The lipids of barley starch are similar in composition to those of wheat starch. Almost equal proportions of LPL and FFA occur in millet, oat, rice, and sorghum starches. The lipid contents of cereal starches also vary with the amylose content. Normal cereal starches contain about 0.6%–1.2% total lipid, which occurs both free and complexed with amylose (below). Waxy maize starch contains only about 0.2% lipid. The amount of lipid also varies with the source. Root, tuber, and legume starches generally contain less lipid than cereal starches. Potato starch contains only about 0.06% lipid, probably largely on the granule surface.

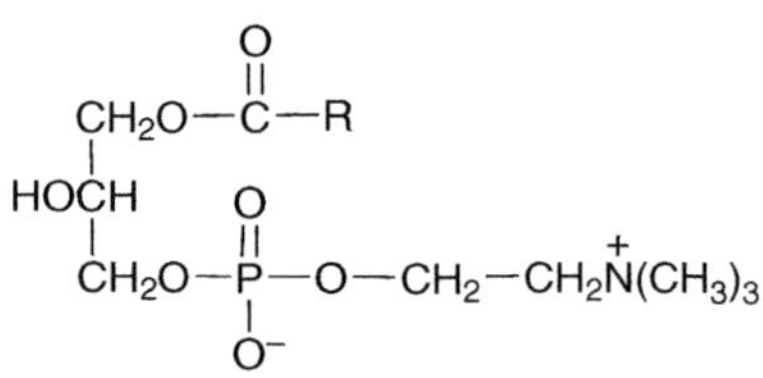

Lysophosphatidycholine

FFA and mono- or diacylglycerols (but not triacylglycerols/triglycerides) can form complexes with amylose by inclusion of one of the fatty acid chains within an amylose single helix (Fig. 6.10) (Section Complexes). This interaction contributes to the low paste viscosity of cereal starches and to their gels being more opaque. Lipids complexed in helices can only be extracted with polar solvents after pasting of the granules.

Protein contamination of starch (corn starch normally contains about 0.35% protein; wheat starch about 0.4%) may be largely residual endosperm storage protein stuck to the surface of granules, but some protein (at least in part as residual synthase enzyme) is entrapped within the granule. Ash (inorganic material) contents of starches are generally 0.1%–0.5%.

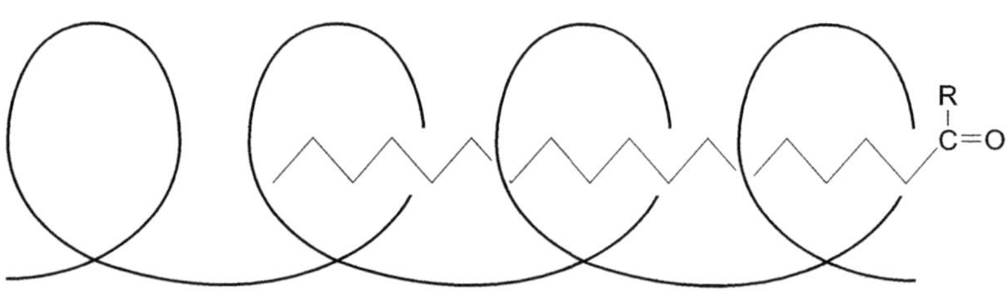

Figure 6.10 A diagrammatic representation of a complex of a molecule of a fatty acid derivative with a segment of an amylose molecule.

Granule types

Corn starch granules, even those from a single plant source, have a diversity of shapes (Figs. 6.1–6.4). Some in the floury center of the corn kernel endosperm are almost spherical. Others are more angular. (When observed using a light microscope [which gives a two-dimensional image], their shapes are those of polygons.) Those in the proteinaceous vitreous (horny) endosperm found near the outside of the kernel may be indented by protein bodies.

Wheat starch granules have a bimodal size distribution due to initiation of granule growth at different times. Distributions vary with wheat types (for example, in hard and soft wheats), but in general, wheat starch is composed of A granules[10] which are large, are lenticular in shape, and have a diameter of more than 10 μm. B granules[10] are small, rather spherical, and about 1–5 μm in diameter. In a typical wheat starch preparation, the large (A) granules constitute 1%–10% (average about 7%) of the granules by number, but 50%–87% (average about 73%) of the starch by weight, while the small (B) granules may constitute 90%–99% (average about 93%) of the total by number, but only 13%–50% (average about 27%) of the starch by weight. The small wheat starch granules contain more lipid (lysolecithin) (Table 6.1).

Rice starch granules, on average, are the smallest of the commercial starches (2–9 μm), although small (B) granules of wheat starch are almost the same size.

The granules in tuber and root starches (such as potato and tapioca/cassava starches) tend to be larger than those of seed starches and are generally less dense and easier to gelatinize (below). Some potato starch granules may be as large as 100 μm along their major axis. Potato and tapioca starch granules swell readily to large sizes.

Banana starch, which can be obtained from green, cull bananas, also has large granules. Amaranth and quinoa starch granules are very small (about 1 μm in diameter).

Starch gelatinization, pasting, pastes, and gels

For practical purposes, uncooked starch granules do not increase the viscosity of water. A 30%–35% slurry of granular starch is easily pumped and stirred, but starch is seldom used in formulations (except as flour in bakery products) at concentrations of more than 7% (usually in concentrations of 2%–5%, the concentration being dependent on the type of starch).

Undamaged starch granules are insoluble in cold water, but undamaged granules can imbibe water reversibly (that is, they can swell slightly and then return to their original size on drying). This reversible range of swelling varies. It increases the size of normal corn starch granules by about 10% and the size of waxy maize starch granules by about 25%. Reversible swelling generally increases with granule diameter, approaching a doubling in size for the largest potato starch granules. The equilibrium

[10] The A and B granule designations of wheat starch are unrelated to the designations of starches that give A- and B-type X-ray diffraction patterns.

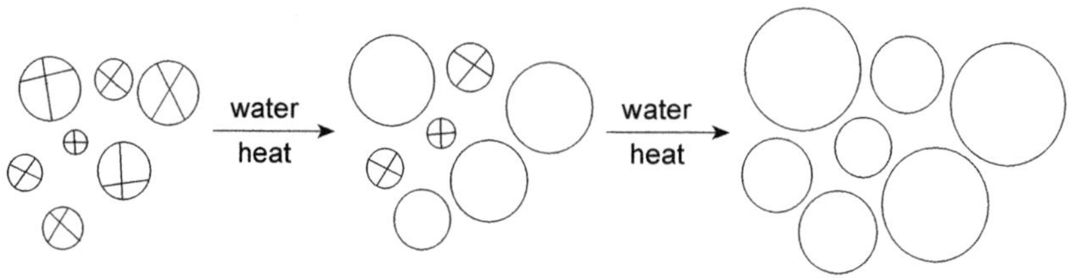

Figure 6.11 A diagrammatic representation of starch granule gelatinization over the temperature range of gelatinization. The diagram indicates that the larger granules in a population of granules swell and lose birefringence at lower temperatures than do with the smaller granules and that, with continued heating, all granules eventually gelatinize.

moisture content of corn starch granules in an aqueous slurry is about 28% and of potato starch granules is about 33%.

When an aqueous slurry of starch granules is heated to temperatures above what is known as the initial gelatinization temperature (see below), dramatic changes occur. As starch granules are heated in excess water, they swell beyond the amount caused by the reversible swelling as hydrogen bonds in amorphous regions are disrupted and water (which is a plasticizer for starch granules) is absorbed. As the temperature continues to increase, more hydration and more swelling occurs in amorphous regions, and hydrated crystallites melt. (Amorphous and crystalline regions appear to be connected, so as amorphous regions swell, the crystallites are undoubtedly distorted so that the starch chains can be, at least partially, hydrated/plasticized, resulting in melting of the crystallites.) This disruption of both amorphous and crystalline structures results in losses of order and the granule properties related to that order, such as birefringence.[7] The process resulting in the irreversible loss of order (that is, resulting in an order–disorder phase transition) within starch granules is called **gelatinization** (Fig. 6.11). Irreversible granule swelling and leaching of amylose occurs during gelatinization. (Some leaching of amylose begins at temperatures about 20° below the **gelatinization temperature [T_{gel}]** [here meaning the initial T_{gel}] due to its location in noncrystalline regions and the fact that it is a smaller linear polymer molecule that can diffuse out of granules.) Each individual granule gelatinizes at a specific and sharp temperature (a melting temperature), but for a population of granules, gelatinization occurs over a **temperature range** because different granules have different gelatinization temperatures.[11] The temperature of initial gelatinization and the range over which gelatinization occurs depend on the type of starch and the method of measurement. Values obtained using a polarizing microscope equipped with a hot stage are the initiation temperature (when the first granule in the field loses birefringence),

[11] Gelatinization of an individual granule occurs at its own specific temperature (like a melting point). However, a population of granules will gelatinize over a range of temperatures. The temperature at which the first granule gelatinizes is called the initiation or onset temperature of gelatinization, and the temperature at which the last granule gelatinizes is called the conclusion or end-point temperature. At temperatures in between the initiation and conclusion temperatures, the percentage of gelatinized granules will be between 0% and 100%.

the midpoint temperature, and the completion or **birefringence end-point temperature** (**BEPT**; when the last granule in the field loses birefringence). Other methods for determining the temperature, extent, and/or heat of gelatinization involve measuring the absorption of heat energy (by differential scanning calorimetry [DSC]), rate of enzyme-catalyzed hydrolysis, loss of turbidity, molecule dissolution, dye adsorption, chemical reactivity, and changes in the X-ray pattern. Of these methods, one of the most sensitive methods for determining the extent of gelatinization, and one easy to perform, is to measure the increase in extent of hydrolysis catalyzed by glucoamylase/amyloglucosidase (Chapter 7) or a mixture of alpha-amylase and glucoamylase, with the produced D-glucose determined quantitatively using glucose oxidase (Chapter 2) because gelatinized starch is much more susceptible to the action of hydrolyzing enzymes than uncooked granules. Gelatinization temperature data for several commercial starches are given in Table 6.1.

Factors other than the type of starch that affect starch swelling and gelatinization include the starch—water ratio, pH, presence or absence of a swelling-inhibiting or swelling-promoting salt, the concentration of the salt, and the presence and concentration of a solute such as sucrose (Chapter 19). In addition, derivatization permanently disrupts hydrogen bonding (Chapter 7), mostly by a steric effect, and makes granule swelling and gelatinization more facile. Finally, for starches within a single species (for example, for starches in different kinds of rice), the gelatinization temperature range appears to be a function of both the proportion of amylopectin short chains and their chain lengths.

Gelatinization of starch is an energy-absorbing process that can be examined by DSC, which measures both the temperature and the enthalpies of gelatinization (Fig. 6.12). Although there is no complete agreement on the interpretation of DSC data and the events that take place during gelatinization of starch granules, the following general picture is widely accepted. Water is a plasticizer for starch. Its mobility-enhancing effect is realized first in the amorphous regions, which are in either a glassy[12] or a rubbery[12] noncrystalline (that is, amorphous) state. When starch granules are heated in the presence of at least 60% water and their glass transition temperature (T_g)[13] is reached, the plasticized glassy regions of the granule undergo a phase transition to a rubbery state. However, the energy absorption peak for this transition is not usually evident. It is believed that melting of the crystalline regions (up to 30% of the total granule) very closely follows the glass transition and thus the enthalpy of the **melting endotherm** (Tm_I)[14] is much larger than that of the glass transition. (An alternative explanation is that T_g has already been reached before the scan has begun.)

[12] A glass is a mechanical solid (a supercooled liquid) capable of supporting its own weight against flow. A rubber is an undercooled liquid that can exhibit viscous flow (Chapter 5).

[13] The glass transition temperature (T_g) of a starch granule is the temperature of melting of a glass (that is, it is the temperature at which a glassy material changes to a rubbery material). T_g of a starch granule is used interchangeably with its gelatinization temperature (T_{gel}) when measured by DSC. It is not clear whether the glass transition is the driving force for gelatinization.

[14] The absorbed energy (enthalpy) is measured by a differential scanning calorimeter. This value is known as the **melting endotherm**.

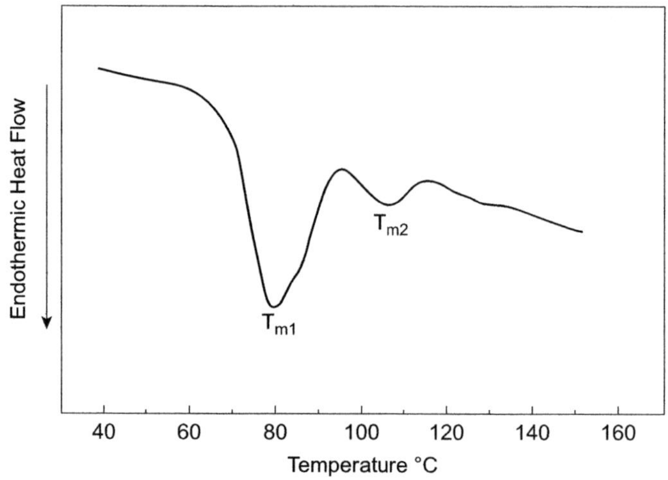

Figure 6.12 A typical differential scanning calorimetry trace of gelatinization of a cereal starch at 50% solids: T_{m1} = the main melting endotherm, T_{m2} = the melting endotherm of amylose-native lipid complexes.

Melting of lipid—amylose complexes, if present (Section Complexes), occurs at much higher temperatures (100—120°C in excess water). The onset of gelatinization also occurs at lower temperatures when damaged[15] granules are present because damaged granules absorb water more easily.

Gelatinization is followed by a process called **pasting**. Swollen, gelatinized starch granules are quite fragile and rather easily disrupted. Continued heating of starch granules in excess water with some shear results in **pasting** (that is, formation of a **paste** [as a hot starch cook is called]), via further granule swelling, additional leaching of soluble components (primarily amylose), and disruption of the granules. A starch **paste** is a viscous mass consisting of a continuous phase (a molecular dispersion/solution) of starch polymer molecules and a discontinuous phase of swollen granules, granule fragments (**granule ghosts**[16] and other remnants), and retrograded molecules (Section Retrogradation and Staling). In other words, cooking of most starches without high shear (as is usually the case) results in a paste that consists of swollen granules, granule ghosts, and fragments of swollen granules (composed mostly of amylopectin) suspended in a solution of mostly amylose molecules. (The swollen granules and granule fragments [that is, the discontinuous phase] reinforce a starch gel much as an aggregate strengthens concrete and glass or carbon fibers reinforce plastics.) Heating a starch gel to temperatures above about 70°C

[15] Granules can become damaged, for example, during milling of wheat to produce flour (during which process the degree of damage is controlled to control the properties of the flour) or in some operations during the milling of a (formerly wet) dried starch cake to produce a fine powder.

[16] The **ghost** is the remnant remaining after cooking starch granules without shear or with only moderate shear. It is at least the remainder of the outer portion of the granule, the inner portion being solubilized. The ghost has the appearance of a hollow shell.

reduces its stiffness by melting the amylopectin crystallites. (Amylose crystallites melt at much higher temperatures.) Complete molecular dispersion is obtained in excess water under conditions of high temperature and high shear—conditions seldom, if ever, encountered in the preparation of food products. Cooling of a hot, corn starch paste can result in a firm, viscoelastic gel[17] that at least partially retains its shape when removed from a container. Mild treatments of granules with acid results in some hydrolysis of glycosidic bonds making the granules easier to cook (paste) (Chapter 7). Reducing the molecular weights of the starch polymers results in a firmer gel when the paste is cooled.

In normal food processing involving heat and moisture, most starch granules undergo both gelatinization and pasting. Gelatinization and pasting of starch granules occurs because, as the temperature of a starch–water suspension increases, molecules within granules vibrate and twist so violently that intermolecular hydrogen bonds are broken, allowing greater penetration of water molecules between polysaccharide chains. As a result, the intermolecular hydrogen bonds between the chains are replaced with hydrogen bonds to water molecules, producing more extensive hydration. Thus, starch polysaccharide molecules become sheathed in layers of water molecules that plasticize them and allow them to move more freely and become more separated from each other. Eventually, molecular segments move to distances and positions that make it impossible for them to return to their original positions and orientations upon dehydration. This causes irreversible swelling of granules to several times their original size.

Cooking behaviors of different starches can be compared with a Rapid Visco Analyzer and/or a Visco-amylograph, both of which record the change in viscosity under low shear rates as a function of temperature (increasing usually from 50°C to 95°C, where it is held for a specified time, then decreased to 50°C) (Fig. 6.13). When a 5%–7% starch suspension is heated with gentle stirring, the granules absorb most of the water, causing them to swell and press against each other, filling the container with a highly viscous paste of highly swollen granules, but highly swollen granules are fragile and easily broken by stirring,[18] resulting in a decrease in viscosity. By the time the peak viscosity is reached, some granules have already ruptured and fragmented due to the slight shear forces. With continued stirring, more granules disintegrate, causing a decrease in viscosity. (The loss in viscosity after the peak viscosity is reached is called **breakdown**. The viscosity reaches a plateau called both the trough viscosity and the hot paste viscosity. (Breakdown [the difference between the peak viscosity and the trough viscosity] is recorded by the instrument.) If the starch contains amylose, it will diffuse from the swollen, gelatinized granules to the external water phase. Some amylopectin may also leach from the swollen granules, but a complete molecular

[17] In starch science, a cooled paste is called a gel, even if it flows. However, technically, even the hot paste is a gel because $G' > G''$ (Chapter 5).

[18] The instruments mentioned measure viscosity by measuring the resistance to movement of a rotating device that provides some stirring (at low shear) in addition to a measure of viscosity.

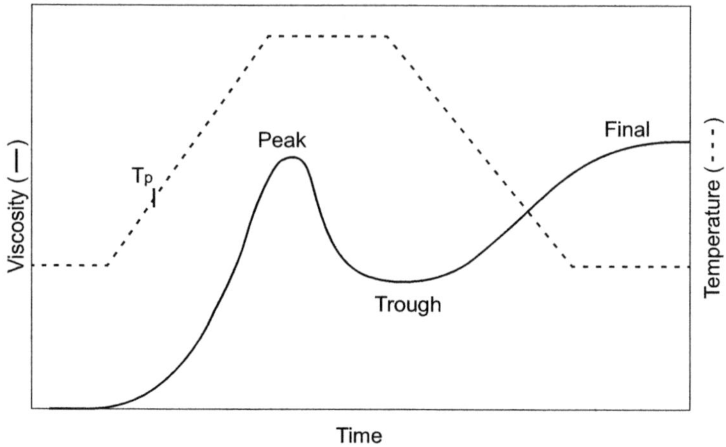

Figure 6.13 A typical pasting and paste property curve of an unmodified starch. The mark (T_p) on the temperature curve marks the pasting temperature, the temperature at which the rapid increase in viscosity begins. The pasting temperature follows the gelatinization temperature by a few degrees. Such curves are used to give indications of the cooking, hot paste, and gel formation characteristics of native and modified starches.

dispersion (solution) is produced only from dilute suspensions subjected to high temperature and high shear[19] (Fig. 6.14). Upon cooling of the hot paste, viscosity increases as it does with all polymer solutions as their temperatures decrease, but with starch, there is an additional factor. As a starch paste cools, some starch molecules begin to reassociate, first increasing opacity, then forming a precipitate or a gel (Chapter 5). This process is called **setback**. (Setback is measured by the instrument as the difference between the final viscosity and the trough viscosity and is an important factor in determining the stability of the formed gel.) The firmness of the gel, which usually is the end result of retrogradation (Section Retrogradation and Staling) depends on the extent of junction zone formation (Chapter 5). Junction zone formation can be either facilitated or hindered by the amount of water present and the presence of other ingredients (fats, proteins, sugars, and acids). Starches with lower amylose (higher amylopectin) contents generally produce more viscous pastes. Starches with higher amylose contents generally produce firmer gels. Rapid cooling in either case generally produces a weak gel exhibiting syneresis,[20] while slow cooling generally results in a firmer gel.

When using instruments of the abovementioned type, viscosities and other characteristics of the pasting curves will vary with concentration, shear rate, temperature profile, and composition of the aqueous phase, so conditions as close as possible to actual use conditions should be used. Box 6.1 lists factors that affect the curve.

[19] High temperatures and/or high shear forces will effect depolymerization (that is, a reduction in molecular size)—especially, and probably exclusively, of the amylopectin molecules.

[20] Syneresis is the expulsion/separation of liquid from a gel caused by shrinking of the gel (Chapter 5).

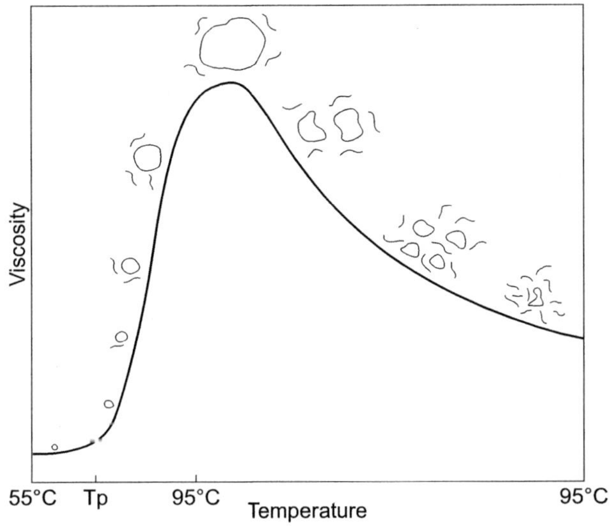

Figure 6.14 Representative pasting curve showing the typical swelling and disintegration that a granule undergoes during the cooking process in relation to viscosity.

Box 6.1 Main Factors That Affect the Cooking Curves of the Rapid Visco Analyzer and the Visco-amylograph

Type of starch (nature of amylose and amylopectin and their ratio)
Types and amounts of any starch modifications (Chapter 7)
Starch–water ratio
Heating rate
Hold temperature and time
Cooling rate
Shear rate
Presence of surfactants/emulsifiers (kind and concentration)
Sugar concentration
pH
Presence of triglycerides (fats and oils)
Presence of certain specific hydrocolloids

Normal corn starch has relatively short chains of amylose and produces chunky (cohesive[21]), opaque gels (opaque because of the presence of crystallites formed by retrogradation) that undergo syneresis.[20] Tuber (potato) and root (tapioca/cassava)

[21] Cohesive refers to the tendency of the molecules to stick together. Chunkiness in gels is an indication of cohesiveness. Cohesive gels are perceived as being difficult to break up in the mouth (Section Gels, Chapter 5).

starches gelatinize more easily than do cereal starches and produce high-viscosity pastes that thin rapidly with moderate shear because their highly swollen granules break easily. They have longer amylose chains and produce gels that are relatively clearer (because of a reduced tendency to retrograde) and are more cohesive (rubbery). Waxy maize starch, of course, has no amylose and also forms clearer gels, but even it will retrograde (that is, become firmer and undergo syneresis[20]) slowly over a period of days or weeks. Waxy maize starch is used more than potato and tapioca starches because it is more resistant/tolerant to acid and shear. High-amylose corn (amylomaize) starch does not cook to any extent at 95°C (Fig. 6.15). Modification (Chapter 7) is employed to overcome defects in starch properties.

Food companies generally prefer the clarity and textures of pastes and gels produced from waxy maize starch, which are more like those of potato and tapioca starches (see below) than they are of normal corn starch, even though they are cohesive[21] and somewhat rubbery. However, waxy maize starch can be easily overcooked to produce stringy pastes. As a result, both normal and waxy maize starches are often modified before use in foods (Chapter 7). High-amylose corn starches produce opaque, strong gels of the kind needed to make gum candies. Potato starch produces clear, viscous pastes. Potato starch is also subject to overcooking (as are all unmodified starches, except high-amylose corn starches, but overcooking is easier with waxy maize, potato, and tapioca/cassava starches) with loss of viscosity; but if used properly, it can be used in concentrations 25%−35% less than required for other starches to give the same viscosity. Tapioca (cassava) starch is much like waxy maize starch and produces clear, cohesive pastes that gel slowly. Normal rice starch produces an opaque gel useful for baby food. Waxy rice starch gels are clear and cohesive.[21] Wheat starch gels are weak and opaque. Generally, what food processors want for many applications are starch products that produce

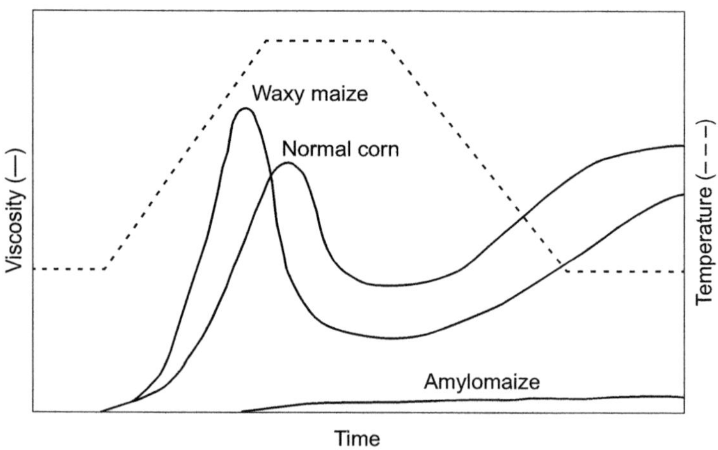

Figure 6.15 Comparison of the pasting and paste property curves of unmodified normal corn/maize, waxy maize, and amylomaize starches.

short-textured, clear, highly viscous pastes and gels. These properties can only be obtained from modified food starches (Chapter 7).

Various added ingredients will affect the shape of the cooking curves, that is, modify the pasting and paste properties of the starch. Surfactants (emulsifiers) (namely, sodium stearoyl 2-lactylate [SSL, a common name used in the food industry rather than a scientific name] and monoglycerides) generally increase the pasting temperature and the overall viscosity. Diacetyl tartaric acid esters of monoglycerides increase the pasting temperature but not the viscosity. Complexation with the starch polymers, primarily with amylose, is undoubtedly involved in the action of surfactants on starches.

Sugar (sucrose), commonly used in products containing starch, generally has little effect at concentrations below 30%, but above sugar concentrations of c. 30%, gelatinization temperatures are dramatically increased and pasting is delayed; so the sugar should be added after cooking the starch to prevent a thin mixture. This phenomenon could be the result of competition for water or due to starch–sucrose interactions, but probably is a result of increasing the molecular weight of the solvent and making it a poorer plasticizer. The presence of various sugars decreases gel strength by interfering with the formation of junction zones between starch polysaccharide molecules.

The presence of acids during cooking always decreases paste viscosity, but usually increases gel strength as a result of depolymerization (reduction in molecular weight/size) of the starch polysaccharides (Chapter 7).

Because starch polysaccharide molecules are for the most part nonionic, salts have little effect on gelatinization and gel formation until their concentration reaches the point that they begin to tie up large amounts of water (thus lowering the amount of water available for hydration of the starch polymer molecules) or by exerting an anti-plasticizing effect. Such concentrations are rarely encountered in food products. Salts and sugars have less of an effect on potato amylopectin and phosphorylated modified food starches (Chapter 7), likely because starch molecules containing ionic groups have a greater water-binding capacity, and in addition, their charges cause the starch molecules to repel each other, at least at the point of derivatization.

The rate of shear, the rate of temperature increase, the hold temperature, the time at that temperature, the rate of cooling, and other factors (Box 6.1) also affect the shapes of cooking curves.

Relationships of starch gelatinization, pasting, pastes, and gels to foods

Starches serve a variety of roles in foods and food production (Box 6.2). Often, but not always, they are used to produce viscous pastes and soft-textured gels. In other cases, for example in producing bakery products, the food material containing starch (in the form of wheat flour) is heated without excess water and/or without shear. The extent of starch gelatinization in baked goods strongly affects product properties, including storage behavior and rate of digestion. In some baked products, many starch granules remain ungelatinized because of the low moisture content of the dough. For

Box 6.2 Some Products Containing Native or Modified Starches

Bakery products
- Breakfast bars
- Brownie mixes
- Cake mixes
- Cereal bars
- Cream cakes
- Cupcake mixes
- Danish fillings
- Frostings
- Glazes
- Pie fillings
- Refrigerated doughs

Breadings for chicken patties
- onion rings, fish products, etc.

Beverages
- Cocktail mixes
- Fruit juices
- Fruit drink mixes

Canned foods
- Canned Spaghetti
- Cream-style corn

Condiments
- Mustard

Confectioneries
- Candies
- Caramel
- Fillings for chocolates
- Marshmallows

Dairy products
- Cheesecakes
- Cheese powders
- Cheese spreads
- Dips
- Low-calorie cheesecakes
- Sour creams
- Yogurts

Desserts
- Mousse mixes
- Pudding mixes

Frozen foods
- Frozen guacamole
- Frozen lasagna
- Frozen pot pies
- Frozen rice
- Frozen yogurt

Gravy mixes

Meat products
- Luncheon meats
- Poultry breast slices
- Sausage

Microwavable meals

Nutritional beverages

Rice noodles

Sandwich spreads

Sauces
- Barbeque
- Cheese
- Cocktail
- Horseradish
- Pizza
- Sweet and sour
- Tarter
- Taco
- Teriyaki

Snacks, puffed Soups, etc.
- Chowders
- Dry soup mixes
- Soups
- Stews

Spoonable salad dressings

Toppings
- Butterscotch
- Marshmallow

Vegetable patties

example, in certain cookies and pie crusts (both high in fat and low in moisture) about 90% of the wheat starch granules remain ungelatinized and close to their native condition. In other products, such as angel food cake and white bread, which are high in moisture, about 96% of the wheat starch granules are gelatinized. Although many become deformed, they are still recognizable as starch granules when the product is examined with a microscope, and the gelatinized granules can be isolated from the bread or cake.

During the dry milling used to produce flour, some starch granules are damaged. Damaged starch increases the absorption of water by the bread dough and produces a weak loaf and, often, a sticky crumb.[22] Starch damage is also undesirable in soft wheat flour used to make cakes and cookies. Damaged granules are more susceptible to hydrolysis by α-amylase (Chapter 7), which is the basis for a method for their detection and measurement.

Water is an active ingredient of foods. It is used to control texture and general physical behavior. It is not the total amount of water, but its availability, that is important. Starches and other carbohydrates absorb, hold, and control the migration of water. Starch granules that are inherent to a foodstuff (for example, those in potatoes) are gelatinized during cooking; but since the cooking is done without shear and the granules are encased within cells, they are not pasted (that is, they remain as swollen, gelatinized, but otherwise intact, granules). Even so, their gelatinization during cooking has a strong influence on the texture of cooked potatoes (firmness, moist or dry mouthfeel, and other characteristics).

Cells of the parenchyma tissue of the edible parts of plants, such as potatoes, contain clusters of starch granules. The cellulosic cell walls of adjoining cells are cemented together, primarily by pectic substances,[23] which are in a structure between plant cells called the middle lamella. As parenchyma tissue is heated, the starch granules in it take up available water, swell, and gelatinize. The water content within parenchyma cells is usually sufficient to plasticize starch granules and aid their gelatinization, but the temperature at which gelatinization occurs is slightly higher than that of the isolated starch. (The higher gelatinization temperature of in situ starch has been attributed to the presence of solutes in the intracellular fluid.) Although gelatinization is complete within the tissue, granule swelling is limited by the surrounding cell wall, but the granules swell (with some leaching of amylose) as much as they can, and in so doing, fill most of the volume of the cells. This granule swelling during heating exerts a measurable internal pressure on the cell walls. Although the magnitude of the swelling pressure is insufficient to rupture the parenchyma cells, they increase in size and become more spherical as a result of the pressure. This phenomenon is referred to as cell "rounding off." It occurs in concert with pectin degradation within the middle lamella by β-elimination (Chapter 15), which results in softening of the tissue. (The softening phenomenon is also observed in

[22] The crumb is the soft inner part of a bread or cake, that is, it is the part inside the crust.

[23] Pectic substances include pectin (Chapter 15) itself and other polysaccharides associated with pectin (namely, rhamnogalacturonans, arabinans, and galactans) (Chapter 17).

tissues that do not contain an appreciable content of starch, such as tomatoes, so the softening seems to be primarily due to pectin degradation within the middle lamella.) Nevertheless, in starch-containing tissues, such as in potato tubers, a high starch content and/or degree of starch granule swelling is associated with a softer, more friable, cooked tissue. It is thought that the cell "rounding off" phenomenon exerts physical pressure on a weakened middle lamella, contributing to cell separation or tissue sloughing. Also, the degree to which gelatinized starch swells to fill the volumes of parenchyma cells is thought to influence the perception of tissue moistness in the mouth. A high starch content and swelling capacity is generally more effective in binding up free moisture within the cooked tissue, producing a corresponding "dry" mouthfeel. In summary, the texture of boiled or baked potatoes and other cooked vegetables is, in large part, a function of gelatinization of the starch granules within parenchyma cells, a process that increases water holding and secondarily effects tissue softening.

Retrogradation and staling

Dramatic changes not only occur when starch granules are heated in excess water but may also occur in the resulting paste when it is cooled, especially if it is made from an unmodified (native) starch. As mentioned earlier in this chapter, cooling a hot starch paste generally produces a firm, viscoelastic gel. In Chapter 5, it was pointed out that gel formation is the result of junction zone formation. In this context, the formation of junction zones can be considered to be the first stage of an attempt by starch polymer molecules to form organized crystallites.

Early users of starch noted that, as starch pastes were cooled and stored, the starch became progressively less soluble. Starch molecules in a dilute solution precipitated, and this insoluble material was difficult to redissolve by heating. They called this return to an insoluble state **retrogradation**. Retrogradation of cooked starch (that is, gelatinized and usually also pasted starch) is a result of crystallization involving both of the two constituent polymers, with amylose undergoing retrogradation at a much more rapid rate than amylopectin. The rate and extent of retrogradation depends on several variables, including the ratio of amylose to amylopectin in the starch and the structures of the amylose and amylopectin molecules, both of which are determined by the botanical source of the starch. Other variables that influence retrogradation are temperature; starch concentration; and the presence and concentration of other ingredients, such as surfactants and salts. Many quality defects in food products, such as loss of viscosity and precipitation in soups and sauces and bread staling, are the results, at least in part, of starch retrogradation.

Staling is the firming of baked products that occurs during storage. Staling is due in part to the gradual transition of amorphous starch to a partially crystalline, retrograded state. Staling of baked goods is noted by an increase in crumb[22] firmness and a loss of product freshness. Staling begins as soon as baking is complete and the product begins to cool. The rate of staling is a function of the product formulation, the baking process, and storage conditions. In many doughs, there is just enough moisture to swell and gelatinize starch granules (while retaining a granule identity),[24] but retrogradation can still occur within those swollen granules in which the amylopectin crystallites are melted.

Staling continues after the product has cooled. Because retrogradation of amylose may be largely complete by the time the baked product has cooled to room temperature, retrogradation of amylopectin is believed to be the predominate process of staling. It has been proposed that over time after cooling the amylopectin network present in soft, fresh bread gradually changes to an extensive, partially crystalline, more permanent network, in which amylopectin crystallites function as junction zones. Also with time, water molecules migrate and, because water molecules are required for formation of the B-type crystallites of retrograded starch, more and more water becomes immobilized. This water of crystallization can no longer contribute to plasticization of either the starch or the gluten. The result is an increase in crumb[22] firmness. Both the degree of retrogradation in, and the tenderness of, cooked rice has been correlated with proportions of different amylopectin branch chain lengths.

Crystallization of partially crystalline polymers such as amylose and amylopectin occurs in three phases. The three processes are nucleation, propagation, and maturation. Crystallization of flexible polymers is a nucleation-limiting process (that is, the crucial first step is formation of crystal nuclei via association of two or more polymer chains). Formation of nuclei is greatest at a temperature just above the glass transition temperature (T_g) or, if the concentration of starch is less than 70%, which is often the case, just above the glass transition temperature of maximally freeze-concentrated starch (T_g') (about 5°C [23°F]). Propagation (growth of nuclei into crystallites) and maturation (perfection of metastable crystallites) is greatest at a temperature just below the melting temperature of the crystallites (T_m), which for normal corn starch is about 68°C (154°F) (the melting point of amylopectin crystallized after extraction from granules). At a temperature just above T_g or T_g', the system is in a rubbery state and the polymer molecules have enough mobility and energy that collisions of properly oriented molecules will associate and form nuclei. As the temperature increases, mobility of the polymer molecules increases so that they can more readily achieve optimal orientation; but of course, for crystallization to occur, the temperature must be below the melting temperature of the crystals. Crystallization of a flexible polymer, held under isothermal (constant temperature) conditions is greatest at a temperature between the optimal temperature for nucleation and the optimal temperature for propagation. For amylopectin molecules, this optimal temperature for crystallization is about 5°C (40°F). This means that staling occurs most readily at the temperature of most refrigerators, but packaged breads usually contain enough antistaling agents that staling is greatly reduced even at this temperature. By cycling between the temperature of highest nucleation rate and the temperature of highest propagation rate, the rate of retrogradation can be greatly accelerated.

[24] The degrees of swelling and gelatinization depend largely on the amount of water in the formulation.

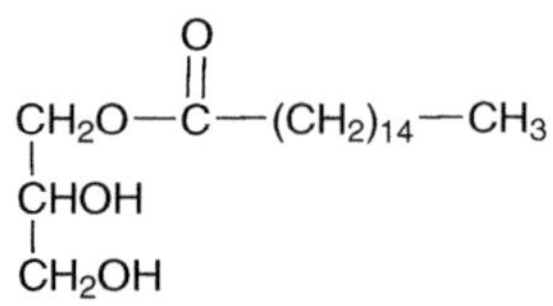

GMP

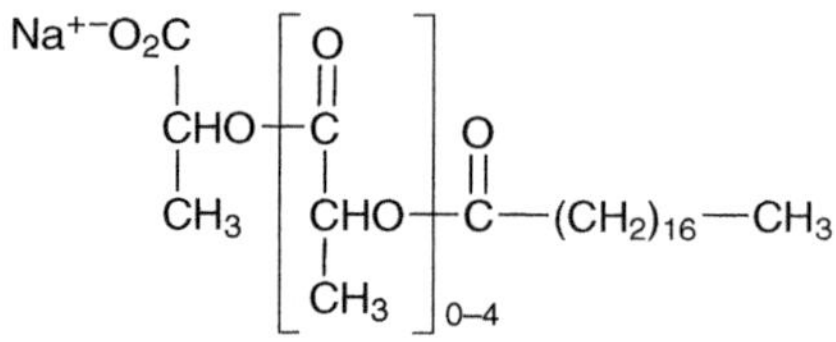

SSL

Such a temperature cycling process is used to accelerate the staling of bread crumb in the production of croutons and to determine the susceptibility (or resistance) of cooked rice to staling.

Most polar lipids with surfactant properties decrease crumb firming, their effectiveness in reducing retrogradation and staling being a function of the structure of the lipid. Compounds such as glycerol monopalmitate, glycerol monostearate, other monoglycerides, derivatives of them, and SSL are incorporated into doughs in part to increase the shelf life of the final product (that is, to slow the process of staling). (Amylose and amylopectin complexes are described in the next section.) Enzymes are also used to reduce/retard staling. (This process is described in Chapter 7)

Complexes

Amylose chains are helical with hydrophobic, lipophilic interiors capable of forming complexes with linear hydrophobic portions of molecules that can fit within the lumen of the helix.[25] When a hot, aqueous dispersion of starch is stirred with 1-butanol, nitropropane, or another slightly polar, slightly water-soluble organic substance and then cooled, amylose complexes crystallize and can be isolated by centrifugation. Only the amylose crystallizes because complexation stabilizes the helix and converts the long, linear molecules into more uniform and more rodlike

[25] As presented in Section Amylose, the inside of an amylose molecule in its natural helical form can be likened to the inside of a nonpolar (hydrophobic) tube because only nonpolar hydrogen atoms line the interior of the helix.

structures that can crystallize. Because only the amylose molecules crystallize, such complexation can be employed to isolate pure amylose.

Iodine (as I_3^-) complexes with amylose and amylopectin molecules. Complexing occurs within the hydrophobic interior of helical segments. The long helical segments of amylose allow chains of poly(I_3^-) to form and produce the blue color that is a diagnostic test for starch. The amylose–iodine complex contains 19% iodine, and the determination of the amount of complexing is used to measure the amount of apparent amylose[1] in a starch. Amylopectin forms reddish-purple complexes with iodine because its branches are too short for the formation of long chains of poly(I_3^-).

Polar lipids (surfactants/emulsifiers and fatty acids) affect starch pastes and starch-based foods in one or more of three ways as a result of complex formation: (1) by affecting starch gelatinization and pasting, (2) by modifying the rheological behavior of the resulting pastes, and (3) by inhibiting the crystallization of starch molecules associated with the retrogradation process. The small amounts of polar lipids in native cereal starches inhibit gelatinization and pasting, but promote retrogradation. The helical complexes formed contain approximately six glucosyl units and six hydrocarbon methylene groups [$-(CH_2)_6-$] of the lipid per helical turn (Fig. 6.10). Specific changes observed upon addition of a lipid depend on its structure, the starch employed, and the product to which it is added; but some generalizations can be stated. Different lipid/surfactants affect the gelatinization and pasting behaviors of a given starch differently, and each lipid/surfactant affects starches from different botanical sources differently. Because complex formation with emulsifiers occurs much more readily with amylose and, as a result, has a much greater effect on amylose molecules than on amylopectin molecules, polar lipids have a much greater affect on normal starches than on waxy (all-amylopectin) starches. Addition of most lipids/surfactants/emulsifiers to starches containing amylose inhibits the processes associated with gelatinization and pasting, but some accelerate the processes and/or cause them to occur at lower temperatures.

Polar lipids can complex at almost any point over the entire length of amylose molecules. Up to 86% of the glucosyl units may be complexed in amylose molecules saturated with a lipid/surfactant. Amylopectin can bind only c. 15% as much lipid per unit weight as can amylose, complexation probably being restricted to the outer chains. The ability of polar lipids to form complexes with amylose and amylopectin is associated with their chain length and degree of unsaturation and the nature of their hydrophilic group. In general, esters of myristic (C_{14}, saturated) and palmitic (C_{16}, saturated) acids are most effective. Complexes between starch molecules (particularly amylose) and fatty acyl compounds are only slowly digested by α-amylase (Chapter 7).

Certain flavor and aroma compounds also complex with starch, resulting in their reduced perception in starchy foods. The binding of such compounds to starch (primarily amylose molecules) appears to be a complex process with competitive, synergistic, and antagonistic effects. However, the major reason all polysaccharides (starches and hydrocolloids) reduce the perception of flavors and aromas is the increased viscosity they impart to aqueous food systems, where increased viscosity retards diffusion of flavor molecules to taste buds and aroma molecules to surfaces where they can escape.

Starch digestion and resistant starch

This book is not a book on the nutritional/physiological aspects of carbohydrates. Nevertheless, a few, very brief comments on digestion of starch are included. Early humans were hunter-gatherers. Their diet consisted primarily of fruits, nuts, beans, and roots. Any starch was consumed in an uncooked form. Pancreatic α-amylase (a digestive enzyme essential for conversion of starch into D-glucose [Chapter 7]) attacks raw (ungelatinized/unpasted) starch granules only very slowly, and the efficiency of starch digestion in the small intestine of early humans was reduced even more because cell walls in the uncooked plant materials were largely intact, blocking access of digestive enzymes to the starch granules within the cells. The slow inefficient digestion of starch resulted in a very inefficient but sustained release of D-glucose (which is good).

Today, starch is largely consumed after cooking of whatever foodstuff it is in (that is, it is consumed in a much more digestible form). The result is more rapid and more complete digestion, elevated blood glucose levels (hyperglycemia), and the accompanying elevated insulin levels (relative to those of our ancient ancestors). This high glycemic response can be advantageous to athletes who practice carbohydrate loading; but for most people, a slow release of D-glucose after a meal would be preferred for health and fitness. Control of blood sugar levels is especially important to those with diabetes mellitus and certain other age-related diseases. So control of the rate and extent of starch digestion is a challenge to the food scientist.

It is now recognized that there are various types of starch with regard to digestibility in any food containing starch. These different forms have been classified into three main groups. Resistant starch (RS) is defined as that portion of starch that does not undergo digestion in the stomach and small intestine (that is, is not broken down into glucose which is absorbed from the small intestine) and may be fermented by the microflora in the large intestine. So RS falls under the definition of dietary fiber (Chapter 17). There are various forms of RS in foods. These different forms are grouped into five general subclasses. RS1 is starch that is physically inaccessible to the α-amylase (Chapter 7) in the small intestine (pancreatic α-amylase) because it is inside plant cells (that is, it is surrounded by plant cell walls) or surrounded by non-digested denatured protein or something else that entraps it. RS2 is uncooked granular starch. (Intact granules [that is, uncooked/unpasted granules] are digested by α-amylase only very slowly.) RS3 is retrograded (crystalline) starch. RS4 consists of granules or portions of granules that are somewhat resistant to digestion because of modification (Chapter 7). RS5 consists of amylose—lipid complexes, which are also resistant to attack by α-amylase. In a practical sense, the concept of RS is only valid for a food that is prepared and consumed in the way it normally would be. For example, retrograded starch (RS3) in a starch-based product that has been stored in a refrigerator will likely be melted when reheated before consumption, converting the RS3 to a highly digestible form. A good example of RS3 is some of the starch in a cooked (boiled or baked) potato that has subsequently been cooled (for example, as used in potato salad) and consumed cold. Another example is the pasta in a pasta salad.

High-amylose corn (amylomaize) starch included in a bread dough, which is subsequently baked and consumed, is only partially digested in the small intestine, so it is an RS and meets the definition of dietary fiber (Chapter 17).

RS that has a measurable dietary fiber content (Chapter 17) of about 40% is produced as a food ingredient. It has a lower water holding content as compared to the parent starch or a chemically modified starch. It is used in extruded breakfast cereals and snacks to improve expansion, crispness, bowl life, eating quality, and shelf life. It may be used in breads (alone or in combination with another source of fiber), muffins, and snack crackers.

There is voluminous literature on carbohydrate nutrition but no consensus among researchers about what is most desirable from a physiological point of view. However, many believe that the most desirable starch ingredient (one that combines both nutritional and food quality aspects) is something between rapidly digesting starch (RDS) and RS, which is known as slowly digesting starch (SDS). An ideal SDS product would be one that would be highly digestible but which would produce blood glucose over an extended period of time, perhaps up to 6 h after ingestion (postprandially). Such products would be classified as low-glycemic-load products and would be beneficial to diabetics and nondiabetics alike. This conclusion is reached from evidence that ingestion of starch of a type that gives a rapid rise in the concentration of glucose in the blood to levels that produce hyperglycemia and the corresponding increase in blood insulin levels (hyperinsulinemia) has a negative long-term impact on human health.

The third type of starch and usually the most prevalent type is RDS. RS, SDS, and RDS are discussed further in Chapter 17.

Manufacture of starches

Corn/maize is the major source of starch because of its abundance and low price. Corn grain cleaned of metal and dust is placed in steep tanks with water and sulfur dioxide (which reduces the pH to 3–4 because of the formation of sulfurous acid[26]) and held at 48–52°C (120–125°F), usually for 20–40 h. Softened grain is then shredded in a mill that does not crush the germ. The germ, consisting of oil and protein, is removed in continuous liquid cyclones.[27] The degerminated ground grain is then ground more

[26] Lactic acid is also present as a result of its production by microorganisms present in the steepwater.

[27] Liquid cyclones or hydroclones are cone-shaped tubes with two outlets, one at the top and one at the bottom, and an inlet port located on the side near the top. A suspension is forced in through the inlet in a way that produces a rotational velocity (a vortex) in the fluid. The centrifugal force produced by the rotational velocity is sufficient to separate particles based on their density, with the heavier particles passing out the bottom of the tube and the lighter ones out the top.

finely to free starch granules. After removing the fibrous seed coat by passing the slurry through a screen, the starch slurry is concentrated in smaller hydroclones, then either pumped into a tank for modification (Chapter 7) or passed to centrifuges and/or filters where the starch is recovered as a cake, which is then dried.

Potato tubers and cassava (manioc, yuca) roots do not require steeping. Potatoes, for example, are put directly through a hammer mill with an aqueous solution of sulfur dioxide. The starch is separated from fiber by use of appropriate screens. Thereafter, the starch is removed in cyclones, washed, and dried.

There is more than one process for the production of wheat starch. In general, a dough is made from wheat flour and water. Starch is separated from the gluten (protein) by kneading the dough in water. The gluten forms a mass, because of adhesion to itself, and is removed. Starch granules in the remaining slurry are collected by centrifugation. Two fractions are obtained. The prime starch (called A starch in the wheat starch industry) contains the large lenticular granules (the A granules) and a few of the small spherical granules (the B granules). A second, lower-value faction (called B starch, tailings starch, and squeegee starch) consists of small granules, damaged granules, and pentosans (Chapter 17) and other cell wall substances.

Other types of commercial starches include starches from arrowroot, mung bean, pea, different types of rice, sago, and sweet potato. All are produced in small amounts. Flours that are high in starch (other than flours from wheat, barley, corn, millet, oat, potato, rice, rye, and sorghum) include banana and pea flours.

Additional resources

General

BeMiller, J., Whistler, R. (Eds.), 2009. Starch: Chemistry and Technology. Academic Press, New York.

Bertoft, E., 2013. On the building block and backbone concepts of amylopectin structure. Cereal Chemistry 90, 294–311.

Bertoft, E., Nilsson, L., 2016. Starch: analytical and structural aspects. In: Eliasson, A.-C. (Ed.), Carbohydrates in Food. CRC Press, Boca Raton, pp. 377–478.

Bertolini, A.C. (Ed.), 2010. Starches: Chracterization. Properties, and Applications. CRC Press, Boca Raton.

Biliaderis, C.G., 1991. The structure and interactions of starch with food components. Canadian Journal of Physiology and Pharmacology 69, 60–78.

Eliasson, A.-C., 2016. Starch: physiochemical and structural aspects. In: Eliasson, A.-C. (Ed.), Carbohydrates in Food. CRC Press, Boca Raton, pp. 479–577.

Huber, A., Praznik, W., 2004. Analysis of molecular characteristics of starch polysaccharides. In: Tomasik, P. (Ed.), Chemical and Functional Properties of Food Saccharides. CRC Press, Boca Raton, pp. 349–369.

Jane, J.-l., Kasemsuwan, T., Leas, S., Zobel, H., Robyt, J.F., 1994. Anthology of starch granule morphology by scanning electron microscopy. Starch/Stärke 46, 121–129.

Jane, J.-l., 2004. Starch: structure and properties. In: Tomasik, P. (Ed.), Chemical and Functional Properties of Food Saccharides. CRC Press, Boca Raton, pp. 81–101.

Nakamura, Y. (Ed.), 2015. Starch: Metabolism and Structure. Springer Japan KK, Tokyo.

Sjöö, M., Nilsson, L. (Eds.), 2018. Starch in Food - Structure, Function, and Applications. Woodhead Publishing, Duxford.

Vamadevan, V., Bertoft, E., 2015. Structure-function relationships of starch components. Starch/Stärke 67, 55–68.

Bread Staling

Goesaert, H., Slade, L., Levine, H., Delcour, J.A., 2009. Amylases and bread firming – an integrated view. Journal of Cereal Science 50, 345–352.

Gray, J.A., BeMiller, J.N., 2003. Bread staling: molecular basis and control. Comprehensive Reviews in Food Science and Food Safety 2, 1–21.

Manufacture of Starch

BeMiller, J., Whistler, R. (Eds.), 2009. Starch: Chemistry and Technology. Academic Press, New York.

Bergthaller, W., 2004. Starch world markets and isolation of starch. In: Tomasik, P. (Ed.), Chemical and Functional Properties of Food Saccharides. CRC Press, Boca Raton, pp. 103–122.

Starches: Conversions, Modifications, and Uses

7

Chapter Outline

Key information and skills that can be obtained from study of this chapter will enable you to

1. Define or identify

converted products	liquefaction
conversion	glucose isomerase
acid-modified starch	high-fructose syrup (HFS)/high-fructose corn syrup (HFCS)
thinning	hydrogenated starch hydrolyzates (HSH)
thin-boiling starch	stabilized/substituted starch
dextrins	starch acetate (acetylated starch)

Continued

Carbohydrate Chemistry for Food Scientists. https://doi.org/10.1016/B978-0-12-812069-9.00007-8

transglycosidation (transglycosylation)	monostarch phosphate (starch phosphate monoester)
dextrose equivalency (DE)	starch octenylsuccinate (OS/OSA starch)
maltodextrins	hydroxypropylstarch (hydroxypropylated starch)
corn syrup solids (glucose syrup solids)	cross-linking (cross-linked starch)
corn syrups/glucose syrups	distarch phosphate (starch phosphate diester)
reversion	oxidized starch
amylase	pregelatinized starch
α-amylase	instant starch
glucoamylase (amyloglucosidase)	cook-up starch
β-amylase	granular cold-water swelling (GCWS) starch
beta-limit dexrtrin	cold-water swelling (CWS) starch
debranching enzyme	heat−moisture treatment (HMT)
cyclodextrins (cycloamyloses)	annealing
clathrates	
guest molecules	

2. Describe the action of each of the following on a starch paste:
   α-amylase
   glucoamylase (amyloglucosidase)
   β-amylase
   isoamylase
   pullulanase
   cyclodextrin glycosyltransferase (cyclodextrin glucanotransferase)
3. Describe the action of glucose isomerase.
4. Using equations, show the reactions involved in making
   distarch phosphate
   hydroxypropylstarch
   starch acetate
   starch octenylsuccinate
5. Describe starch "stabilization."
6. Describe the most important reason for making stabilized starch.
7. Using a diagram, describe the effect of cross-linking on cooking characteristics of a starch, including the effect of the degree of cross-linking.
8. Explain how pregelatinized and cold-water-swelling starches are made and the differences between the two.
9. List three reasons for using a pregelatinized starch in a formulation.

Products of hydrolysis (products of conversion)

Starch molecules, like other polysaccharide molecules, are depolymerized by hot acids (Chapters 1 and 4), the reaction becoming faster as the temperature is increased. Products of hydrolysis are known as **converted products** or products of **conversion**. The least converted starch products are acid-modified starches.

To produce acid-modified starch products, either hydrochloric acid is sprayed onto well-mixed starch or stirred moist starch is treated with hydrogen chloride gas and the mixture is heated until the desired degree of hydrolysis (which is very slight in any case) is obtained or the pH is lowered after modification (section on Modified Food Starches in this chapter) and the slurry is held at the modification temperature until the desired degree of hydrolysis is obtained. The acid is then neutralized, and the product is recovered by filtration or centrifugation, washed, and dried. This process is called **thinning**. The products, called **acid-modified** or **thin-boiling starches**,[1] are still granular but break up (cook out in industrial jargon) easily and largely dissolve when heated in water. Even though only a few glycosidic bonds are cleaved, the starch granules disintegrate much more easily during heating in water. (Hydrolysis occurs more or less randomly to produce both linear and branched molecular fragments.) Thin-boiling form gels with improved clarity and increased gel strength, but they provide less solution viscosity as compared to the native parent starch. Thin-boiling starches are used in gum candies such as jelly beans, jujubes, orange slices, and spearmint leaves, in molds for depositing candies, and in processed cheese loaves. Acid-modified corn starch is used to prepare especially strong and fast-setting gels. Derivatized starch products (section on Modified Starches in this chapter) may also be thinned.

A little more extensive modification with acid produces **dextrins**. (Dextrin is a generic term applied to a variety of products obtained by partial hydrolysis of starch,[2] but in practice, the term dextrin is applied to products obtained by heating [pyroconversion of] "dry" acidified starch.) There are generally five steps in making a dextrin (acidification, predrying at a low temperature, heating, cooling, rehumidifying). Three types of reactions occur during dextrinization of a starch: (1) hydrolysis of glycosidic bonds, (2) **transglycosylation (transglycosidation)**,[3] and (3) reversion (Chapter 3). Hydrolysis depolymerizes the starch polymer molecules.

Transglycosylation makes more highly branched polymer structures. The relative amounts of these three reactions depends on the moisture content of the "dry" starch

[1] In the starch industry, they are also called fluid starches and fluidity starches.

[2] The term dextrin is also applied to products obtained by partial hydrolysis of other polysaccharides (for example, cellodextrins from cellulose), but less commonly.

[3] **Transglycosidation (transglycosylation)** refers to the transfer of a portion of a polysaccharide chain from the monosaccharide unit to which it is attached to another unit, creating a branch (Fig. 7.1). When starch is the substrate for dextrinization, the new linkages formed are not necessarily, and probably not, the same as those in the original amylose and amylopectin molecules (that is, they will be linkages other than the $(1 \rightarrow 4)$ and $(1 \rightarrow 6)$ linkages) and some of the D-glucopyranosyl units at the reducing end of the transferred chain will have the β configuration.

(which usually is the starch's normal moisture content [about 13% for cereal starches and about 18% for potato and tapioca starches]) and the temperature of the process. As an acidified starch ("dry", but with some moisture present) is heated, glycosidic bonds undergo hydrolysis, some water is consumed in the reaction, and a rise in temperature occurs. After the temperature surpasses 100°C, dehydration occurs and transglycosylation and reversion begins. The resulting dextrin is more highly branched than the native starch.

Because they produce solutions of lower viscosity (as compared to the native parent starch), dextrins are used at higher concentrations than are the parent starches. They have film-forming and adhesive properties; they are used in coatings in products such as candies and pan-coated roasted nuts. They are also used as fillers, as encapsulating agents, and as carriers of flavors (especially spray-dried flavors). They increase crispness and browning in certain bakery products.

Dextrins are classified by their cold-water solubility and color (white or yellow) but mainly by their **dextrose equivalency (DE)**, which is the reducing power of the substance as a percentage of the reducing power of pure D-glucose (dextrose), so by definition D-glucose has a DE of 100. Amylose and amylopectin have no measurable reducing power (that is, their DE is zero), but each hydrolytic cleavage of a polysaccharide chain produces a new reducing end; so the DE value increases with the degree of hydrolysis. DE values are related to the degree of polymerization (DP) through the following equation: $DE = 100 \div DP$,[4] with both DE and DP being average values for a population of molecules. Thus, the DE of a hydrolysis product is inversely proportional to its average molecular weight (that is, the higher the DE value the lower the molecular weight). Products with low DE values (relatively high molecular weights) retain large amounts of linear, long chain fragments, allowing them to form strong gels.

A specially made dextrin in which the digestible carbohydrate has been removed is marketed as soluble dietary fiber (Chapter 17). In this case, a starch is first converted into a dextrin by heating it in such a way as to enhance transglycosylation and introduce linkages other than the normal α-(1 $\rightarrow$ 4) and α-(1 $\rightarrow$ 6) linkages. The pyrodextrin is treated with amylases (amylases being specific for the normal starch polysaccharide linkages [section on Hydrolyzing Enzymes in this chapter]) to remove most of the digestible portions leaving, for the most part, a nondigestible product (which by definition makes it dietary fiber). Although many naturally occurring linkages remain, the product as a whole is largely indigestible by the enzymes of the small intestine and analyzes as about 85% dietary fiber.

Hydrolysis of hot starch dispersions with either an acid or an enzyme(s) to DE values of less than 20 (average $DP > 5$) produces **maltodextrins**.[5] Maltodextrins are officially defined as "purified, nutritive mixtures of saccharide polymers obtained by partial hydrolysis of edible starch." Most maltodextrin molecules are

[4] Likewise, $DP = 100 \div DE$. (DP = degree of polymerization.)

[5] Malto- and amylo- are prefixes used in naming substances related to starch. Dextrin is a generic term applied to depolymerization products of polysaccharides, but is most often applied to starch.

maltooligosaccharides (DP 2−20), so maltodextrin products are mixtures of maltooligosaccharides, which mixtures have average DE values of less than 20. The DE values of the products are mostly in the range of 5−15, so their average DP values are mostly in the 7−20 range. The term nutritive refers to the fact that they can be digested and used as an energy source by humans.

Maltodextrins made by acid-catalyzed hydrolysis have a high quantity of linear chains that are able to partially crystallize. To prevent haze formation in their solutions and to produce maltodextrins of low hygroscopicity and high water solubility, acid-catalyzed hydrolysis to DE 5−10 is often followed by enzyme-catalyzed hydrolysis with an α-amylase (alpha-amylase) (section on Hydrolyzing Enzymes in this chapter). Maltodextrins are also produced by use of an α-amylase alone and with a combination of an α-amylase and a debranching enzyme (section on Hydrolyzing Enzymes in this chapter).

Maltodextrins can be produced from any starch. They are generally recognized as safe (GRAS) food ingredients. Maltodextrins are often used to increase the solids content of foods. The main attributes of maltodextrins are good dispersibility and solubility; provision of body, smooth texture, and a pleasant mouthfeel; bland flavor; moisture control; crystallization inhibition; film formation; easy digestibility; easy spray-drying; flavor and aroma encapsulation; and low to no sweetness. As the DE of maltodextrins increases (average DP decreases), the following properties increase: browning efficiency, freezing point depression, hygroscopicity, osmolality, solubility, and sweetness. The following properties increase as the DE decreases (average DP increases): ability to provide body, binding capability, crystal growth inhibition, and viscosity. Some typical food product uses of maltodextrins are given in Box 7.1.

Maltodextrins that have been prepared so that they form microcrystals that are about the size of oil or fat droplets (about 1 μm) are used as fat replacers (sparers/mimics/mimetics) (Chapter 17). These products are used, for example, to reduce the fat in cakes and cream fillings. They can also be used to prepare reduced-fat, low-fat, and fat-free frozen desserts; low-fat salad dressings; low-fat processed cheese; and similar products.

Hydrolysis to DE 20−35 (average DP 3−5) gives mixtures of molecules that, when dried, are called **corn syrup solids** (in the United States) or **glucose syrup solids**. Syrup solids dissolve rapidly and are mildly sweet. Syrup solids are used in many of the same applications as are maltodextrins. Syrup solids are often preferred for infant formulas, coffee whiteners, and whipped toppings. They are also recommended for frozen foods and desserts (as cryoprotectants),[6] meat products, and processed cheese products, along with maltodextrins.

Continued hydrolysis of starch with an acid and/or enzymes produces a mixture of D-glucose, maltose, other maltooligosaccharides, and a few products that result from these molecules recombining (forming new glycosidic bonds). The recombination process is called **reversion** (Chapter 1); a small amount of reversion occurs because of the high concentrations at which the conversions are done. The products

[6] A cryoprotectant is a substance that protects against freezing damage.

Box 7.1 Some Applications of Maltodextrins

Agglomerating agents (for water-soluble gums)

Binders
- Frozen meat analogues
- Granola bars

Bulking agents/carriers
- Artificial sweeteners
- Dry beverage mixes
- Dry milk-flavoring mixes
- Dry sauce mixes
- Dry soup mixes
- Gravies
- Spice blends
- Spoonable salad dressings

Coatings
- Dry roasted peanuts (oxygen barrier)
- Panned candies

Crystallization inhibitors
- Confections
- Frozen foods and desserts

Dried flavors (spray-dried and extruded)
- Dry beverage mixes
- Dry roasted peanuts

Fat replacers, sparers, and mimics
- Bakery fillings
- Cream cheese
- Fat-free confections
- Frozen desserts
- Margarines
- Salad dressings
- Whipped toppings
- Yogurt

Moisture control
- Fruit leather
- Granola bars

Nutritional beverages
- Adult formulas
- Dietetic products
- Infant formulas

Processing aids
- Cheese powder
- Extruded products

of this extensive hydrolysis are syrups known as **glucose syrups** (called **corn syrups** in the United States). Glucose syrups are purified, concentrated, aqueous solutions of nutritive saccharides with a DE value of greater than 20 obtained from a food-grade starch.

Glucose syrups may be made by one of three processes: (1) acid conversion (pH about 2, temperature above 100°C), (2) acid−enzyme conversion (an acid-converted hydrolyzate is treated with one or more starch hydrolyzing enzymes [amylases, the choice of which depends on the desired saccharide profile of the finished syrup] [section on Hydrolyzing Enzymes in this chapter]), (3) enzyme−enzyme conversion. In the latter process, a starch suspension is pasted, and the paste is liquefied using a thermostable α-amylase in a single step. The product is then treated with one or more other enzymes depending on the desired saccharide profile. Saccharide compositions of typical syrups produced in the United States are given in Table 7.1.

Lower-DE syrups (DE less than about 43) contain higher proportions of maltooligosaccharides. They are used in icings and glazes because they impart both body and flexibility, and the oligosaccharides prevent crystallization so that shelf life is

Table 7.1 Saccharide compositions of typical corn syrups

		% Saccharides, carbohydrate basis						
Type of syrup	**DE**	**Mono-**	**Di-**	**Tri-**	**Tetra-**	**Penta-**	**Hexa-**	**Higher**
Acid converted	27	9	9	8	7	7	6	54
	36	14	12	10	9	8	7	40
	42	20	14	12	9	8	7	30
	55	31	18	12	10	7	5	17
Acid–enzyme converted	43	8	40[a]	15	7	2	2	26
	49	9	52[a]	15	1	2	2	19
	65	39	31	7	5	4	3	11
	70	47	27	5	5	4	3	9
	95	93	3	1	1	-	-	2

[a]High-maltose syrups.

extended. As seen in Table 7.1, syrups with intermediate DE values (DE 43–49) are usually high-maltose syrups. Higher DE syrups (DE more than about 64) contain greater amounts of D-glucose. They are humectants and are used in the production of soft cookies.

Glucose syrups are produced in enormous quantities from starch (section on D-Glucose and D-Fructose Production in this chapter). The most common glucose syrup has a DE of 42. Syrups are stable because crystallization does not occur easily in such complex mixtures and they are sold in concentrations of sufficient osmolality (about 70% solids) to prevent growth of ordinary microorganisms. Waffle and pancake syrup is a glucose syrup that has been colored with caramel coloring (Chapter 18) and flavored with maple flavoring. High-maltose syrups are the starting material for production of maltitol (Chapters 2, 3, and 19).

Uses of glucose syrups in foods are widespread. They are used largely as humectants and to provide body[7] and bulk[8] and with various degrees of sweetness—their sweetness generally being moderate. Products with moderate sweetness are used in foods requiring body, but not high sweetness, such as sauces, toppings, and some confections and sweet bakery products. Products with low sweetness provide body to ice cream, canned fruits and vegetables, bakery products, and some confections. Other properties are also important. Viscosity is important as it relates to the handling characteristics of the syrup and to the texture of the food product. Glucose syrups may be used to reduce water activity, lower the freezing point, increase the osmotic pressure,

[7] Body refers to the organoleptic attributes (often referred to as mouthfeel) due to the rheological properties (Chapter 5) of a food or beverage.

[8] Bulk refers to the solids content that contributes to the texture and palatability of a food product.

and increase the chewiness of a food product. They inhibit crystallization of sucrose. In ice cream, they provide resistance to meltdown.[9] They increase the shelf life of hard candies and peanut brittle by preventing crystallization and providing resistance to changes in moisture. They provide fermentable substrates for bakery products and contribute to browning (Chapter 18). Glucose syrups are used in a plethora of other applications from bakery fillings to salad dressings.

Syrups can be made from any starch. They can also be made from flours. For example, brown rice syrup is made from brown rice (Chapter 19). Partial hydrolysis of oat and related flours is presented in Chapter 17.

Hydrolyzing enzymes

Amylase is the generic term for enzymes that catalyze the hydrolysis of starch. Three to four amylases are used for industrial hydrolysis of starch to D-glucose (dextrose). Bacterial and fungal alpha-amylases (α-amylases) and glucoamylase readily attack damaged and cooked starch granules. (These enzymes will attack ungelatinized corn starch granules, but only very slowly. The result is an increase in granule porosity (that is, they appear to drill holes in the granules). However, they act many orders of magnitude faster and much more completely on cooked [pasted] starch.)

alpha-Amylase is an endoenzyme[10] that cleaves both amylose and amylopectin molecules internally. α-Amylases from different sources have different action patterns and produce different product mixtures. One α-amylase commonly used industrially produces large amounts of maltopentaose (G5)[11] through maltononaose (G9). A typical digest after long incubation will contain G6-G12 > G5 >> G3 >> G2 > G4 > G1. The larger oligosaccharides may be singly, doubly, or triply branched via (1 → 6) linkages, as α-amylase acts only on the (1 → 4) linkages of starch. alpha-Amylase does not attack double-helical starch polymer segments or polymer segments complexed with a polar lipid (that is, stabilized single-helical segments). Hence, the starch has to be pasted first to destroy the granule structure and to unwind helical structures.

[9] Meltdown refers to the rate of loss of structure and shape when ice cream is warmed to room temperature. When ice cream is warmed, two things happen: (1) the ice crystals melt, and (2) the foam structure collapses. The ice crystals can melt without a complete loss of shape due to stability of the foam provided by the milk proteins and emulsified fat globules, the degree of which stability being what is indicated by the meltdown test. Ideally, an ice cream should melt rather quickly, and the melt should have characteristics similar to those of the original mix before freezing.

[10] In this context, an **endoenzyme** is an enzyme that attacks polymeric molecules somewhere other than a chain end, that is, somewhere in the middle of the chain. (The term can also refer to an intracellular enzyme.)

[11] G5 is a shorthand designation for a maltooligosaccharide containing five α-D- glucopyranosyl units. Other maltooligosaccharides are similarly designated.

Glucoamylase (also known as **amyloglucosidase**) is an exoenzyme[12] that, in combination with an α-amylase, is used commercially for producing D-glucose (dextrose) syrups and crystalline D-glucose. It sequentially releases single D-glucosyl units from the nonreducing ends of amylose and amylopectin molecules, even those joined through (1 → 6) bonds (bonds which become available as amylopectin molecules are eroded down to branch points). Consequently, the enzyme can completely hydrolyze starch to D-glucose. Commercial end products, however, contain only about 95% D-glucose because the reaction in concentrated solutions reaches an equilibrium with a small reverse reaction producing about 1% each of maltose and isomaltose and 2%–3% of higher oligosaccharides.

Beta-Amylase (β-amylase) is an exoenzyme[12] that releases maltose sequentially from the nonreducing ends of amylose and amylopectin. It cannot cleave the (1 → 6)-linkages at branch points. Thus, in the case of amylopectin, it leaves a pruned molecule termed a limit dextrin—specifically a **β-limit dextrin**. Because there may be an even or an odd number of D-glucopyranosyl units in the outer branches of amylopectin, there will be two or three D-glucopyranosyl units left attached as a branch (A chain) and one or two D-glucopyranosyl units left beyond the branch in the B chain to which the A chain is attached. To produce maltose, starch is hydrolyzed with a thermostable[13] α-amylase or with hydrochloric acid to a DE of about 20; then the hydrolyzate is treated with β-amylase. Maltose is the starting material for the production of maltitol (Chapters 2, 3, and 19).

There are several **debranching enzymes** that specifically catalyze hydrolysis of the (1 → 6)-linkages of amylopectin, producing numerous linear, lower-molecular-weight molecules (the branch chains). One such enzyme is called isoamylase; another is pullulanase.

Cyclodextrin glycosyltransferase (glucanotransferase) is an enzyme that forms rings of (1 → 4)-linked α-D-glucopyranosyl units from amylose and amylopectin. The enzyme can form six-, seven-, and eight-membered cyclic maltooligosaccharides (Fig. 7.1) by means of an intramolecular transfer (cyclization). As the normal helical conformation of a linear portion of a starch molecule contains 6–7 glucosyl units per turn of the helix, transfer of a glycosidic bond from one that joins adjacent segments of a spiral to one lying beside it forms a circular structure. These products, originally called "Schardinger dextrins" after their discoverer, are now known as **cyclodextrins** or **cycloamyloses.** The six-, seven-, and eight-membered rings are respectively, alpha-, beta-, and gamma-cyclodextrins. γ-Cyclodextrin, which has the largest ring, is the most soluble. β-Cyclodextrin, the intermediate-size compound, is the least soluble because of an extensive band of intramolecular hydrogen bonds covering the entire outer surface.

Cyclodextrin molecules are shaped like a hollowed-out rubber stopper, that is, the molecules are tapered at one end and have a large hole in their centers. Because they

[12] In this context, an exoenzyme is an enzyme that attacks polymeric molecules at one of the chain ends (that is, in the case of polysaccharides, at either the reducing end or the nonreducing ends). (The term can also refer to an extracellular enzyme.)

[13] A thermostable (or thermally stable) amylase in this context is one whose activity will not be completely destroyed at the temperature of boiling water for at least several minutes.

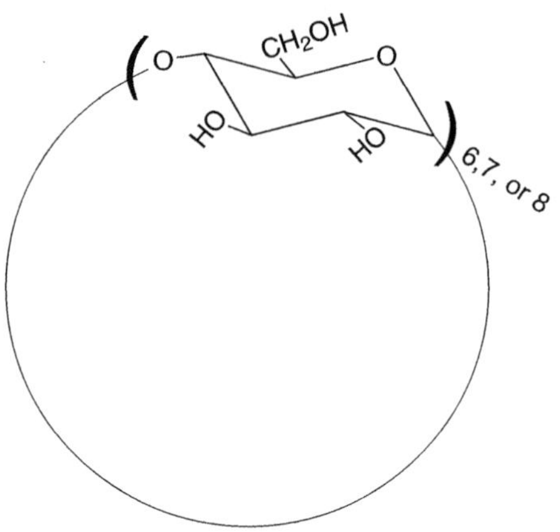

Figure 7.1 Structures of the three cyclodextrin molecules.

are derived from a turn of a helix, their hydroxyl groups are all projecting outward from the truncated donut structure, and the cavity inside the ring is hydrophobic (lipophilic), just as the inside of an amylose helix is. As a result, cyclodextrins have the ability to complex with hydrophobic substances, which is the basis for their usefulness. Such inclusion complexes are termed **clathrates**, and the complexed nonpolar molecules are known as **guest molecules**. This complexation is useful for fixing and protecting aromas and flavors from thermal decomposition, oxidation, volatilization, and/or light-induced degradation. In this way, volatile essential oils can be converted into dry powders in which the flavoring or aromatic substance is protected from light and oxygen but is readily released when the complex is added to an aqueous system because of the water solubility of the cyclodextrin. The size of the cavity in cyclodextrin molecules increases as the number of α-D-glucopyranosyl units increases (that is, the order of cavity sizes is $\alpha < \beta < \delta$). All three cyclodextrins may be used as an ingredient in foods, but essentially only β-cyclodextrin is used because of its lower cost and its functionality. It is used mostly in powdered essential oils. Approved food applications for complexes with flavor and aroma compounds include dry mixes for baked goods, beverages, and soups; flavored coffee and teas; savory snacks and crackers; breakfast cereals; chewing gum; tableted candies; processed cheese products; gelatin desserts; and puddings. Insoluble polymeric beads of cyclodextrins have been shown to be useful for removal of bitter components from citrus juices.

Many other enzymes that act on starch (amylases) have been discovered in bacteria, archaebacteria, and fungi and characterized. Some have been commercialized. Certainly, many more remain to be discovered and perhaps used to produce new ingredients or to be used in food processing, such as in baking.

Enzymes in baking

Bacterial and fungal α-amylase preparations with intermediate thermostability are added during the preparation of doughs, primarily to reduce the rate of staling of the finished product. These enzymes have optimal activity at temperatures slightly above the gelatinization temperature of wheat starch but are then inactivated as the temperature approaches 100°C. The mechanism of the antistaling effect of these enzymes has been discussed and debated to great length. A probable mechanism for the staling process was presented in Chapter 6. It has been proposed that the effect of amylases in retarding or preventing staling is a result of their limiting the formation and strength of the amylopectin network and the immobilization of water molecules. One type of an α-amylase weakens the network by cutting the chains connecting the crystallites. Another type, decreases the degree of crystallization by shortening the branch chains to the point that they cannot participate in crystallite formation.

Another polysaccharidase, xylanase (hemicellulase) is used to depolymerize the arabinoxylan of wheat flour (Chapter 17) and to indirectly produce a more elastic and flexible gluten network.

D-Glucose and D-fructose production

In the United States, corn syrup is the major source of D-glucose and D-fructose. A glucose syrup is made by first passing a slurry of starch in water containing a thermally stable[13] α-amylase through a jet cooker, where rapid gelatinization and enzyme-catalyzed hydrolysis (**liquefaction**)[14] occur. After cooling the solution to 55–60°C (130–140°F), glucoamylase is added and hydrolysis is continued. When hydrolysis is complete, the syrup is clarified, decolorized, and concentrated. When crystalline D-glucose (dextrose) or its monohydrate is desired, the conversion is continued until as close to DE 100 as possible is reached. Then, seed crystals are added.

Crystalline D-glucose is used in many of the same products that utilize glucose syrups when a dry form of the sugar is desired. In addition, when it is used as part of the sugar coating of donuts and in sandwich cookie and sugar wafer fillings, it imparts a desirable cooling sensation in the mouth because of its slight negative heat of solution. D-Glucose is the starting material for the production of sorbitol (Chapter 2) and via sorbitol ascorbic acid (vitamin C) (Chapter 2).

For production of D-fructose, a solution of D-glucose is passed through a column containing bound **glucose isomerase** (an enzyme that catalyzes isomerization of D-glucose to D-fructose, producing an equilibrium mixture of approximately 58% D-glucose and 42% D-fructose). Higher concentrations of D-fructose are required to make the **high fructose syrup (HFS)**, often called **High fructose corn syrup (HFCS)** in the United States. To make HFS/HFCS, the isomerized syrup is passed through a bed of cation-exchange resin in the calcium salt form. The resin binds D-fructose, which is then eluted to provide a fraction enriched in D-fructose. This fraction is added to the 42% fructose solution to produce syrups with higher concentrations of D-fructose. D-Fructose is not easily crystallized, but crystalline D-fructose can be obtained

[14] **Liquefaction** refers to the rapid reduction in viscosity that occurs when a starch paste is acted on by an α-amylase.

from the enriched solution and is available commercially. It is about 20% sweeter than sucrose; so for equal sweetness, fructose has about 20% less calories as compared to sucrose Chapter 19). D-Fructose gives a rapid, short, and relatively intense sweet taste. Because other flavors are not masked by it, it is used in fruit-flavored products such as sorbets.

While HFS/HFCS is almost always able to replace sucrose or invert sugar (Chapter 3) on about an equal dry solids weight basis, the nature of the specific food system will affect the exact replacement ratio. HFS cannot replace sucrose in products such as certain cookies, certain crackers, and chocolate, where crystallization of the sugar contributes to the texture (Chapter 19).

Two major types of HFSs are produced. HFS 42 (HFCS 42) contains about 42% D-fructose, the remainder being essentially D-glucose. HFS 42 is used as the sweetener in many fruit-flavored beverages, in the doughs used in production of most yeast-raised breads and rolls, and in the batters used in production of cakes. The sugars in HFS are fermentable (about 96%). Yeast utilizes D-glucose preferentially; so at the end of the proof period, about one-third of the HFS will have been fermented and the D-fructose:D-glucose ratio in the dough will have changed from 45:55 to 70:30. The D-fructose results in a slight sweetness, which is desired in some breads. HFS is blended with sucrose and glucose syrups in fruit canning. Dairy products (namely, chocolate milk, eggnog, frozen desserts, frozen novelties, ice cream, milk-based nutrition drinks, and yogurts) and nondairy products (such as barbecue sauces, cake frostings, fruit fillings, and jams) may also contain HFS 42. Most cola drinks are made with HFS 55, which contains about 55% D-fructose.

Hydrogenated starch hydrolyzates

Hydrogenation (reduction) of starch hydrolyzates produces what are known as **hydrogenated starch hydrolyzates (HSH)**. These products contain mixtures of reduced saccharides, including sorbitol, maltitol, and reduced higher maltooligosaccharides. A common one is made from a high-maltose syrup and, thus, maltitol is the predominant component in it. While they can be used as bulk sugar replacers and texture modifiers, their most prevalent use is in food products for diabetics and in sugarless candy and chewing gums (Chapter 19). In general, HSH products are slightly sweet, nonreducing, and noncariogenic and behave as humectants and cryoprotectants.[6]

Different types of starch conversion products

A company providing starch conversion products as food ingredients will usually offer several versions each of thin-boiling products, food dextrins, maltodextrins, glucose syrups, corn syrup solids, dextrose, and HFS. In the case of thin-boiling products, dextrins, and maltodextrins, different versions could be based on different base starches, different degrees of conversion, or different conversion processes. In the case of glucose syrups and corn syrup solids, the different versions could be based on different degrees of conversion (DE values) or different conversion processes. In the case of HFS, the primary difference is in the D-fructose:D-glucose ratio.

Oxidized starches (because they too are depolymerized during the oxidation process) are also classified as converted starches (section on Oxidized Starches in this chapter).

Modified food starches

Starch ingredients can provide bulk[8] body,[7] and improved texture and mouthfeel. Native (that is, unmodified starches) find limited use as food ingredients (other than as components of flours, such as wheat, corn, potato, rice, rye, and sorghum flours). Isolated native corn starch is generally only used to dust molds for jelly-type confections, for moisture control (for example, in salt), to help dissipate heat during grinding of sugar to produce powdered sugar, and for dusting of marshmallows and chewing gum. Products made with cooked unmodified starches have a very short shelf life and must be consumed within 4–6 h. An exception is the use of a high-amylose starch to produce the strong gels of gum candies.

For the majority of applications, food processors prefer starches with better behavioral characteristics than provided by native starches. For example, Upon cooking slurries of cereal starches (with the exception of waxy maize starch), weak-bodied, cohesive, rubbery pastes and undesirable gels are produced. Native waxy maize, potato, and tapioca starches produce more desirable pastes and gels, but their properties still are not ideal for most applications. However, the functional properties of starches can be improved dramatically by small amounts of modification. Modifications are done to enhance the positive attributes of the starch, to reduce or eliminate its negative attributes, and to introduce specific functionalities. Examples are modifications done to increase the ability of the paste produced by cooking to withstand the heat, shear, and acid associated with processing conditions, to make the product in which starch is used as an ingredient more stable to storage conditions, and to introduce specific functionalities, such as emulsifying properties. Modified food starches are abundant, functional, useful, and low-cost food ingredients—generally macroingredients.

Modifications of starch can be accomplished chemically, physically, or genetically. (Only the first two are covered in this book—with emphasis on the first because of much greater use of chemically modified starches in foods.) Types of chemical modifications that are made (singly or in combinations) to improve the properties of starches are cross-linking of polymer chains, noncrosslinking derivatization (often called stabilization or substitution), depolymerization (covered in the previous section), and pregelatinization. Some specific property improvements that can be obtained are the following:

- reduction in the energy required for cooking
- modification of cooking characteristics
- increased solubility
- increased or decreased paste viscosity
- increased freeze-thaw stability of pastes

- enhancement of paste clarity
- increased paste sheen
- inhibition of gel formation
- enhancement of gel formation and gel strength
- reduction of gel syneresis
- improvement of interaction with other substances
- improvement in stabilizing properties
- enhancement of film formation
- improvement in water resistance of films
- reduction in paste cohesiveness
- increased stability of the granules and pastes to low pH, heat, and shear

Covered in this section are chemical modifications and modified food starches used in the United States. (Others that may be used in other countries are also mentioned.) Those modifications covered in some detail are the following:

1. Stabilized starches
   - **a.** Starch acetates (starch esters)
   - **b.** Monostarch phosphates (starch esters)
   - **c.** Starch octenylsuccinates (starch esters)
   - **d.** Hydroxypropyl starches (starch ethers)
2. Cross-linked starches
   - **a.** Distarch phosphate (starch diesters)
   - **b.** Distarch adipate (starch diesters)
3. Cross-linked and stabilized starches
   - **a.** Hydroxypropylated distarch phosphate
   - **b.** Phosphorylated distarch phosphate
   - **c.** Acetylated distarch phosphate
   - **d.** Acetylated distarch adipate
4. Oxidized starches

Any starch can be chemically modified. Modification is practiced significantly on normal corn and waxy maize starches and to lesser extents on potato, wheat, and tapioca/cassava starches.

The most important commercial derivatives of starch are those in which only a very few of the hydroxyl groups are reacted. Normally ester or ether groups are attached at very low degrees of substitution (DS) values (Chapter 5). DS values are often less than 0.1 and generally in the range 0.002–0.2. Thus, generally on average, there is one substituent group on every 500 to 5 D-glucopyranosyl units. Uncooked, derivatized granules cannot be distinguished from uncooked, unmodified granules when examined with a microscope, but the small levels of derivatization dramatically change the properties of starches and greatly extend their usefulness and broaden their applications. Inserted groups, such as hydroxypropyl ether groups, restrict intermolecular associations and the ability of starch chains to form junction zones. Thus, gels of these derivatized starches remain stable and do not undergo syneresis[15] as readily or as extensively as do gels of unmodified starches. Such starch products that are esterified

[15] Syneresis is the expulsion/separation of liquid from a gel caused by shrinking of the gel (Chapter 5).

or etherified with monofunctional reagents and resist interchain associations are called stabilized starches (section on Stabilized Starches in this chapter). Use of difunctional reagents produces cross-linked starches (section on Cross-linked Starches in this chapter). Modified food starches are often both cross-linked and stabilized. They may also be thinned.

Starches are modified in an aqueous slurry. For esterification or etherification, a starch slurry of 30%–45% solids from wet milling (in the case of corn starches) is introduced into a stirred reaction tank. Sodium sulfate (or sodium chloride) is added to 10%–30% concentration to inhibit gelatinization. The pH is adjusted (usually with sodium hydroxide) to pH 8–12, the exact value depending on the reaction to be conducted. The chemical reagent is added. Temperature is controlled (often to 49°C [120°F] to prevent pasting so that the derivatized starch can be recovered in the form of granules). Following reaction to the desired degree of substitution, the starch is recovered by filtration or centrifugation, washed, and dried.

Chemical reactions both currently allowed and used to produce modified food starches in the United States are as follows:

1. esterification with acetic anhydride, succinic anhydride, the mixed anhydride of acetic and adipic acids, 2-octenylsuccinic anhydride, phosphoryl chloride, sodium trimetaphosphate, sodium tripolyphosphate, or monosodium orthophosphate
2. etherification with propylene oxide
3. treatment with hydrochloric and sulfuric acids
4. bleaching with sodium hypochlorite (Also allowed is bleaching with hydrogen peroxide, peracetic acid, and potassium permanganate.)
5. oxidation with sodium hypochlorite
6. various combinations of these reactions

Other reactions may be allowed and used in other countries.

On food labels, chemically modified starch ingredients must be labeled modified food starch or food starch modified (Table 7.2).

Stabilized starches

Derivatization of starches with monofunctional reagents reduces intermolecular associations, which results in reductions in gelation, gel opacity, and precipitation. Hence, such derivatization is sometimes called stabilization, and the products are known as **stabilized starches** (also known as **substituted starches**). Pastes of unmodified normal corn starch produce opaque, cohesive,[16] rubbery, and syneresing[15] gels. Pastes made from native waxy maize starch are relatively clear and have little tendency to gel at room temperature, making waxy maize starch the preferred base starch for most modified food starches. However, waxy maize starch pastes become cloudy and chunky and exhibit syneresis[15] when stored under refrigerator or freezer

[16] Cohesive refers to the tendency of the molecules to stick together. Chunky gels are an indication of cohesiveness. Cohesive gels are perceived as being difficult to break up in the mouth (section on Gels, Chapter 5).

Table 7.2 Starch modifications and conversions[a]

Treatment	Products
Etherification	Hydroxypropyl starches
Esterification	Starch acetates Starch adipates[b] Distarch phosphates[b] Monostarch phosphates Starch octenylsuccinates
Oxidation	Oxidized starches
Low-degree conversions[c]	Dextrins Thin-boiling starches
High-degree conversions[c]	Cyclodextrins Dextrose Fructose Glucose syrups High-fructose syrups Maltodextrins
Heat	Granular cold-water swelling starches Pregelatinized starches

[a]For food use only.
[b]Cross-linked starches.
[c]With enzymes and/or acids.

conditions. The most common derivatives employed for starch "stabilization" are the acetate ester and the hydroxypropyl ether. Lesser amounts of monostarch phosphate ester are used.

Starch esters

Starch granules in an alkaline slurry are acetylated by treatment with acetic anhydride.[17] The product of acetylation is known as a **starch acetate** or **acetylated starch**. Acetylation of starch to the maximum allowed in foods (DS 0.09) lowers the gelatinization and pasting temperatures, improves paste clarity, and provides stability to retrogradation and freeze-thaw cycling of food products made from them. Acetylation results in a higher peak viscosity than is obtained using the native/unmodified parent starch, and the viscosity of the cooled paste is less than that obtained from the unmodified starch, an indication of improved stability (that

[17] The equation below for the reaction of starch molecules with acetic anhydride shows that, after treatment with alkali (NaOH is essentially always used), the starch molecules are converted into the alkoxide (Starch-O^-) form. Only a fraction of the hydroxyl groups are actually converted into alkoxy groups, but it is these groups that react with the reagent. This applies to other dervatization reactions of polysaccharides presented in this and other chapters.

is, a reduced amount of retrogradation). In other countries, starch in an alkaline slurry may be acetylated by treatment with vinyl acetate (a transesterification reaction). (This reaction is not used in the United States because acetaldehyde is a by-product and must be washed out of the product.) Starch is hydroxypropylated (section on Starch Ethers in this chapter) for essentially the same purpose (that is, for stability of the paste produced by cooking in excess water).

Starch phosphate monoesters are made by drying starch in the presence of monosodium orthophosphate or sodium tripolyphosphate. **Monostarch phosphates** produce clear, stable pastes with a long, cohesive texture that have freeze-thaw stability. Paste viscosity is generally high and can be controlled by varying the concentration of reagent, time of reaction, temperature, and pH. Cooking a monostarch phosphate in the presence of salt reduces the viscosity of the paste. Phosphate esterification lowers the gelatinization and pasting temperatures. Less monostarch phosphate is probably used as a modified food starch as compared to the amounts of hydroxypropylated and acetylated starches.

Preparation of an octenylsuccinate ester is a way to attach a hydrocarbon chain to starch molecules. The hydrocarbon chain is first attached to succinic anhydride. The product is then reacted with starch granules in an alkaline slurry to make a monoester. The product of this reaction (namely, low-DS **starch 2-octenylsuccinate**) when

$$\text{Starch}-\text{OH} \xrightarrow{\text{OH}^-} \text{Starch}-\text{O}^- + \text{Ac}_2\text{O} \longrightarrow \text{Starch}-\text{OAc} + \text{HOAc}$$

$$\text{Ac} = -\overset{\overset{\displaystyle \text{O}}{\|}}{\text{C}}-\text{CH}_3$$

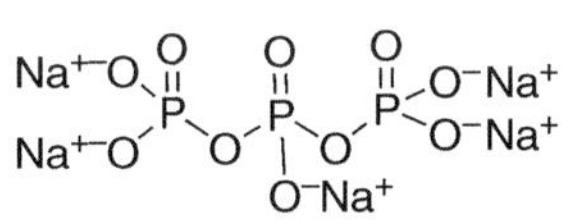

Sodium tripolyphosphate

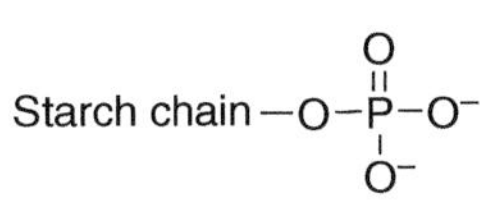

Monostarch phosphate

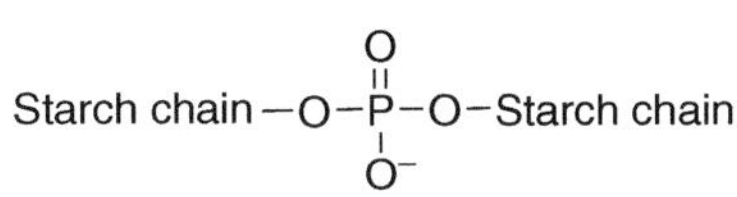

Distarch phosphate
(Cross-linked starch)

cooked in water, forms stable dispersions because the bulky substituent group keeps molecules separated. More importantly, because of the hydrophobicity of the alkenyl group, starch 2-octenylsuccinate molecules concentrate at the interface of oil-in-water dispersions, making them emulsifiers. Starch 2-octenylsuccinate is also an emulsion stabilizer because of its polymeric nature. These combination emulsifiers and emulsion stabilizers can be used in a variety of food applications where emulsion stability is needed, such as in pourable dressings and flavored beverages. The presence of the aliphatic chain tends to give the starch derivative a sensory perception of fattiness, so it is possible to use the derivative as a partial replacement for fat in certain foods. Starch 2-octenylsuccinate ester products are also referred to as lipophilic starches and **OSA** or **OS starches**.

Starch ethers

Hydroxypropylstarch is prepared by reacting starch granules in an alkaline slurry with propylene oxide to produce low levels of etherification (DS 0.02–0.2). Its properties are similar to those of starch acetate because the appended groups also produce "bumps" along the polymer chains, making it a stabilized starch. **Hydroxy-propylated starch** has lower gelatinization and pasting temperatures (as compared to the native/unmodified parent starch), that is, it is easier to cook and it produces clear pastes that do not retrograde and which better withstand freezing and thawing.

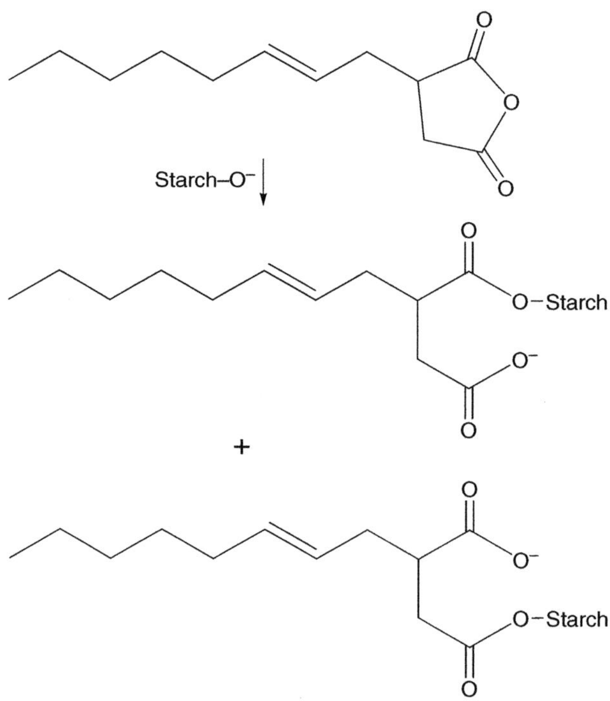

Hydroxypropyl starches are more extensively used than are acetylated starches because the ether linkage is completely stable, whereas the ester linkage is not. They are used as thickeners and extenders. Lower-molecular weight products are good coffee whiteners. To improve viscosity under acidic conditions, acetylated and hydroxypropylated starches are often also cross-linked with phosphate diester groups (next section).

Cross-linked starches

Many modified food starch products are cross-linked. **Cross-linking** occurs when a starch is reacted with a difunctional reagent that reacts with hydroxyl groups on different starch polysaccharide molecules and, thus, joins (cross-link) the molecules in granules.

Most cross-linking is accomplished by producing **distarch phosphate** (also known as **starch phosphate diester**). Phosphate ester cross-linking occurs when starch granules in an alkaline slurry are reacted with phosphoryl chloride ($POCl_3$; commonly called phosphorus oxychloride), sodium trimetaphosphate, or a mixture of sodium trimetaphosphate and sodium tripolyphosphate. ($POCl_3$ is so very reactive that it will react with hydroxyl groups without them being in the alkoxy form. When it is used as the cross-linking reagent, the function of the alkali is to neutralize the HCl by-product.)

$$2\ \text{Starch–O}^- \text{ chains} + \text{X–R–X} \longrightarrow \text{Starch chain–O–R–O–Starch chain} + 2\text{X}^-$$

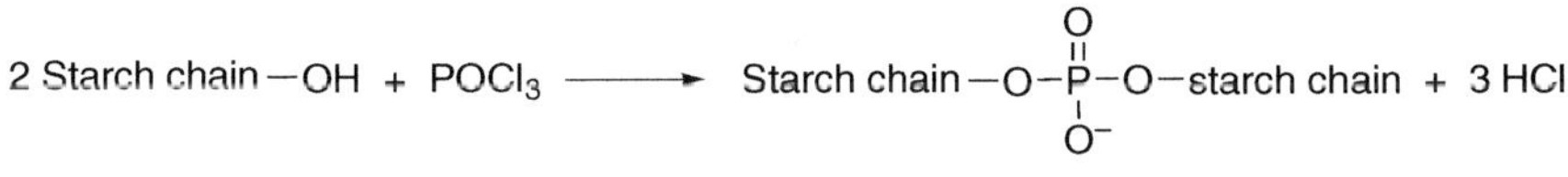

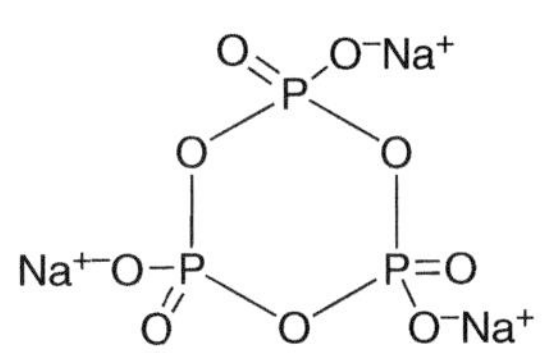

Sodium trimetaphosphate (STMP)

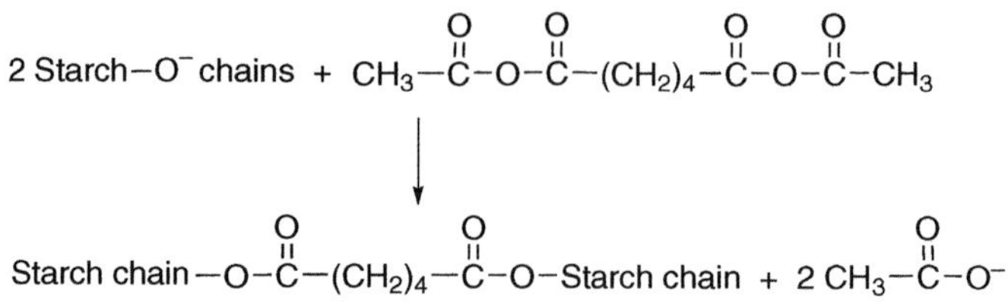

The only other cross-linked food starch product produced and used in the United States is the distarch ester of adipic acid, which as a carboxylic acid ester is less stable than the phosphate diester cross-link.

In some countries other than the United States, reaction with a reagent (commonly known as epichlorohydrin) which produces a diether cross-link is used. (This reaction is voluntarily not used in the United States because of the carcinogenic potential of the reagent [not the product].)

The joining/cross-linking of starch chains strengthens the granule and reduces both the rate and the degree of granule swelling and subsequent disintegration during cooking in water. Thus, cross-linked starch granules are less sensitive (more stable) to processing conditions (high temperature; extended cooking times; low pH; high shear during mixing, homogenization, and/or pumping) than are native granules. Hot pastes of cross-linked starches are more viscous, heavier bodied, shorter textured, and more stable during extended cooking, exposure to low pH conditions, or when subjected to severe agitation. Only a small amount of cross-linking is required to produce an effect. (Most cross-linked food starches contain less than one cross-link per 1000 α-D-glucopyranosyl units.)

With relatively low levels of cross-linking, granule swelling increases as the amount of cross-linking increases because increased granule stability means that granules can swell more without disintegrating. Increased granule swelling means that the volume occupied by swollen granules and, hence, paste viscosity increases. For example, treatment of starch with only 0.0025% of sodium trimetaphosphate[18] reduces both the rate and the degree of granule swelling, greatly increases paste stability, changes the pasting curve dramatically (Fig. 7.2), and changes the textural characteristics of the paste. At relatively high levels of cross-linking, the opposite occurs (that is, granule swelling decreases as the amount of cross-linking increases)

[18] If the reaction were 100% efficient, this amount of reagent would provide a DS of about 1.3×10^{-5} or 1 crosslink for 77,000 D-glucopyranosyl units, but the reaction is undoubtedly much less efficient.

because granule swelling becomes increasingly restricted to the point that the volume occupied by swollen granules in hot pastes becomes less than it is in the native starch. Thus, treatment with 0.08% trimetaphosphate produces granules whose swelling is so restricted that no peak is observed in the cooking curve (Rapid Visco Analyzer or Brabender Visco/amylo/graph). Also as the number of cross-links increases, the granules become more and more stable to physical conditions and acidity and less and less dispersible by cooking. Energy requirements to reach maximum swelling and viscosity are also increased. Even though hydrolysis of glycosidic bonds occurs during heating under acidic (low pH) conditions, the tying together of starch molecular chains by means of phosphate cross-links holds granules together and provides large molecules and elevated viscosity.

Cross-linked starches that form stable, high-viscosity pastes when dispersions are subjected to high shear, high temperature, and/or low pH are particularly useful in continuous processes. High levels of cross-linking are also useful for extruded products. Storage-stable thickening following retort sterilization of canned foods is also provided by cross-linked starches; for their reduced rate of gelatinization and swelling allows the low initial viscosity of the cans' contents to remain long enough to facilitate the heat transfer and temperature rise required to provide uniform sterilization. (A high viscosity would restrict the heat transfer required for sterilization.) Ultimately, however, when the temperature gets high enough, granule swelling develops the desired viscosity, texture, and other properties of the final product. Cross-linked starches are used in products such as batter mixes, canned soups,

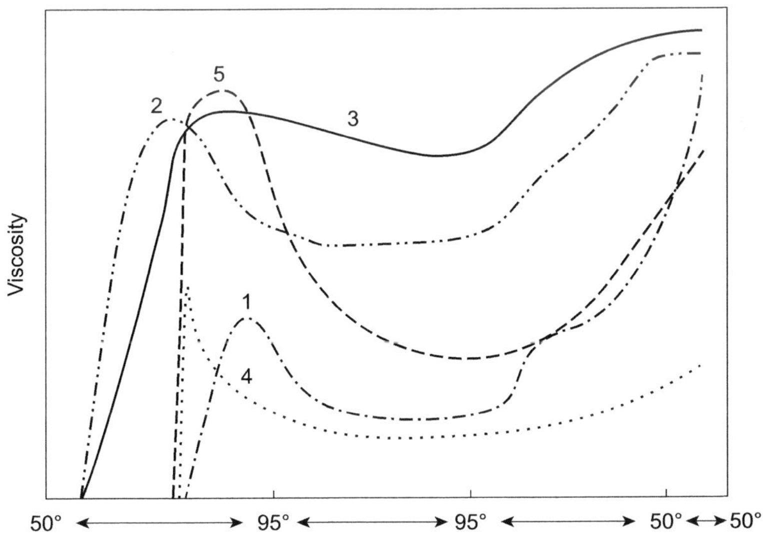

Figure 7.2 Typical Rapid Visco Analyzer and Brabender Visco/amylo/graph curves: 1 = normal corn starch, 2 = stabilized normal corn starch, 3 = moderately cross-linked and stabilized normal corn starch, 4 = waxy maize starch, 5 = potato starch.

canned gravies, canned puddings, cream-style canned corn, and spoonable salad dressings.

Cross-linking of waxy maize starch gives a clear paste that is sufficiently rigid that fillings in cut sections of pies hold their shape. Neither will the filling boil out when heated.

Starches that are both cross-linked and stabilized are used in canned, frozen, baked, and dry foods. They are used if the system is acidic (salad dressings and tomato-based sauces), is to be retorted (soups, sauces, gravies), when the process requires hot-fill application (pie fillings, bakery glazes), and in aseptically filled products (puddings, cheese sauces). They provide stability during long-term storage to products such as baby foods and fruit and pie fillings in cans and jars, frozen fruit pies, pot pies, and gravies.

Oxidized starches

Bleaching and sterilization of starches is often done with small amounts of sodium hypochlorite, hydrogen peroxide, peracetic acid, or potassium permanganate. When starch is treated with greater amounts of an oxidizing agent (oxidant), the result is a product known as **oxidized starch**. Sodium hypochlorite is the primary oxidant used in the United States. It oxidizes C6 to an aldehydic group and C2 and C3 to keto groups. Cleavage of the C—C bond between C2 and C3 to produce two aldehydic groups also occurs (Fig. 7.3). Some of the aldehydic groups on C2, C3, and/or C6 are further oxidized to carboxylate groups. Carbonyl (either aldehydic or keto) groups at either C2, C3, or C6 become a part of a β-alkoxy carbonyl system and, because sodium hypochlorite solutions are quite alkaline, β-elimination and depolymerization occur (Chapter 4). Therefore, because of the depolymerization as with acid-thinned starches, the energy required to cook oxidized starch and the hot-paste viscosity are reduced and these starches are also classified as converted starches. Batters used for coating fried foods made with hypochlorite-oxidized starch exhibit improved adhesion. Oxidized starch is whiter and produces more stable pastes that have a reduced tendency to form a gel and to retrograde (as compared to the parent starch).

In some other countries, starches are oxidized with hydrogen peroxide and copper(II) sulfate, which results in depolymerization without formation of carbonyl or carboxylate groups.

Pregelatinized and cold-water swelling starch products

If starch has been pasted and dried without excessive molecular reassociation, it can be dissolved in water at room temperature. Such precooked, cold-water soluble products, known industrially as **instant** or **pregelatinized starches**,[19] are made in two ways. In

[19] Unfortunately, many in the starch industry refer to them as "pregels", although they have nothing to do with gels. Gelatinization and gelation are quite different phenomena.

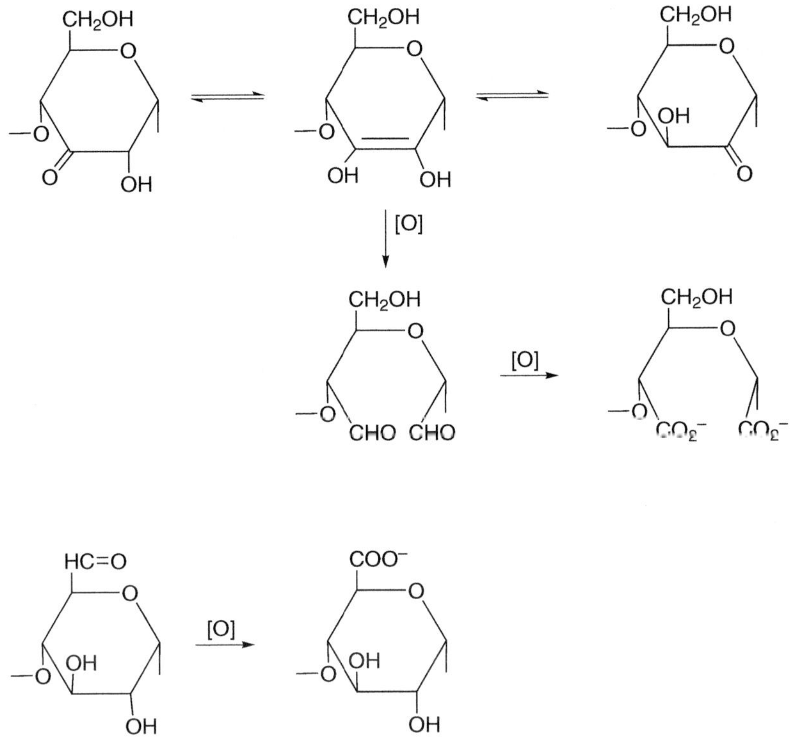

Figure 7.3 Modified glucosyl units produced via reaction of a starch with hypochlorite ions. The first product is a carbonyl (aldehydic or keto) group. When an aldehydic group is formed, it is rather easily oxidized to a carboxylate group. As explained in Chapter 4, because the reaction conditions are alkaline, units containing carbonyl groups are points of chain cleavage via beta-elimination.

one, a starch—water slurry is applied to a steam-heated roll or into the nip between two nearly touching and counter-rotating, steam-heated rolls where the starch is rapidly gelatinized, pasted, and dried. The dry film is scraped from the roll and ground. The resulting product should contain few intact granules, except when the starting material is a highly cross-linked starch or a high-amylose starch. Pregelatinized products are also prepared using extruders. In both cases, the dried product needs to be ground (powdered) to the desired mesh size.

Because pregelatinized starches are precooked, they hydrate rapidly and can be used without cooking, producing dispersions without lumps if prepared correctly. In general, pregelatinized starches behave similarly to water-soluble gums, so solutions/dispersions of them need to be prepared in a similar manner (Chapter 5). Some pregelatinized starches are designed to produce some graininess or pulpiness (desired in some products) when dispersed in water. Many pregelatinized starches are used in dry mixes, such as instant pudding mixes. They disperse readily with high-shear stirring (not usually encountered in the home kitchen) or when mixed with sugar or other dry ingredients (for use in the home). Both chemically modified and unmodified starches can be used to make pregelatinized starches. If modified starches are used, the properties introduced by the modification(s) are also exhibited by the pregelatinized products; thus, paste properties, such as stability to acid, shear, and freeze-thaw cycles, may also be characteristics of pregelatinized starches. A pregelatinized, slightly cross-linked starch is useful in instant soups, pizza toppings, and extruded snacks. It also has application in limited-moisture systems such as in soft cookies and bakery fillings.

Pregelatinized starch products are used when (1) a source of heat is not available, (2) there is no step in the process in which sufficient heat will cook the nonpregelatinized starch, or (3) heat cannot be applied because of the thermal lability of one or more of the other ingredients. They are especially useful in dry mix products designed for use in the home. They are also used to provide functionality in the final product. Mixes for moist cakes take advantage of their high water absorption. Because pregelatinized starches do not develop the same gel firmness as **cook-up starches**,[20] they produce smoother puddings and cream fillings. (Most dispersions of pregelatinized starch products do develop additional viscosity when heated.) Pregelatinized starch products contribute texture and moisture management to snack foods and viscosity and moisture management to frostings and toppings.

Intact granules that swell extensively in the presence of room temperature water are made by heating normal corn starch in 75%−90% ethanol to 150−175°C (300−345°F) for 0.5−2.0 h or by very quickly heating a starch slurry in a special spray-drying nozzle. (The droplets containing gelatinized granules are then dried in the spray drier.) These products are known as **granular cold-water-swelling (GCWS) starches** or simply as **cold-water-swelling (CWS) starches**. The granules in them are gelatinized, but their granular form is maintained (as opposed to traditional pregelatinized products). They are also categorized industrially as "instant starches". (The product described first in this section, while historically called pregelatinized starch, is actually prepasted starch.) GCWS/CWS products are also often called cold-water soluble starches, although pregelatinized starch is actually more soluble in room temperature water.

GCWS/CWS starch products hydrate rapidly. They produce the functionalities of cook-up starches without heating. A special property is that, when they are dispersed in sugar or glucose syrups by rapid stirring, as in a Waring blender, and the resulting dispersion is poured into molds, it will set to a rigid gel that can be sliced easily to make a gum candy.[21] The ability to swell in unheated water is also useful for making desserts and in muffin batters that contain particles, such as blueberries, that would settle to the bottom when the batter thins by heating (during baking before the

[20] Starches that are neither pregelatinized or GCWS/CWS and, therefore, need to be cooked to generate a paste are known as **cook-up starches**.

[21] Some of the intermolecular hydrogen bonding within starch granules is disrupted in the preparation of GCWS starch products. When these granules are placed in water, hydration and swelling is rapid. In this condition, retrogradation reintroduces insolubility and results in aqueous dispersions of poor quality. However, when the treated granules are dispersed in a sugar solution, retrogradation is minimal.

temperature reaches the gelatinization temperature of the wheat starch and its thickening effect is realized). A further modification of the process converts a mixture of normal corn and waxy maize starch to an agglomerate that readily produces high viscosity when added to water. Waxy maize starch alone on heating in aqueous ethanol losses its granule structure and, on addition to hot water, dissolves to produce, on cooling, a hard candy-like product. GCWS starches can be made from either native or modified normal starches.

Other physical modifications

Processes for physical modifications of starches are usually divided into those that are thermal treatments and those that are not, but the demarcation is not always clear. Many physical modifications have been explored—primarily to avoid the requirement to label chemically modified starch as modified food starch or food starch modified. Other reasons for wanting a physically modified starch that provides desirable functionalities include a desire to reduce or eliminate process effluents, reduce costs, and increase amounts of slowly digesting and resistant starch (RS) (Chapters 6 and 17).

Two thermal treatments (pregelatinization and production of GCWS) were discussed above. Other thermal treatments include HMT, annealing, and heating of dry starch. Nonthermal treatments include ultrasound, milling, high-pressure, and pulsed electric field treatments. The modifications to starches effected by these and other nonthermal treatments have largely been studied because the treatments are employed or have been investigated as nonthermal processes for the pasteurization of food products, and it was of interest to know what changes were made to ingredients during processing. They are not covered in this book.

Heat–moisture treatments

Heat-moisture treatments (HMT) consist of heating moistened starch granules. The temperatures to which a starch is heated are higher than its normal gelatinization temperature, but the granules do not gelatinize because the moisture content is low. A wide range of conditions have been used with different starches. Moisture contents are usually between 20% and 30%; temperatures used are usually between 80 and 140°C (175–285°F), and the length of heating varies between 1 and 24 h. (The source of heat may also vary, such as from conventional to microwave.) Because of the different starch sources and the wide range of process variables, a variety of products with a variety of properties can be made, and it is not possible to describe the properties of HMT starches in a way that applies to all products. Granules usually remain intact during HMT. HMT products often exhibit increases in pasting temperature and hot-paste (trough) viscosity and decreases in granule swelling, peak viscosity, breakdown, and solubility (that is, the products often have characteristics similar to those of a lightly cross-linked starch).

HMT starches usually contain slightly to moderately more thermostable slowly digesting starch (SDS) and RS (Chapters 6 and 17). The natures of the products vary as the treatment conditions (especially the moisture content) varies. Treatment of the starch with acid before HMT or conducting the HMT under slightly acidic conditions effects depolymerization and increases the amount of thermostable RS. Addition of a fatty acid that will complex with amylose (especially lauric acid) before HMT increases the amounts of thermostable SDS and RS.

It is not known with certainty what changes take place in granules during HMT, but it has been logically proposed that the plasticizing effect of the water present and the thermal energy allows double helices to move within crystallites so as to form more ordered and more closely packed structures (that is, HMT disrupts the least stable structures and allows the growth and/or more perfect alignment of the more stable native crystallites). HMT does bring about changes in crystal type. It partially or completely changes B-type crystals to A-type crystals.

Annealing

Annealing is another hydrothermal process that changes the physical properties of starches. **Annealing** occurs when starch granules in an excess of water (more than 40%) are held at a temperature below the initiation/onset temperature of gelatinization for periods of minutes to days. Again (as with HMT), because of the many possible combinations of the type of starch, the temperature, and the holding time, many different products with diverse properties can be made, and it is not possible to describe the properties of annealed starches in a way that applies to all products. However, again granules remain intact. And again, the properties of annealed starches are usually similar to those of a lightly cross-linked starch, that is, there is usually an increase in the gelatinization temperature and a reduction in swelling power, solubility, and breakdown. Both HMT and annealed starch products usually produce firmer gels. Only slight or no increases in thermostable SDS and RS values have been reported.

Again, it is not known with certainty what changes take place in granules during annealing, but again, it has been logically proposed that annealing effects increased alignment of double helices and perfection of crystallites. So the mechanisms of these hydrothermal processes seem to be similar. It is somewhat certain that, during annealing, molecular rearrangements occur within both amorphous and crystalline regions of granules and that the changes within the crystallites primarily result in their perfection, rather than changes in the amount of crystalline material or in the crystal type, making that at least one difference between annealing and HMT in terms of mechanism.

Annealing has been done both as single- and multistep processes. Both annealing and HMT have been done before and after each other, prior to and following chemical modification, and following acid thinning.

Heating dry starch

Products with acid-, shear-, and heat-tolerance characteristics similar to those of chemically cross-linked starches with a low to modest degree of cross-linking are prepared by heating a starch with less than 15% moisture at temperatures above 100°C (212°F), but below the temperature at which thermal degradation occurs. Making the starch alkaline and drying to less than 1% moisture facilitates the changes in properties.

Multiple modifications

Modified food starches are tailor-made for specific applications. Many modified food starches are made from different native starches by a combination of cross-linking and stabilization (introduction of monosubstituent groups). Acid-thinning and/or pregelatinization may also be employed. These widely used products can have a wide diversity of functional properties and attributes and are made for use with specific processes and in specific products (Table 7.3, Box 7.2).

Using modified food starch

In choosing a modified food starch to use in making a processed food product, the functionalites demanded of the starch ingredient must be identified. Questions to be asked are likely to include, "What are the critical processing parameters?", "Are there any formulation issues, including interactions with other ingredients?", "What are the packaging, storage, and shelf-life requirements?", "Under what conditions will the consumer reconstitute and/or use the product?", etc. Then, obtaining optimum functionality of any modified food starch depends on its proper cooking, as any starch can be both undercooked and overcooked.

Blends of starches and hydrocolloids

A starch product is often the first choice of a food product developer needing thickening, bulk, or body. Starch products usually have the advantage of lower cost and easier handling and processing, as compared to hydrocolloids, but starch is not the ideal thickening, gelling, and/or texturizing agent for all applications. Blends of two or more hydrocolloids and blends of a native or modified food starch and a hydrocolloid are often employed to take advantage of either two or more individual attributes or attributes produced by synergistic interactions and/or to reduce the cost of ingredients. Starch-hydrocolloid combinations are mentioned in the chapters on cellulose and cellulose derivatives (Chapter 8), galactomannans (Chapter 9), xanthan (Chapter 11), gellan (Chapter 12), carrageenans (Chapter 13), and alginates (Chapter 14). An example is described here. ι-Carrageenan (Chapter 13) produces a weak gel at a concentration of 0.2% with calcium ions. When the proper modified waxy maize starch product is used in combination with an ι-carrageenan product, the gel texture becomes much less elastic (that is, more rigid). Because the carrageenan molecules are probably mostly excluded from starch granules because of their molecular size, they are probably confined to the aqueous phase outside

Table 7.3 Some attributes imparted by chemical modifications of starches

Modification	Attributes
Acid-modified (thinned)	Increased solubility Lower gelatinization and pasting temperatures (easier cooking) Lower hot-paste viscosity Increased paste clarity Increased gel strength
Dextrinized	Imparts crispness High water solubility (can make high-solid solutions without high viscosity) Increased browning Film formation
Stabilized	Lower gelatinization and pasting temperatures (easier cooking) Increased product stability at refrigerator temperatures Improved freeze-thaw stability of product Decreased product setback Greater product clarity Increased steam-table stability Increased moisture control More easily redispersed when pregelatinized Often nongelling
Cross-linked	Increased stability to heat Increased stability to low pH/acid Increased shear resistance Increased cooking temperature (delayed pasting) Decreased product setback Increased or decreased paste viscosity Increased body Salvelike texture
Cross-linked and stabilized	Lower gelatinization and pasting temperatures, but increased paste viscosity Other attributes of combined cross-linking and stabilization A variety of textures and rheological properties
Octenylsuccinylation	Emulsifying properties Emulsion-stabilizing properties Encapsulation
Hypochlorite-oxidized	Whiter Lower gelatinization and pasting temperatures (easier to cook) High water solubility (can make high-solid solutions without high viscosity) Makes softer, clearer gels Greater adhesion of pastes More sterile
Pregelatinized and granular cold-water swelling starch	Thickening without cooking If otherwise modified, other attributes of the modified starch

Box 7.2 Some food product functionalities that can be controlled by selection of the proper modified starch

Acid stability
Adhesion
Clarity
Color
Emulsion stability
Film formation
Flavor release
Heat stability
Hot and cold viscosity
Hydration rate
Moisture retention
Mouthfeel
Oil migration
Physical state (liquid, semisolid, solid)
Shear stability
Sheen
Shelf stability
Tackiness
Temperature required to cook
Texture/consistency

the swollen starch granules, where their concentration increases as the granules take up water and swell. Because the setting and melting temperatures of an ι-carrageenan (or any other) gel are not concentration dependant, they do not change. Also, the ι-carrageenan solution (at temperatures above the melting temperature of its gel) has a lubricating effect and reduces the friction between swollen granules. This reduces the viscosity and improves heat exchange during processing. So the ι-carrageenan modifies properties of the modified food starch, and the modified food starch modifies properties of the ι-carrageenan.

Encapsulation

Some benefits that can be derived from use of encapsulated food ingredients are given in Table 7.4. Several processes are used to encapsulate ingredients—the most widely used being spray-drying. Considerations in choosing the encapsulating process include the nature of the end product and the process in which the encapsulated ingredient will be used (which will dictate the appropriate encapsulating substance), the characteristics of the substance or material to be encapsulated, and economics. Together, these factors influence selection of the most appropriate encapsulating process.

Carbohydrates are popular encapsulating agents. Gum arabic (Chapter 16) is widely used carbohydrate for flavor oil encapsulation via spray drying. Octenylsuccinylated starch products designed to replace gum arabic are quite effective alternatives. Use of β-cyclodextrin is limited by its cost. Maltodextrins and syrup solids are much less expensive, but less effective, alternatives.

Table 7.4 Some benefits of encapsulation

Functionality	Examples of encapsulated substances
Protection against moisture oxidation reactions with other ingredients temperature volatilization	Aromas Bioactives Flavors Natural colors Nutraceuticals Sweeteners Vitamins
Improve handling and dispersibility by converting to a dry powder	Natural colors Oils Vitamins
Delayed dissolution (timed release)	Acidulants (meat products) Enzymes Flavors (chewing gum) Leavening agents Minerals Salt (bakery products) Sweeteners Vitamins
Reduction of off-flavors	Minerals Nutraceuticals Vitamins

Additional resources

Conversions and Products of Conversion

Blanchard, P.H., Katz, F.R., 1995. Starch hydrolysates. In: Stephen, A.M. (Ed.), Food Polysaccharides. Marcel Dekker, New York.

Brumm, P.J., 1998. Enzymatic production of dextrose. Cereal Foods World 43, 804−807.

Buck, A.W., 2016. High fructose corn syrup. In: O'Brien-Nabors, L. (Ed.), Alternative Sweeteners, fourth ed. CRC Press, Boca Raton, pp. 403−422.

Chronakis, I.S., 1998. On the molecular characteristics, compositional properties, and structural-functional mechanisms of maltodextrins: a review. Critical Reviews Food Science 38, 599−637.

Hull, P., 2010. *Glucose Syrups: Technology and Applications*. Wiley-Blackwell, Oxford.

Schenck, F.W., Hebeda, R.E. (Eds.), 1991. Starch Hydrolysis Products: Worldwide Technology, Production and Application. VCH Publishers, New York.

Schenck, F.W., 1996. Solid starch hydrolysates. Cereal Foods World 41, 388−390.

Schenck, F.W., 1996. Solid starch hydrolysates. Starch/Stärke 48, 188−190.

White, J.S., 2016. Crystalline fructose. In: O'Brien-Nabors, L. (Ed.), Alternative Sweeteners, fourth ed. CRC Press, Boca Raton, pp. 379−402.

Cyclodextrins

Szejtli, J., 2004. Cyclodextrins. In: Tomasik, P. (Ed.), Chemical and Functional Properties of Food Saccharides. CRC Press, Boca Raton, pp. 271–289.

Szente, L., Szejtli, J., 2004. Cyclodextrins as food ingredients. Trends in Food Science & Technology 15, 137–142.

Use of Amylases in Baking

Goesaert, H., Slade, L., Levine, H., Delcour, J.A., 2009. Amylases and bread firming – an integrated view. Journal of Cereal Science 50, 345–352.

Chemically Modified Starches

Chen, Y.-F., Kaur, L., Singh, J., 2018. Chemical modification of starch. In: Sjöö, M., Nilsson, L. (Eds.), Starch in Food, second ed. Woodhead Publishing, Duxford, pp. 223–253.

Wurzburg, O.B. (Ed.), 1986. Modified Starches: Properties and Uses. CRC Press, Boca Raton.

Physically Modified Starches

BeMiller, J.N., 2018. Physical modification of starch. In: Sjöö, M., Nilsson, L. (Eds.), Starch in Food, second ed. Woodhead Publishing, Duxford, pp. 223–253.

BeMiller, J.N., Huber, K.C., 2015. Physical modification of food starch functionalities. Annual Review of Food Science and Technology 6, 19–69.

Encapsulation

Dziezak, J.D., April 1988. Microencapsulation and encapsulated ingredients. Food Technology 136–153.

Jin, Y., Li, J.Z., Nik, A.M., 2018. Starch-based microencapsulation. In: Sjöö, M., Nilsson, L. (Eds.), Starch in Food, second ed. Woodhead Publishing, Duxford, pp. 660–690.

Reineccius, G.A., March 1991. Carbohydrates for flavor encapsulation. Food Technology 144–149.

Cellulose and Cellulose-Based Hydrocolloids

8

Chapter Outline

Key information and skills that can be obtained from study of this chapter will enable you to

1. Define, describe, and/or identify

powdered cellulose	hydroxypropylmethylcelluloses (HPMCs)
microcrystalline celluloses (MCCs)	gelation temperature (gel point)
carboxymethylcelluloses (CMCs)	hydroxypropylcelluloses (HPCs)
methylcelluloses (MCs)	

2. Describe the (1) preparation, (2) properties, and (3) uses of powdered cellulose.
3. Describe the properties that allow MCC to function (1) as a foam stabilizer, (2) as an emulsion stabilizer, (3) as a fat extender, (4) in dietetic foods, (5) as a heat-stable gel former, (6) in controlling ice crystal growth, and (7) as a carrier.
4. Describe the difference between powdered MCC and colloidal MCC.
5. Describe and discuss the concept behind derivatization of cellulose to change it into a water-soluble hydrocolloid.
6. List and describe the five principal variables that control the properties of cellulose derivatives.
7. Describe the general characteristics of water-soluble cellulose derivatives.
8. Describe the property differences between uniformly and nonuniformly substituted CMC and explain the differences in molecular terms.

Carbohydrate Chemistry for Food Scientists. https://doi.org/10.1016/B978-0-12-812069-9.00008-X

9. Describe and discuss interactions of CMC with cations, including proteins (polycations).
10. Write chemical equations and give conditions for chemical modification of cellulose to make (1) a CMC, (2) a MC, (3) a HPC, and (4) a HPMC using a shorthand designation for cellulose.
11. Give the probable explanation for thermal gelation of HPMC solutions.
12. Explain, on a molecular basis, the effectiveness of HPMC in nondairy whipped toppings.

Introduction

Because cellulose is the principal cell wall component of higher plants, it is the most abundant organic compound and the most abundant carbohydrate on Earth.[1] Approximately one-half the mass of perennial plants and one-third the mass of annual plants are cellulose. Cellulose is present as the principal cell wall material in all vegetables and fruits consumed by humans. It is a high-molecular weight (MW), linear, insoluble homopolymer of β-D-glucopyranosyl units joined by (1→4) glycosidic linkages. (The structure below shows the structure as repeating cellobiosyl units to show that every other unit is inverted [as compared with the units ahead and behind it], giving the molecule a flat ribbon-like shape/conformation.) Because of their linearity and stereoregular nature, cellulose molecules associate over extended regions, forming polycrystalline, fibrous[2] bundles, where the molecular chains in crystalline regions are held together by numerous hydrogen bonds. Cellulose is insoluble except in a few special solvents that can disrupt its intermolecular hydrogen bonds. However, certain derivatives of cellulose are water soluble and important as hydrocolloids (see Sections Carboxymethylcelluloses, Methylcelluloses and hydroxypropylmethylcelluloses).

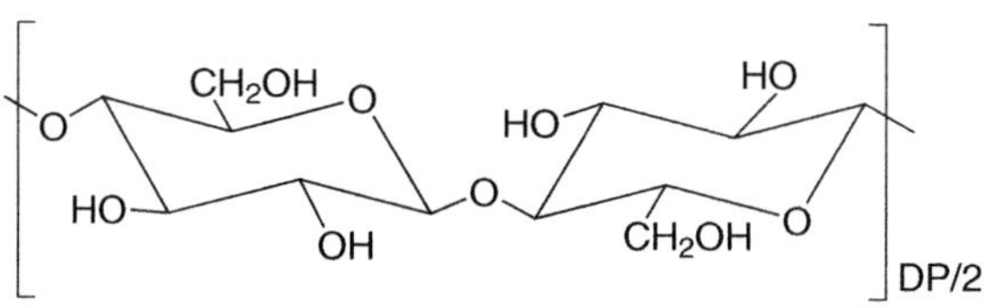

Cellulose

Cellulose

Cellulose and its modified forms serve as dietary fiber (Chapter 17) because no forms of cellulose are digested by humans and, thus, none are either significant energy or carbon sources for humans. Dietary fiber does, however, serve important physiological functions (Chapter 17).

The chemical structure of cellulose is the same regardless of the source, but cellulose preparations differ in the physical arrangement of the polymer molecules (their fibrillar

[1] When cellulose is considered as a polymer of D-glucose, D-glucose can be said to be the most abundant carbohydrate and even the most abundant organic compound on Earth.

[2] The term fibrous indicates that the lengths of the bundles of molecules are far greater than their widths.

structure) and, therefore, properties from source to source. The commercial source of cellulose is wood pulp or cotton linters.[3] Because cotton fibers are at least 93% cellulose, cotton linters need only a treatment with a hot sodium hydroxide solution to remove very small amounts of protein, pectic substances,[4] and wax to obtain high-quality cellulose. Wood chips (about 50% cellulose, 30% hemicellulose [Chapter 17], and 20% lignin) are subjected to a pulping (delignification) process to solubilize and remove the latter two components. To effect pulping, wood chips are digested with an alkaline solution of sodium sulfide (the kraft or sulfate process), calcium bisulfite in the presence of sulfur dioxide (the bisulfite process), or sodium hydroxide alone (the soda process). The latter process is the primary source of cellulose used to make water-soluble derivatives for the food industry. This pulp is further purified by treatment with alkali and an alkaline solution of sodium hypochlorite to remove additional amounts of hemicellulose and color and the remaining traces of lignin.

Powdered celluloses

A purified cellulose powder obtained by special pulping of wood is available as a food ingredient. A measure of the quality of cellulose is its content of alpha-cellulose (that portion which is insoluble in 18% sodium hydroxide solution). Highly purified pulp with an alpha-cellulose content of more than 99% can be considered to be pure $(1 \rightarrow 4)$-linked β-glucan. Chemical purity is not required for food use because cellulosic cell wall materials are components of all fruits, vegetables, flours, meals, and brans. However, to make the **powdered cellulose** used in food, pulp that has been produced by delignification of wood chips is purified and bleached until it meets the specifications of the *Food Chemicals Codex.* Then, particle size can be reduced by milling. Original fibers may vary in length from 500 to 4000 μm and in width from 5 to 350 μm. After size reduction, products with average fiber lengths in the 22–120 μm range are obtained. Larger particles absorb more water and provide more bulk.[5] (Depending on the fiber length, cellulose powder can absorb and hold 3.5–10 times its weight of water.) Smaller particle sizes produce food products with smoother texture (as compared with products made with larger average particle sizes). The powdered cellulose used in foods has negligible flavor, color, and microbial contamination.

Powdered cellulose may be added to bread to provide noncaloric bulk. Reduced-calorie[6] baked goods containing powdered cellulose have an increased content of dietary fiber (Chapter 17) and stay moist longer. Powdered cellulose reduces caking

[3] Cotton linters are the short fibers remaining on cotton seeds after the long fibers are removed for textile manufacturing.

[4] Pectic substances are polysaccharides associated with pectin (Chapter 15) in the middle lamella between plant cells.

[5] Bulk refers to the solids content that contributes to the texture and palatability of a food product.

[6] There is no official definition of reduced calorie. Here, it means containing relatively fewer calories than a comparable product. To be considered a low-calorie food product, the product must contain not more than 40 Kcal per a standard serving.

of grated and shredded cheeses. In barbecue sauces, and salad dressings, powdered cellulose contributes to texture and increases cling and viscosity. Use of smaller particle size powdered cellulose in icings and fillings for baked goods and confections reduces their caloric value with minimal change in texture. Powdered cellulose of at least 110 μm particle size increases the viscosities of solutions of guar gum (Chapter 9), sodium CMC (see Section Carboxymethylcelluloses), and xanthan (Chapter 11) because interactions of these hydrocolloids with cellulose result in network formation. Many of these same attributes are also provided by MCCs (next section). In addition, a large volume of use of powdered cellulose is for filtration of beers, juices, and wines, with or without addition of diatomaceous earth in the filter cake.

Other products in this category are microfibrillated and microreticulated celluloses. Bacterial (namely, *Acetobacter xylinum*) cellulose gel (called nata) is a favorite dessert in the Philippines. *A. xylinum* cellulose, a noncrystalline cellulose, has been investigated for other food applications (Table 8.1).

Microcrystalline celluloses

Microcrystlline celluloses (MCC) are made by hydrolysis of purified wood pulp (alpha-cellulose pulp), followed by separation of the cellulose microcrystals. Cellulose molecules are fairly rigid, completely linear chains of (1→4)-linked β-D-glucopyranosyl units that easily align in long junction zones because of their flat, ribbon-like conformation produced by every other monomeric unit being inverted with respect

Table 8.1 Characteristic properties of cellulosics

Product	Major characteristics
Cellulose powders	Provide noncaloric bulk, retain moisture
Microcrystalline celluloses	Stabilize foams, stabilize emulsions, replace fats and oils, form thixotropic gels, improve adhesion (cling), provide freeze-thaw stability, provide anticaking activity in grated and shredded cheese
Carboxymethylcelluloses	Thicken, retard ice crystal growth, form films, bind and hold water, protect colloids, act as processing aids, act as humectants, retard sugar crystallization, prevent syneresis, stabilize proteins, control batter viscosity
Methylcelluloses and hydroxypropylmethylcelluloses	Exhibit thermal gelation, reduce the amount of fat required, provide lubricity, form and stabilize emulsions, form and stabilize foams, form films, provide freeze-thaw stability, provide adhesion and binding, bind and hold water, provide pseudoplastic rheology, are nonionic, are compatible with sugar

to the units preceding and following it. (Cotton linter cellulose has an average degree of polymerization (DP) of about 3000. Wood cellulose molecules have average DP of about 500 to about 1500, depending on the tree source.) These long chains do not fit together over their entire lengths, rather cellulose molecules in microfibrils pass in and out of crystalline regions. The ends of crystalline regions are simply the result of divergence of cellulose chains away from order into more random arrangements, but all chains do not leave the crystallites at the same place. When purified wood pulp is treated with acid, hydronium (H_3O^+) ions most easily penetrate the noncrystalline (amorphous) regions and effect hydrolytic cleavage of the glucan chains. Hydrolysis of the cellulose chains in the amorphous regions releases individual crystallites, which have chains protruding from their ends; hence, the crystallites are sometimes called fringed crystallites. The released crystallites increase in size because the protruding chains that constitute the fringes at either end of the crystallites now have greater freedom of motion and can align in crystalline order and enlarge the crystallite (Fig. 8.1).

Two types of MCC are produced. One is powdered MCC. Powdered MCC is a spray-dried product. Spray-drying produces aggregates of microcrystals that are porous and sponge-like. Average particle sizes in different products generally range from about 20 μm to about 90 μm. A particle 30 μm in diameter typically contains about 600 million microcrystals of about 0.1 μm in length. Powdered MCC is used as an anticaking agent and flavor carrier for grated and shredded cheese. It is used to make reduced-calorie[6] and/or high-fiber bakery products and as an extrusion aid for expanded snacks and restructured, frozen, French fried potatoes. The second type of MCC is termed colloidal MCC. It is water dispersible and has functional properties somewhat similar to those of hydrocolloids, although it (like powdered MCC) is insoluble in water. To make colloidal MCC, considerable mechanical energy is applied after hydrolysis to tear apart the weakened microfibrils and provide a major proportion

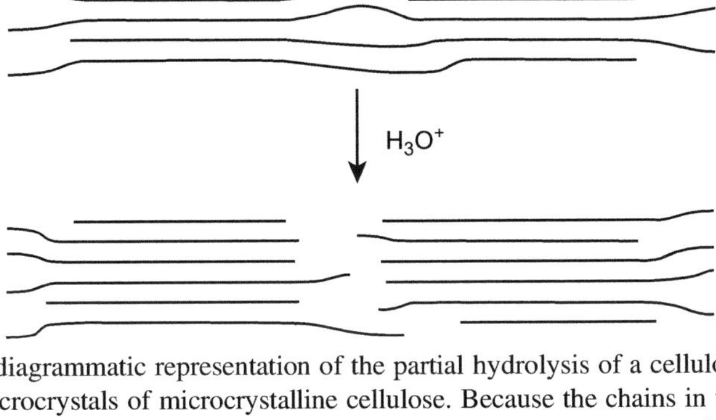

Figure 8.1 A diagrammatic representation of the partial hydrolysis of a cellulose fiber to produce the microcrystals of microcrystalline cellulose. Because the chains in the amorphous regions have greater freedom of motion after cleavage, the crystals can grow in size and are actually larger in the product than they were in the original pulp.

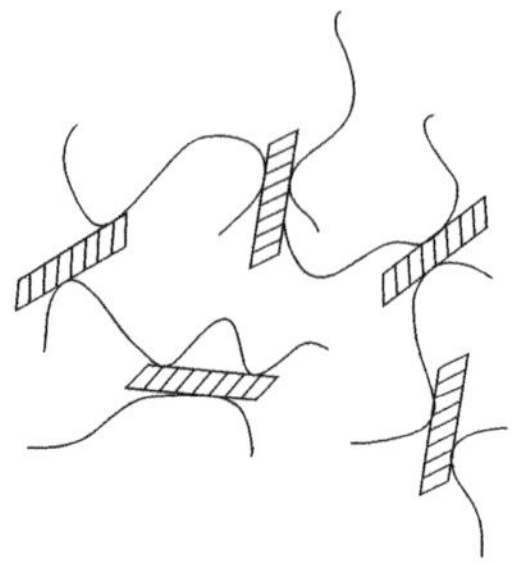

Figure 8.2 A diagrammatic representation of the tying together of cellulose microcrystals with sodium carboxymethylcellulose (not drawn to scale).

of colloidal-sized aggregates (<0.2 μm in diameter). To prevent rebonding of the aggregates during drying, sodium CMC (see Section Carboxymethylcelluloses) or an alginate (Chapter 14) may be added. These anionic hydrocolloids bind to the surfaces of the microcrystals, aid in their redispersion, and act as barriers to reassociation by giving the particles a stabilizing negative charge from the ionized carboxyl groups of the adhering hydrocolloid molecules. They also leash the particles together after dispersion (Fig. 8.2).

As already mentioned, MCC products do not dissolve in water. Viscosity development is dependent on shear to disperse the particles.[7] During high shear in water, particle dispersion progresses from being one of partially hydrated particles, to being one of fully swollen particles, to being a partial dispersion of microcrystals, to being a fully dispersed colloidal suspension of microcrystals. The viscosity of the dispersion continues to increase over 12–24 h as the dispersed microcrystals hydrate and swell. This increase in viscosity can be as much as ten-fold. The nature of the suspension varies with various factors, including the particular MCC product used, the temperature, and the presence of other ingredients; but generally, it forms a thixotropic gel (Chapter 5). The rheology and stability of the system can be modified through the use of hydrocolloids. Xanthan (Chapter 11) works best as a protective colloid; CMCs and HPMCs (see Section Methylcelluloses and hydroxypropylmethylcelluloses) are also effective. Xanthan is the only hydrocolloid that is effective in very acidic systems (for example, pH 3). In general, as with hydrocolloids, MCC should be added to the water first, using as much shear as possible for as long as possible to give as complete a dispersion as possible. If the pH is below 4, a protective colloid is required.

Both types of MCC are relatively stable to heat and acids. Their major functions are listed in Table 8.2. Both are dispersed by shear rather than dissolved with heat as are starch and some hydrocolloids.

[7] The particles are agglomerates of microcrystals.

Table 8.2 Major functions of microcrystalline celluloses

Functionality	Comments
Stabilize foams, even during high-temperature processing	Collects at the air–water interface and strengthens the interfacial film. Produces same foam stiffness at a reduced fat concentration.
Stabilize emulsions, even during high-temperature processing	Collects at the oil–water interface and strengthens the interfacial film.
Form heat-stable gels	Gives soft, thixotropic gels with salvelike textures (at 1.0%–1.2% concentration) via formation of networks of hydrated microcrystals rather than intermolecular junction zones. Because microcrystalline cellulose (MCC) is insoluble, the gels do not melt. MCC can be used to make dressings that are stable to retorting. MCC will also form gels from fruit syrups and sugar syrups and toppings such as marshmallow and butterscotch.
Stabilize heated starch (Chapter 6) and high-methoxyl pectin (Chapter 15) gels	MCC crystals do not melt.
Provide freeze-thaw stability/ heat shock[a] improvement	MCC controls ice crystal growth by collecting at the ice–water interface.
Fat and oil extender and replacer	Particle sizes of <0.2 μm mimic fat and oil droplets (Chapter 17), providing creaminess/richness in low-fat foods. MCC is nondigestible and noncaloric, so it can be used to make low-calorie food products.
Carrier for flavors	Porous, sponge-like aggregates absorb flavor oils and similar substances.
Improves adhesion	Enhances cling in batters and breadings and starch-based sauces.
Modifies texture	Modifies the functionality of other texturizers. MCC is a nonnutritive filler/bodying agent.
Holds water	Prevents boil out when used in bakery fillings. Prevents moisture migration in icings.
Prevents caking	Absorbs surface oil and prevents adhesion in shredded and grated cheeses.
Extrusion aid	Gives a finer cell structure and greater crispness when used to make products such as puffed corn curls.

[a]Heat shock is the formation of a grainy/sandy mouthfeel due to the formation of relatively large ice crystals as ice cream slightly melts and refreezes (due, for example, to the defrost cycle of a home freezer).

Table 8.3 Typical products containing microcrystalline celluloses, carboxymethylcellulose, methylcellulose, and hydroxypropylmethylcellulose

Microcrystalline celluloses
Barbecue sauces, cereal bars, frozen cheese lasagna, frozen guacamole, fruit drink mixes, grated cheese, liquid diet products, low-calorie mayonnaise, low-calorie spoonable salad dressing, marshmallow topping, no-oil pourable salad dressing, sandwich spread, shredded cheese, whipped topping
Carboxymethylcelluloses
Cake mixes, cheesecakes, cupcake mixes, diet dairy drink mixes, dips, frozen chicken sticks, frozen chicken and turkey nuggets, frozen desserts, frozen gravy, fruit drink mixes, grated and shredded cheese, hot cocoa mixes, ice creams, instant cappuccino, low-calorie cheesecakes, low-calorie mayonnaise, marshmallow topping, pancake syrups, pasta salad mix, protein shakes, snack cakes, sour cream, vegetable patties
Methylcelluloses
Bakery products, barbecue sauce, breakfast bars, frozen desserts and novelties, fruit drink mixes, fruit juices, meatless franks, onion rings, pancake mix, sandwich spread, twice-baked potatoes, vegetable patties
Hydroxypropylmethylcelluloses
Cream cheese, honey-roasted peanuts, low-calorie cheesecake, salad dressing mix, sandwich spread, whipped toppings and mixes

Colloidal MCC is the type most commonly used in preparing food products. The largest volume is used in frozen desserts. In frozen desserts, it controls ice crystal size, improves foam stability (reduces the rate of meltdown),[8] improves extrusion, and increases body[9] and stiffness. It is a common ingredient in reduced-fat ice cream and ice cream-like products. Colloidal MCC is used in ice cream at concentrations of 0.2%–0.4% with care to disperse it as small particles, some less than 0.2 μm in diameter. In salad dressings, it improves pasteurizability, spreadability, stability, and uniformity. It is not an emulsifier, but it collects at interfaces in oil-in-water emulsions, where it acts to thicken the mixture and to improve mouthfeel. It improves the cling of sauces—even at elevated temperatures. MCC can add pulpiness to products like low-solids tomato sauce. With starch, it improves thickening with less masking of flavor. It improves the body, texture, and stability of a 20%–27% vegetable-fat whipped topping (Table 8.3).

[8] Ice cream is a frozen, fat-stabilized foam. Meltdown is the collapse of the fat-stabilized foam structure of an ice cream after the temperature of the ice cream rises above the temperature at which the ice crystals melt.

[9] Body refers to the quality of a food or beverage; it is the sum of rheological properties (Chapter 5) that contribute to mouthfeel.

[10] It is also the sodium salt of the glycolic acid ether of cellulose.

Coprocessed mixes of MCC and guar gum (Chapter 9) and MCC and an alginate (Chapter 14) are particle aggregates that can mimic the rheology of fat-in-water emulsions and contribute fatlike mouthfeel to non- and low-fat products (Chapter 17).

Ingredient labels on food products containing MCC may read MCC or cellulose gel.

Modified cellulose products

The principle behind converting cellulose into water-soluble products (hydrocolloids) is to separate the molecules in a cellulose fiber, then derivatize some of their hydroxyl groups so that the molecules cannot reassociate through intermolecular hydrogen bonding over enough of their length to bring about insolublization. Swelling of cellulose fibers to separate the cellulose molecules is accomplished by treating a highly purified wood pulp (or cotton linters) with 18% sodium hydroxide solution. This treatment forms "alkali cellulose". In alkali cellulose, many of the hydroxyl groups (-OH) have been converted into alkoxy groups ($-O^-$), which are quite nucleophilic. (The alkoxy groups also give the molecules negative charges which repel each other and swell the fibers, destroying some of their crystallinity and making the molecules more accessible to derivatizing reagents.) (Note that a much more alkaline system is used than is used for the derivatization of starches.) If an oxidant, such as oxygen, is present, some hydroxyl groups are oxidized to carbonyl groups, and under the alkaline conditions, a beta-elimination reaction (Chapter 4) occurs and the cellulose is depolymerized (that is, reduced in average molecular size). So various reduced viscosity types are made by aging the alkali cellulose (that is, holding it until the desired average MW is achieved). All modified cellulose products used in foods are cellulose ethers made by reaction of alkali cellulose with an alkyl halide (to make CMC and MC) or an epoxide (to make a hydroxyalkylcellulose, namely, HPC) (Fig. 8.3).

Variables in cellulose derivatives are (1) the amount of etherification, (2) the nature of the substituent group, (3) the average MW of the derivatized cellulose molecules, (4) the distribution of the substituent groups on O2, O3, and O6 of the β-D-glucopyranosyl units, and (5) the distribution of the substituent groups along the cellulose chain.

The average MW (average DP) of a cellulose derivative determines the viscosity type. The largest molecules produce the highest viscosity (Chapter 5). For example, viscosities of 1.0% solutions of CMC at room temperature can range from less than 10 mPa s (cps) for a low-MW product to about 10,000 mPa s for a high-MW product. The nature and number of substituent groups determines solubility. Distribution of substituent groups determines the rheology type (of importance only with CMC).

General characteristics of water-soluble cellulose derivatives are as follows:

- Wide range of viscosity types available
- High-viscosity types available
- High purity
- pH stable
- Tasteless

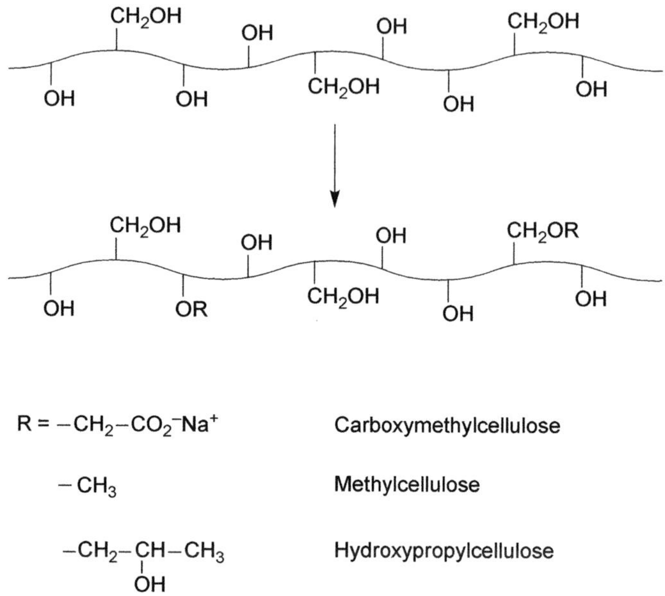

Figure 8.3 Representation of the introduction of carboxymethyl, methyl, and hydroxypropyl ether groups into a cellulose molecule. In the substituted molecule (bottom one with R groups), there are two substituent groups at an O2 position and one at an O6 position. There are three substituent groups per nine hydroxyl groups (three glucopyranosyl units) and, thus, a DS of 1.0. The diagram indicates that a polysaccharide composed only of neutral hexopyranosyl units (like cellulose) is a molecule of only primary and secondary hydroxyl groups available for substitution.

- Odorless
- Uniform quality
- Form clear solutions
- Film forming
- Noncaloric

Carboxymethylcelluloses

Carboxymethylcellulose (CMC), the sodium salt of the carboxymethyl ether of cellulose,[10] is widely used as a hydrocolloid. For production of CMC, alkali cellulose is reacted with the sodium salt of chloroacetic acid. (Alkali cellulose is shown below as being 100% ionized, but may not be. In the case of cellulose, a much stronger solution of sodium hydroxide is used, so there are undoubtedly more hydroxyl groups in the alkoxy ($R{-}O^-$) form than in the case of starch.

[11] Monodisperse means that the polymer molecules are present as individual hydrated molecules and not as agglomerates.

$$\text{Cell–OH} \xrightarrow{OH^-} \text{Cell–O}^- + ClH_2CO_2^-Na^+ \longrightarrow \text{Cell–O–}CH_2\text{–}CO_2^-Na^+ + NaCl$$

Carboxymethylcellulose (CMC)

Most food-grade sodium CMC products have degree of substitution (DS) values in the 0.7—0.8 range, but some products may have a DS as low as 0.4. (For water solubility, CMC must have a DS of at least 0.4.) The usual CMCs of DS 0.7—0.8 hydrate rapidly because they are quite ionic. Therefore, to make a smooth solution (that is, one without lumps), one of the means of dispersing it described in Chapter 5 must be followed. As the DS of CMC increases, salt tolerance, hygroscopicity, and alcohol tolerance increase and its thixotropic nature (see below) decreases.

Substituent groups are found at O2, O3, and O6 of D-glucopyranosyl units in the approximate proportion of 2.1:1.0:1.6. The degree of uniformity of derivatization along the cellulose chain determines the behaviors of its solutions. Nonuniformly derivatized CMC is a good example of a hydrocolloid that produces solutions with thixotropic rheology. Nonuniform distribution leaves unsubstituted stretches of molecules available for junction zone formation with unsubstituted regions of other CMC molecules, producing thixotropy. As explained in Chapter 5, thixotropic fluids have the characteristics of a weak gel at rest. Why a CMC with nonuniform substitution behaves this way and undergoes shear thinning is diagrammed in Fig. 8.4. Nonuniformly derivatized CMC produces solutions with thixotropic rheology (that is, with hysteresis as apposed to instantaneous, reversible shear thinning [pseudoplasticity]) (Chapter 5) because (1) the molecules with shorter contacts (contacts held together by the fewest hydrogen bonds) disengage from junction zones first, while those with longer contacts require more energy and time for pulling away from associations of molecules, and (2) the relatively short "naked regions" require time to "find each other" and reestablish a junction after shear is removed. So there is a time dependency to both shear thinning and the subsequent rethickening when shear is removed. Uniform derivatization

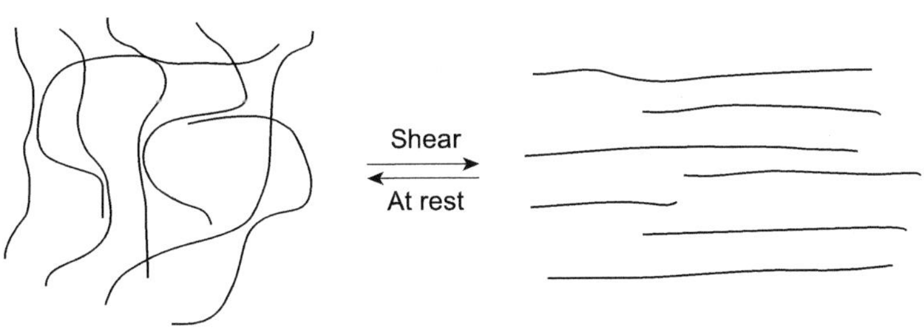

Figure 8.4 A diagrammatic explanation of why a solution of nonuniformly substituted carboxymethylcellulose is a weak gel at rest and undergoes shear thinning when stirred, pumped, or swallowed (that is, is thixotropic). The diagram depicts a situation where sufficient shear is applied to (A) break all associations and (B) to align the dissociated molecules in a fully extended, linear arrangement. Neither will completely occur at low rates of shear.

produces molecules with no long unsubstituted stretches. Because they do not contain "naked regions" that can bind to the "naked regions" of other CMC molecules, products with uniform substitution will form smooth, stable solutions. The "naked regions" of nonuniformly substituted CMC allow binding to powdered cellulose and MCC. Nonuniformly substituted CMC also interacts with certain hydrocolloids such as galactomannans (Chapter 9). Maximum viscosity is produced when CMC and guar gum are used together in a CMC to guar gum ratio of about 1:3.

CMC is available in a wide range of viscosity types, and as already discussed, solution viscosity depends mainly on the average MW of the water-soluble polysaccharide (Chapter 5). Solutions of higher MW (higher viscosity) products also exhibit greater pseudoplasticity or thixotropy. Because it is a highly ionic hydrocolloid, CMC's solubility and the viscosity of its solutions are affected by salts. As explained in Chapter 5 (Fig. 5.13), CMC should be dissolved in water before other solutes are added (not dispersed in a salt solution) to take full advantage of its properties. Effects of salts on solutions of CMC are a function of the type (DS and viscosity type) of the CMC, the type and concentration of the salt, and the pH. In general, monovalent cations form soluble CMC salts; divalent cations produce hazy dispersions; and salts of trivalent cations are insoluble (that is, aluminum ions will bring about precipitation, but aluminum ions are not encountered in foods).

Because CMC consists of long, fairly stiff molecules bearing negative charges, its molecules in solution are stretched out due to electrostatic repulsion of chain segments (that is, chain folding is restricted because any folding would bring the carboxylate groups closer together where their negative charges would repel each other). In addition, because the molecules repel each other, monodisperse,[11] highly viscous, stable solutions result. Lowering the pH to less than 4 represses ionization of the carboxyl groups so that some lose their charge ($\text{-COO}^- \rightarrow \text{-COOH}$). Molecular association then occurs and viscosity increases for a time, but viscosity cannot be maintained long term at pH values of less than 4 because of hydrolysis (depolymerization). At pH 3, insolubilization (precipitation) occurs due to extensive loss of the negative charge, changing the anionic polymer to a neutral polymer, which allows chain associations to occur. CMC undergoes hydrolysis with loss of viscosity when acidic solutions of it are retorted.

CMC is widely used in food products to absorb and hold water, to control crystal growth, to thicken, as a binder, to increase shelf life, and to provide desired texture or body. Its largest volume use is in the preparation of dry pet foods that form their own gravy when warm water is added. Its second largest use is in the preparation of ice cream, sherbets, and other frozen desserts. It is the primary stabilizer in most ice cream products, where it is used to prevent growth of ice crystals (see Chapter 13). Keeping ice crystals small maintains the smooth, creamy consistency of the product. CMC also controls sugar crystal size in fondants.

CMC is used when proteins must be stabilized, such as in yogurt, fruit, soy, and other acidic drinks containing protein. CMC, a polyvalent anion, can stabilize protein dispersions, especially near their isoelectric pH value where they are least soluble, because the protein molecules will have multiple positive charges and can bind to CMC molecules. Using CMC, milk products can be stabilized against casein precipitation when milk is acidified because CMC forms stable soluble complexes with casein at pH values in the range 3–6 where casein is insoluble.

Because of its rapid hydration, it affects hydration of other dry mix components. It is both a binder and an extrusion aid in the preparation of pet foods and other extruded products. In baked goods, like cakes, CMC is added to adjust the consistency of the batter, to increase product volume, to improve the quality of the finished product, to provide moisture retention, and to prolong freshness. Some of the many applications of CMC are given in Tables 8.2 and 8.4.

There is no limit to the concentration of CMC that may be used in food products, but it should be used with good manufacturing practices in mind. Ingredient labels on products containing CMC may read sodium carboxymethylcellulose, sodium carboxymethyl cellulose, carboxymethylcellulose, carboxymethyl cellulose, CMC, sodium CMC, or cellulose gum.

Methylcelluloses and hydroxypropylmethylcelluloses

When alkali cellulose is treated with methyl chloride, the protons of some of the hydroxyl groups of cellulose are replaced by methyl groups (that is, methyl ether groups are introduced). The products are **methylcelluloses (MC).** Most members of this class of hydrocolloids also contain hydroxypropyl ether groups (cellulose-O-CH_2-CHOH-CH_3). **Hydroxypropylmethylcelluloses (HPMC)** are made by reacting alkali cellulose with both propylene oxide and methyl chloride. The DS of commercial MC range from 1.1 to 2.2. The hydroxypropyl MS (moles of substitution [Chapters 4 and 7]) levels in commercial HPMCs are in the range 0.02–0.3. At this low degree of derivatization, it is unlikely that much, if any, formation of poly(propylene oxide) [poly(-propylene glycol)] chains occurs. Both members of this hydrocolloid family (that is, both MCs and HPMCs, are generally referred to simply as MCs) (Scheme 8.3).

MC products are available in a wide range of viscosity types, with the viscosity of 2% solutions varying from less than 10 to more than 250,000 mPa s. MC products are cold water soluble because the protrusions along the chains prevent intermolecular reassociation. They are not readily hydrolyzed by cellulase and, hence, are somewhat resistant to microbial attack.

A unique and most important property of the MC and HPMC is that of reversible thermal gelation. Their solutions gel when heated; the gels reliquefy when cooled. While a few ether groups spread along cellulose molecules enhance water solubility by prevention of intermolecular associations, they also decrease chain hydration by replacing water-binding hydroxyl groups with less polar ether groups. The ether groups, while responsible for water solubility by blocking interchain associations, restrict hydration of the chains to the point that MC are on the borderline of water solubility. (Even so MC hydrate rapidly.) In solutions at room temperature, MC and HPMC chains are hydrated and there is little polymer–polymer interaction, even through the hydrophobic ether groups. However, when an aqueous solution is heated, the water molecules solvating the polymer molecules achieve sufficient kinetic energy that many dissociate from the chain, allowing intermolecular van der Waals associations (hydrophobic associations) between the ether groups and gelation to occur

Table 8.4 Typical applications of carboxymethylcelluloses

Product types	Functions
Beverage mixes	Alcohol tolerance
Cake, doughnut, and related mixes	Batter thickener Humectant (improves texture and extends shelf life) Increases volume (film former)
Cheese spreads	Protective colloid
Dietetic foods	Thickener Bodying agent
Dressings	Thickener
Dry pet food	Makes gravy when water is added
Dry-powder fruit drink mixes and salad dressings	Suspending aid Rapid hydration
Extruded products	Lubricant Binder Film former Processing aid
Fondants	Inhibits sugar crystal growth
Frozen and dried egg white	Protein stabilizer
Hot cocoa mixes	Thickener
Ice cream and other frozen dessert products	Inhibits ice crystal growth Improves mouthfeel, body and texture Stabilizes casein
Icings and frostings	Inhibits sugar crystal growth Prevents moisture loss.
Meat emulsions	Texturizer Binder
Milk products	Protein stabilizer
Pie fillings	Improves texture Prevents syneresis
Puddings	Inhibits sugar crystal growth
Sauces	Suspending aid
Syrups	Transparent thickener for low-calorie pancake syrup
Toppings	Holds moisture Prevents syneresis

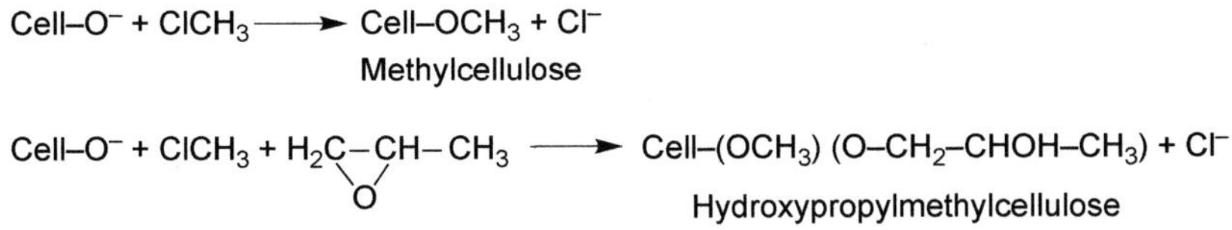

when the incipient **gelation temperature (gel point)** is reached. Gel strength continues to increase to a maximum as the temperature is held above the gelling temperature, but the process is completely reversible. The gelation temperature and gel strength are determined by the specific type of MC. Reducing the temperature once again brings about rehydration and solubility. Thus, the thermal gelation is reversible. This property is the basis for many of the applications of MC and HPMC (Table 8.5).

An interesting aspect of this property is that MC and HPMC are insoluble in hot water because these and other polysaccharides are insoluble under conditions at which they form gels; so MC (the generic term that includes HPMC) can be dispersed in hot water without thickening or clumping. Then, the stirred dispersion can be cooled to give a clear, smooth solution. In fact, the rate and degree of solubility are inversely related to the temperature of the water.

Table 8.5 Typical applications of methyl- and hydroxypropylmethylcelluloses

Product types	Functions
Acidic foods	pH stability
Bakery products	Gas entrainment (film formation)
Batters and coatings	Provides a barrier to fat absorption via thermal gelation and film formation, provides hot cling (adhesion)
Deep-fried foods	Reduction of oil absorption via thermal gelation and film formation
Extruded products	Lubricant, binder
Frozen foods, desserts, and whipped toppings	Provides freeze-thaw stability, maintains texture and viscosity, inhibits phase separation and syneresis
High-sugar products	Compatibility
Meat analogues	Adhesion and binding, lubricant
Pie fillings	Reduces boilover, prevents water migration from the filling to the crust
Salad dressings	Emulsifier and emulsion stabilizer (inhibits phase separation)
Structured foods	Adhesion binding
Whipped toppings	Foam stabilizer, provides freeze-thaw stability

Surface activity is another property of MC (generic) products. They share their emulsifying and emulsion stabilization properties with gum arabic (Chapter 16), propylene glycol alginate (Chapter 14), and starch octenylsuccinate (Chapter 7). Being derivatives of linear molecules, they also form films, and MC (generic) products add lubricity during processing. Low-MW MC (generic) interact synergistically with hydroxypropylstarch (Chapter 7).

MC (generic) products can be used to reduce the amount of fat in food products via two mechanisms: (1) they impart fatlike properties so that the fat content of a product can be reduced, and (2) they reduce absorption of fat by products being fried. When foods to be battered and deep fried, such as onion rings, are formulated with a MC product in the batter, fat absorption is reduced (because the gel structure produced by thermal gelation provides a barrier to penetration of fat and oil) and adhesion is improved (because MC products form continuous films and have adhesive properties). MCs also function as binders in certain structured foods because of their thermal gelation, and MC products impart richness (creaminess), provide structure and body, and generate and stabilize foams. In baked products, their thermal gelation gives greater gas retention during baking, and they aid in moisture retention, improve tenderness, and extend shelf life.

Other attributes of MC (generic) products are the provision of hot cling (via thermal gelation), reduction in boilover (via thermal gelation), prevention of moisture migration, maintenance of texture (inhibition of phase separations, such as syneresis) during freeze-thaw cycling, both emulsification and emulsion stabilization, and tolerance to high concentrations of sugar. The ability to provide lubricity is also useful in making products by extrusion.

A major use of HPMC products is in nondairy whipped toppings. Its interfacial activity stabilizes foams and imparts better whipping characteristics, and its emulsifying property prevents phase separation. It provides freeze-thaw stability because it becomes more and more soluble as the temperature is lowered. MC creates a denser foam as compared with HPMC.

Acceptable label designations for MC are methylcellulose, methyl cellulose, and modified cellulose; modified vegetable gum is used occasionally. Acceptable label designations for HPMC are hydroxypropylmethylcellulose, hydroxypropyl methylcellulose, hydroxypropyl methyl cellulose, modified cellulose, and propylene glycol ether of MC; carbohydrate gum is used occasionally.

Hydroxypropylcelluloses (HPC) are made by reacting alkali cellulose with propylene oxide only. Food-grade HPC products are water soluble as long as the water temperature is below about 40°C (105°F). (Like the MC and HPMC, HPC is insoluble in hot water). As HPC solutions made at a lower temperature are heated to a temperature above about 40°C (105°F), the HPC molecules precipitate rather than form a gel, but the principle is the same as in gelation of HPMC and MC. Because HPC molecules are nonionic, they are compatible with most salts at low concentrations, but higher concentrations of some salts (for example, 5%—10%) and sucrose (more than 10%) lower the precipitation temperature. Hydoxypropylcelluloses are on the edge between water solubility and water insolubility, so anything that competes with them for the water molecules needed to hydrate the polymer molecules and keep them in solution will effect precipitation.

It requires much more derivatization to make cellulose molecules water soluble than it does to stabilize starch molecules, so while the MS of hydroxypropylated starch does not often exceed 0.20, commercial HPC products have MS values of more than 2.0. In fact, the legal limit for food use is MS 4.6, which is on average 4.6 hydroxypropyl groups per glucosyl unit. A molecule of HPC with MS 4.6 will contain 62% by weight of hydroxypropyl groups. (This is an excellent example of why MS, rather than DS, is used when hydroxypropylation is involved.) To achieve MS values greater than 3, some propylene oxide molecules have to react with the newly added hydroxyl groups on the added hydroxypropyl ether groups (that is, chaining has to occur). In fact, the chaining reaction begins long before MS 3.0 is reached because the hydroxyl group on the hydroxypropyl ether group is more reactive than any of the hydroxyl groups on glucosyl units and, as already pointed out, the hydroxyl group on C3 is not very reactive.

Like the more highly substituted HPMC, HPC products are thermoplastic. Like other cellulose derivatives, HPC can be cast into transparent, water-soluble, edible films.

The label designation may be HPC or hydroxypropyl cellulose.

Ethylmethylcellulose

Cellulose containing both ethyl and methyl ether groups is an approved food ingredient in the European Community.

Regenerated cellulose

A food-related use of cellulose is in the manufacture of regenerated cellulose. Cellulose can be dissolved in a solution of sodium hydroxide containing carbon disulfide. The soluble molecule is actually the xanthate ester of cellulose (Cellulose-O-CS_2^-), which is called viscose. If this solution is forced through a hole into a bath of dilute sulfuric acid and sodium sulfate, the cellulose is regenerated and a rayon fiber results. If the viscose solution is forced through a slit into the regenerating bath, a thin, clear, transparent sheet called cellophane is made. Cellophane is used as a packaging material and as a wrapping material for home use. And the regenerated cellulose can be made in the form of tubes that are used as casings in the manufacture of several processed meat products. First, the meat is comminuted (chopped into very small pieces). Other ingredients, such as spices, are added. If the product being made is wieners/franks/hot dogs, etc., the emulsion is stuffed into a casing of the proper diameter, which is twisted when the desired length of the wiener is reached. The string of wieners is cooked and smoked. Then the casing is slit and peeled off to produce skinless wieners/franks/hot dogs, etc. Other casing-free sausages and luncheon meat products are made in a similar way.

Additional resources

Cash, M.J., Caputo, S.J., 2009. Cellulose derivatives. In: Imeson, A. (Ed.), Food Stabilisers, Thickeners and Gelling Agents. Wiley Blackwell, Oxford, pp. 95–115.
Krawczyk, G., Venables, A., Tuason, D., 2009. Microcrystalline cellulose. In: Phillips, G.O., Williams, P. (Eds.), Handbook of Hydrocolloids. CRC Press, Boca Raton, pp. 741–759.
Murray, J.C.F., 2009. Cellulosics. In: Phillips, G.O., Williams, P.A. (Eds.), Handbook of Hydrocolloids. CRC Press, Boca Raton, pp. 710–723.

Guar, Locust Bean, Tara, and Cassia Gums

9

Chapter Outline

Key information and skills that can be obtained from study of this chapter will enable you to

1. Define or identify:
   galactomannan
   guaran
   guar gum
   locust bean gum (LBG)
   carob gum
2. Describe a galactomannan.
3. Describe the natures of the commercial guar and LBG based on how they are prepared.
4. Give the idealized repeat unit structure of guaran using both (a) Haworth structures (b) and shorthand notations.
5. Describe the difference between guar gum and guaran.
6. Describe the two structural differences between guaran and the galactomannan of LBG.
7. Describe the process and give the rationale for making "synthetic LBG."
8. Describe the difference in solubilities of guar and LBG.
9. Give a molecular explanation for the differences in solubility of guar and LBG polysaccharides.
10. Compare the synergistic interactions of guar and LBG with xanthan.
11. Give the molecular basis for the synergistic interaction of LBG with xanthan and/or carrageenan using diagrams.

Carbohydrate Chemistry for Food Scientists. https://doi.org/10.1016/B978-0-12-812069-9.00009-1

Sources, natures, and structures of guar and locust bean gums

Guar and locust bean gums are the most widely used of this family of hydrocolloids (the galactomannans; see below) for both food and nonfood uses. Both hydrocolloids produce high viscosities; 1% solutions of commercial guar gums usually have viscosities of 4000–7500 mPa·s, although 1% solutions of very high quality guar gum can produce viscosities as high as 10,000 mPa·s.

Both hydrocolloids are the ground endosperm of seeds, rather than isolated polysaccharides; so they are actually flours. The two hydrocolloids/flours are prepared in similar ways. Clean seeds are passed through a special type of mill that breaks them in half. (The halves are called splits.) The germ is then removed from the splits. The degerminated splits are heated to aid in removal of the seed coat, which is done using another special mill. The resulting degerminated, dehusked splits can be directly ground to a flour or soaked in water and then flaked, extruded, or ground. Soaking in water before milling generally produces a gum of a higher viscosity grade. An extract can also be produced. To make an extract, the guar or locust bean gum is dispersed in hot water. After the dispersion is filtered, the dissolved polysaccharide is precipitated with 2-propanol (isopropanol). The precipitate is dried, ground, sieved, and perhaps blended for quality assurance purposes. The extracted material contains much less protein, fiber, ash, and lipid than the usual materials but is seldom, if ever, used commercially.

The polysaccharide of each gum (that is, the main component of both endosperms) belongs to a class of polysaccharides known as **galactomannans**. Galactomannans have a main chain of (1 → 4)-linked β-D-mannopyranosyl units, with some of the mannopyranosyl units bearing single α-D-galactopyranosyl units attached to O6. Figs. 9.1 and 9.2 give idealized structures of **guaran** (the polysaccharide/galactomannan of guar gum).[1] They show 50% substitution and regular substitution, while the actual substitution in a high-quality guar gum is about 56% (Man:Gal ratio of 1.8:1) and nonuniform.

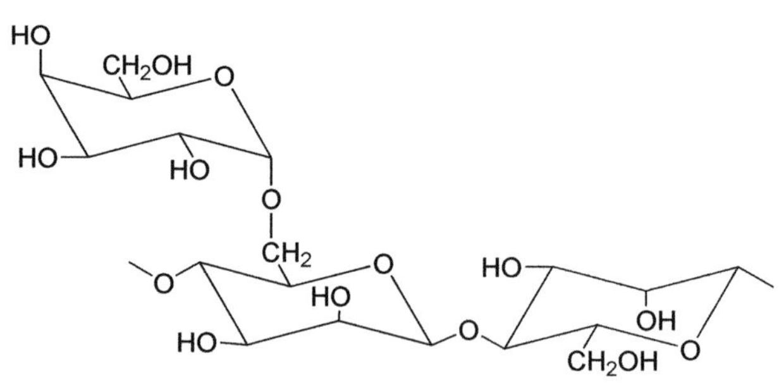

Figure 9.1 A representative segment of a galactomannan molecule (see also Chapter 4).

[1] Guaran was used to describe determination of polysaccharide structures in Chapter 4.

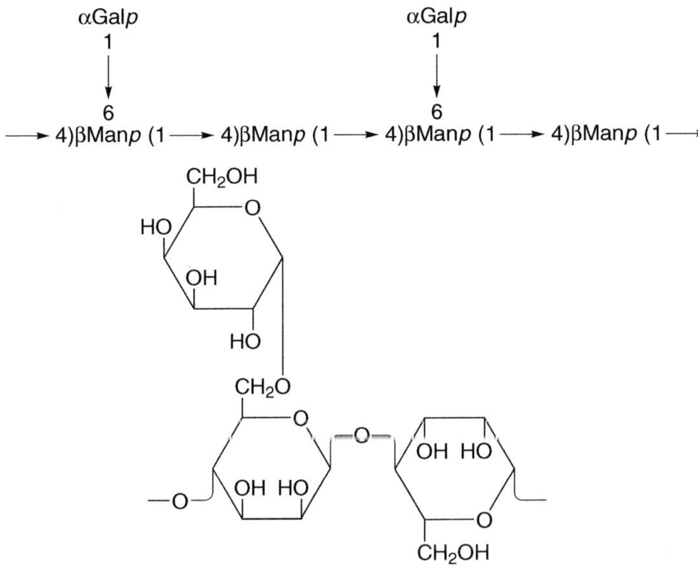

Figure 9.2 Idealized structure of guaran, the polysaccharide of guar gum: (top) shorthand structure, (bottom) Haworth representation. This is the same trisaccharide unit shown in Fig. 9.1.

Guar plants look rather like soybean plants; they have an erect stem with seeds in numerous pods in clusters. As a result, guar plants are sometimes called cluster bean plants. Commercial **guar gum** is the ground endosperm of guar seeds.

Commercial **Locust bean gum (LBG)** (also called **carob gum**) is the ground endosperm of locust bean (carob) seeds. Locust bean seeds are obtained from the large seed pods of locust (carob) trees, which grow around the Mediterranean Sea where they were planted by early traders who used the uniformly sized seeds for measurement of weight. (This weight terminology has carried over to the present measurement of diamonds in karats; originally one karat was the weight of one carob seed.) Pods dropped in the fall by locust trees are collected and taken to kibblers, who separate the seeds for milling and grind the empty pod to a powder that may be used for extending or replacing cocoa. Commercial LBG is the ground endosperm of locust bean/carob seeds. Typical compositions of both of these endosperm powders (commercial guar and LBG) are 73%–87% polysaccharide, 8%–14% moisture, 3%–6% protein, 1%–4% fiber, 0.8%–2.0% ash, and 0.5%–1.0% lipid. Both gums are available as powders of various grades, qualities, and mesh sizes.

Locust bean galactomannan has fewer side units (branches) than does guaran. It has stretches averaging about 80 underivatized β-D-mannopyranosyl units (so-called "naked regions"), alternating with sections averaging about 25 β-D-mannopyranosyl units long, in which almost every main chain unit has an α-D-galactopyranosyl unit glycosidically connected to its O6 position. On average, in the polysaccharide of a high-quality LBG, only about 26% of the main chain units carry a D-galactosyl unit (Man:Gal ratio of 3.9:1).

LBG is more expensive than guar gum because of a more restricted source. Adulteration with less expensive guar gum is a problem because they are difficult to distinguish from each other. However, as explained in the next section, they are used for rather different purposes, and it is important to use unadulterated LBG. Because the gums are flours, they contain distinguishing proteins. Three general ways have been developed to determine the degree of adulteration of LBG with guar gum. One involves capillary electrophoresis of a protein extract to examine for proteins specific to the guar seed; another uses antibodies raised to guar endosperm proteins in immunoassays, and the third employs DNA identification. A "synthetic LBG" is made by treating guar gum with an α-galactosidase to remove enough of the α-D-galactopyranosyl units to give some open stretches of unsubstituted β-D-mannopyranosyl units (naked regions), which gives the product properties similar (but not identical) to those of LBG. Because LBG is almost always used in combination with other hydrocolloids which form junction zones with the unsubstituted sections of its molecules, the presence of these naked stretches with enough substitution to make the product soluble is essential. Commercial products may contain as little as 13% D-galactose (Man:Gal ratio of 6.7:1); they always contain less D-galactosyl units than the about 23% found in high-quality LBG.

Properties of guar and locust bean gums

Because of the differences in structures of the polysaccharides, guar gum and LBG have different physical properties, even though both galactomannans are composed of long, rather rigid chains that have a large hydrodynamic volume (Chapter 5), thus providing high solution viscosity. Guar gum at low concentrations especially produces high viscosity solutions with pseudoplastic rheology, so it is classified as a "thickening hydrocolloid" (Table 5.2).

In general, galactomannans, which are nonionic, hydrate more slowly than do ionic hydrocolloids such as carboxymethylcellulose (CMC) (Chapter 8). Powders with different mesh sizes and rates of hydration are available. Guar gum hydrates and builds viscosity when dispersed in room temperature water. Although different mesh sizes hydrate and produce viscosity at different rates, generally about half of the final viscosity is obtained in 10–15 min and almost full viscosity in 1 h in unheated dispersions. A higher viscosity will be obtained after first heating a dispersion of guar gum (made in room temperature water), then cooling it. Guar gum is about 70% soluble in cold water and about 80% soluble in hot water (because other guar seed endosperm components are insoluble).

Guar gum and LBG (because they are neutral molecules) are compatible with most other food substances, including other biopolymers, but because guaran has its galactosyl units fairly evenly spaced along the backbone chain, there are few locations along the molecules that are suitable for significant junction zone formation. Guar gum does, however, exhibit interactions with starches (Chapter 6), cellulose (Chapter 8), agar,

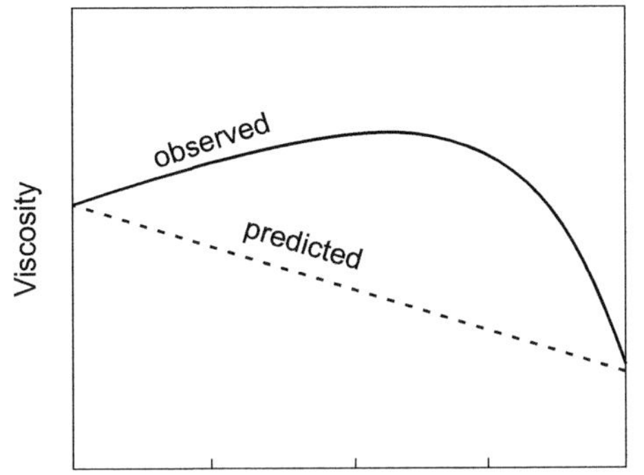

Figure 9.3 A graph indicating the synergistic interaction between guar gum and xanthan to give viscosities greater than that predicted by summing the two viscosities, indicating that they do not act independently of each other.

κ-carrageenan (Chapter 13), and xanthan (Chapter 11). Interactions result in synergistic[2] increases in viscosity. For example, when a combination of guar gum and xanthan is used, the observed viscosity is greater than would be predicted if the molecules of guar gum and the molecules of xanthan acted independently, indicating a synergistic intermolecular interaction (Fig. 9.3). The xanthan:guar gum ratio required for maximum viscosity varies with ionic strength; from about 1:1 in distilled water to about 9:1 in hard water, but it is usually about 7:3. The intermolecular interaction of guar gum with microcrystalline cellulose has been used to make a fat mimetic (Chapters 8).

Because guar gum consists of long, soluble, rather rigid molecules, its solutions are pseudoplastic, and it is able to reduce fluid friction during pumping of aqueous liquids, thereby reducing energy requirements. Guar gum solutions tolerate salt well, but high concentrations of calcium salts effect precipitation, as do many polyvalent cationic salts.

Because LBG molecules contain "naked" regions (that is, pure β-(1 → 4)-mannan segments) that fit together in much the same way as do cellulose molecules (which are β-(1 → 4)-glucan molecules—the only difference being the orientation of the hydroxyl group on C2), LBG is only slightly soluble in room temperature water; most particles only partially swell in room temperature water. Heating an aqueous dispersion of LBG to about 85°C (185°F) is required for good dissolution (Fig. 9.4). Only about 35% of LBG is soluble in cold water; at least 85% is soluble in hot water. LBG potentially produces the most viscosity of the commercial galactomannans, but the viscosity achieved is a function of the temperature reached during its dissolution.

[2] A synergistic interaction is an interaction between two substances that produces a value for a property that is greater than what would be predicted by adding the values for the property produced by the two substances acting independently.

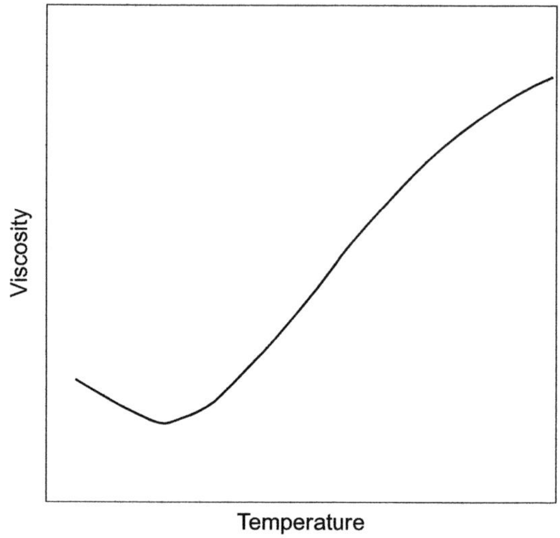

Figure 9.4 A graphical representation of the dissolution of locust bean gum (LBG) in water. A small portion of the hydrocolloid is usually soluble in room temperature water. The solution that forms undergoes a reduction in viscosity as the temperature rises (as do other solutions of hydrocolloids). As the temperature increases further, more and more LBG molecules go into solution, resulting in an increase in viscosity.

LBG solutions do not gel. However, LBG, with its long "naked chain" regions, can form junction zones and produce thixotropic solutions. LBG interacts with xanthan (Chapter 11) (Fig. 9.5) and κ-carrageenan (Chapter 13) helices and forms rigid, elastic gels with these two hydrocolloids (that is, hot solutions of mixtures of LBG with κ-carrageenan [Chapter 13] or xanthan [Chapter 11] gel when cooled). For maximum interaction between LBG and the other hydrocolloids, the hydrated mixtures need to be heated to temperatures above 60°C (140°F) to at least partially dissolve the LBG. With κ-carrageenan, not only is an increased gel strength (over that obtained with κ-carrageenan, which is a gelling hydrocolloid, alone) realized, the texture of the gel is also modified. LBG makes κ-carrageenan gels less brittle and more elastic by replacing carrageenan–carrageenan junctions with weaker carrageenan–LBG junctions. Syneresis[3] is also reduced. LBG has little effect on an iota-carrageenan gel. When a hot solution containing xanthan (a nongelling hydrocolloid) and LBG (another nongelling hydrocolloid) is cooled, a gel is formed because of junction zone formation between the two polysaccharides. Xanthan may form junction zones with LBG when it is in either a single-helical (Fig. 11.4) or a double-helical (Fig. 9.6) form.). And LBG molecules also interact with bare portions of cellulose derivatives (Chapter 7) to form junctions and to produce a synergistic[3] increase in viscosity.

[3] Syneresis is the expulsion/separation of liquid from a gel caused by shrinking of the gel (Chapter 5).

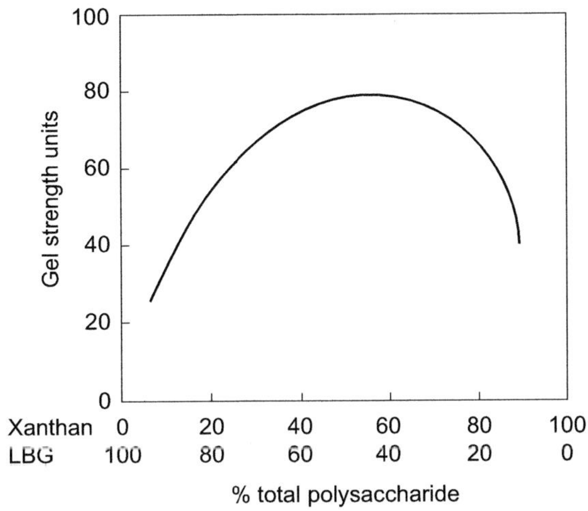

Figure 9.5 Gel strengths of the gels formed by mixtures of locust bean gum and xanthan, each by themselves nongelling hydrocolloids.

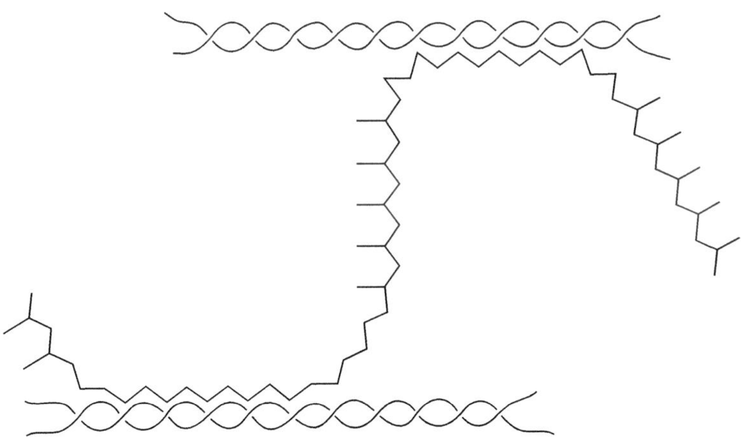

Figure 9.6 A representation of the hypothesized interaction of a locust bean gum molecule with double-helical segments of carrageenans and/or xanthan to form a three-dimensional structure and a gel.

In summary, combinations of LBG and xanthan or κ-carrageenan are used in many human and pet foods. Guar gum is often used with such combinations to soften the resulting chewy gel texture, except in pet foods where chewiness is less of a problem. Addition of guar gum to a xanthan–LBG gel will improve its freeze-thaw stability.

Galactomannans are acid sensitive, as are many other polysaccharides, and will lose viscosity at pH values below 3.5, especially when the temperature is raised.

Uses of guar and locust bean gums

Guar gum

In common with several other hydrocolloids, galactomannans serve to thicken and control water. Especially, guar gum is used in this way because of its low cost. (Applications of LBG, which is more expensive, most often make use of its synergistic actions with other gums.)

Guar gum is an economical thickening agent for many food products. Its main applications are in bakery products, dairy products, pet foods, prepared meals, and sauces (Box 9.1). It is frequently used in combination with other hydrocolloids, particularly in dairy products. For example, it is used in ice cream and other frozen desserts as a stabilizer; LBG, CMC (Chapter 8), and κ-type carrageenans (Chapter 13) are also common ice cream ingredients. The primary function of guar gum is to bind water. It also aids in preventing ice crystal growth by immobilizing water released by slight melting (provides freeze-thaw stability), improves mouthfeel, reduces the chewiness produced by a combination of κ-carrageenan and LBG, and slows meltdown[4] (Chapters 8 and 13). It also is used to thicken dry mix beverages, instant hot cereals, and instant soups. It is more slowly hydrated than products like CMC and pregelatinized starches, which is an advantage in preventing lumping in these dry mix products.

Guar gum improves mixing and recipe tolerance in bakery products such as specialty breads, cakes, and donuts. In mixes, guar gum thickens the batter rather quickly and promotes uniform hydration of the flour. Guar gum promotes a higher level of moisture retention after baking, thereby improving freshness (that is, reducing staling) and overall quality for a longer period. It is commonly used (often in combination with other hydrocolloids) in pizza and tortilla doughs. It acts as a plasticizer in flour and water doughs, improving kneading and rolling. It also improves the strength and pliability of the finished product, even after freezing and heating. Guar gum reduces water migration into pizza crust and maintains its crispness.

Guar gum minimizes syneresis[3] in processed cheese, icings, and pie fillings. Significant amounts of guar gum are used in canned and intermediate moisture pet foods because of its high viscosity and heat stability.

A special cold-water-soluble form that hydrates more rapidly is used in instant milk shake products. Two other special forms of guar gum are also available. One is a product with reduced flavor and odor. The other is a hydrolyzed product that is marketed as soluble dietary fiber (Chapter 17). A patent has been issued for esterifying a low-molecular weight (low-viscosity) type of guar gum by reaction with a reagent like octenylsuccinic anhydride to produce a product that behaves like starch octenylsuccinate (Chapter 7) in being both an emulsifier and an emulsion stabilizer.

[4] Meltdown is the collapse of the fat-stabilized foam structure of an ice cream after the temperature of the ice cream rises above the temperature at which the ice crystals melt.

Box 9.1 Typical Products Containing Galactomannans

Guar gum

- Bakery Products
  - Bagels
  - Brownie and brownie mixes
  - Cakes and cake mixes
  - Cheesecake, including low-calorie
  - Corn bread mix
  - Donuts
  - Low-calorie cheesecake
  - Pizza and pizza dough mix
  - Tortillas
- Beverages
  - Chocolate drink
  - Frozen lemonade
  - Nutritional drinks
  - Protein shakes
  - Soy beverages
- Breakfast bars
- Chicken strips
- Condiments
  - Horseradish sauce
  - Pickle relish
  - Salsa
- Dairy products
  - Cheese spreads
  - Cold pack cheese
  - Cottage cheese, including low-fat
  - Cream cheese
  - Dips
  - Eggnog
  - Frozen cheese products
    - Cheese
    - Cheesecake
    - Lasagna
  - Frozen desserts
  - Frozen novelties
  - Frozen yogurt
  - Ice creams, including fat-free
  - Milk shakes
  - Processed cheese
  - Sour cream
  - Whipped topping
  - Yogurt, low-fat
- Dry mixes
  - Soup
  - Pasta salad
  - Pudding

Locust bean gum

- Bakery products
  - Cakes
  - Cheesecake
  - Donuts
  - Icings
- Beverages
  - Fruit drinks
- Candy, chewy
- Cereal bars
- Chicken strips
- Dairy products
  - Cheese dressings
  - Cheese spreads
  - Cottage cheese, including low-fat
  - Cream cheese
  - Dips
  - Dry mixes
    - Pasta
    - Soup
  - Flan
  - Frozen cheese products
    - Cheesecake
  - Frozen desserts
  - Frozen novelties
  - Frozen yogurt
  - Ice creams, including fat-free
  - Milk shakes
  - Sour cream
  - Whipped toppings
  - Yogurt
  - Dry mixes
    - Soup
- Frozen meals
- Fruit fillings
- Gelled desserts
- Gel sticks and snacks
- Low-sugar fruit spreads and jams
- Meal, ready-to-eat (MRE), microwaveable
- Salad dressings, low-calorie
- Sauces
  - Barbecue
  - Pasta
  - Sloppy Joe

Continued

Box 9.1 Typical Products Containing Galactomannans—cont'd

Guar gum	Locust bean gum
Egg substitutes	
Frozen foods (other than dairy)	
Chicken strips	
Fish sticks	
Onion rings	
Turkey nuggets	
Twice-baked potatoes	
Hot cereals, instant	
Low-sugar fruit spreads	
MRE, microwaveable	
Pet foods	
Ready-to-eat cereals	
Sauces	
Barbeque	
Pasta	
Pizza	
Sloppy Joe	
Steak	
Vegetable	
Salad dressings, pourable	
Low-calorie	
Spanish olive pimento stuffing	
Vegetable patties	

Locust bean gum

About 85% of LBG is used in dairy and frozen food products. It is not often used alone, but almost always in combination with other hydrocolloids such as CMC, carrageenan, xanthan, and guar gum in ice cream and related products (see Chapter 13), where it provides heat-shock resistance,[5] smooth meltdown,[4] and desirable texture and chewiness. It does increase whey off,[6] especially in soft-serve ice cream, so it is always used with in combination with κ-carrageenan to prevent that from happening. A typical use level is 0.05%–0.25%.

In soft cheese, such as cream cheese, LBG speeds coagulation, increases curd yield, makes curd separation easier, and provides shear stability. It improves texture (enhances resilience and body), holds moisture, and prevents syneresis[3] (of LBG) in the finished cheese. It provides excellent spreadability and stability to cheese spreads and may be used in imitation sour cream, sour cream–based dips, and yogurt.

[5] The heat shock test is a specific test that measures the stability of the ice cream to repeated freeze-thaw treatments (−20°C and 20°C [−4°F and 68°F] for prescribed times.

[6] Whey off is the separation of whey (that is, phase separation) in an ice cream mix.

LBG can be used as a binder and stabilizing agent in processed comminuted meats (such as sausages) to improve homogeneity and texture. It has a lubricating effect on the meat mix, resulting in easier extrusion and stuffing. Its ability to bind water results in improved yield and longer shelf life. LBG is widely used in combination with xanthan or a κ-type carrageenan to form the typical gel structure of canned pet foods and to provide retort stability.

Like guar gum, LBG can be used in barbecue sauces, salad dressings, and dry mixes for soups and gravies. It can provide mouthfeel to marshmallows and meringue toppings and control water adsorption and dough and batter rheology in the production of breads, cakes, and biscuits. It enhances the crispness of rehydrated freeze-dried products such as celery.

Products

Typical products containing guar and LBGs are given in Box 9.1. It is obvious from Box 9.1 that guar gum is more widely and extensively used than is LBG. Its maximum specified use in foods (US) is 2.0%, with the range generally being 0.35%–2.0%. The maximum specified use of LBG in foods is 0.8%, with the range generally being 0.15%–0.8%.

Appropriate label designations are guar gum, locust bean gum, and carob gum.

Tara gum

Tara gum is the ground endosperm seeds of the tara shrub. Commercial tara gum comes from South America. Its Man:Gal ratio is in between those of guar gum and LBG, and its properties reflect this composition. It forms gels with a κ-type carrageenan and xanthan, but weaker gels than are formed with LBG. As with guar gum and LBG, higher viscosities are obtained after heating aqueous suspensions, a phenomenon most marked with LBG. Tara gum is also in the middle of the three galactomannans with regards to producing viscosity, with 1% solutions having viscosities of 3500–5000 mPa·s.

Tara gum is used in dairy products, namely, in ice cream, frozen yogurt, and cream cheese. The properties of tara gum are closer to those of guar gum than to those of LBG, particularly with regard to its efficacy in preventing syneresis,[3] so it is not surprising that it is not extensively used.

Cassia gum

Another galactomannan hydrocolloid is obtained from the seeds of *Cassia obtusifolia* (also known as *Senna obtusifolia*)—an annual legume which grows wild in Africa, Asia, Central America, North America, Oceania, and South America. The seed endosperm is reported to contain more than 75% galactomannan. Little is known about its structure or properties. Its galactose–mannose ratio has been reported to be 3.6:1 to about 5:1, which would make its composition most like that of LBG. However, it has also been reported that the galactosyl branches are more randomly distributed along the mannan backbone than they are in LBG, which would give cassia gum quite different properties than those of LBG. When whole seeds were extracted with hot water, about half of the extracted polysaccharides were galactomannan in composition and half had the composition of poly(galacturonic acid) (Chapter 15). It was assumed that the galactomannan came from the endosperm and the poly(galacturonic acid) from the hull. Cassia gum is approved as a thickening and gelling agent for human and canned pet food in the European Union and Japan.

Additional resources

Dierckx, S., Dewettinck, K., 2001. Seed gums. In: Vandamme, E.J., De Baets, S., Steinbüchel, A. (Eds.), Biopolymers, Polysaccharides II, Polysaccharides from Eukaryotes, Vol. 6. Wiley-VCH, Weinheim, pp. 321–343.

Fox, J.E., 1997. Seed gums. In: Imeson, A. (Ed.), Thickening and Gelling Agents for Food, second ed. Blackie Academic & Professional, London, pp. 262–283.

Maier, H., Anderson, M., Karl, C., Magnuson, K., Whistler, R.L., 1993. Guar, locust bean, tara, and fenugreek gums. In: Whistler, R.L., BeMiller, J.N. (Eds.), Industrial Gums, third ed. Academic Press, San Diego, pp. 181–226.

Mathur, N.K., 2011. Industrial Galactomannan Polysaccharides. CRC Press, Boca Raton.

Reid, J.S.G., Edwards, M.E., 1995. Galactomannans and other cell wall storage polysaccharides in seeds. In: Stephen, A.M. (Ed.), Food Polysaccharides and Their Applications. Marcel Dekker, New York, pp. 155–186.

Wielinga, W.C., 2009. Galactomannans. In: Phillips, G.O., Williams, P.A. (Eds.), Handbook of Hydrocolloids, second ed. CRC Press, Boca Raton, pp. 228–251.

Inulin and Konjac Glucomannan

10

Chapter Outline

Key information and skills that can be obtained from study of this chapter will enable you to

1. Describe and give complete and shorthand structures of a representative inulin molecule.
2. Discuss the molecular size of inulin and the reasons for its polydispersity.
3. Explain the instability of inulin in low pH conditions.
4. Describe inulin's most important functional property.
5. Define and/or describe fructooligosaccharides (FOS).
6. Describe the key health benefit of inulin and FOS.
7. Describe the structures of native and deacylated konjac glucomannan.
8. Describe the nature of konjac glucomannan flour.
9. Describe the two ways to form thermally reversible gels containing konjac glucomannan.
10. Describe the five ways to form thermally irreversible gels containing konjac glucomannan.
11. Explain why gels cannot be formed from native konjac glucomannan alone.

Introduction

Polysaccharides that are storage substances in roots and underground stems (tubers) (including starches such as potato, cassava/tapioca, and arrowroot starches) are used as ingredients in processed foods. Two of the nonstarch polysaccharides that are storage substances are inulin and konjac glucomannan—two quite different polysaccharides.

Carbohydrate Chemistry for Food Scientists. https://doi.org/10.1016/B978-0-12-812069-9.00010-8

Inulin

Structure and stability

Inulin is composed of β-D-fructofuranosyl units linked 2 → 1. The polymer chains are often, but not always (because of degradation—either natural or during isolation), terminated at the reducing end with a sucrose unit. Inulin's degree of polymerization (DP) rarely, if ever, exceeds 60. It occurs in plants together with fructooligosaccharides, giving an overall DP range of 2–60.

$$[\beta Fru\mathit{f}\text{-}(2 \rightarrow 1)]_{n<60}\text{-}[\alpha Glc\mathit{p}]_{0-1}$$

This structure (with a terminal sucrose unit at the reducing end) is the same as that given in Fig. 3.7, with the exception that the chain of fructofuranosyl units is longer (DPs of the most abundant molecules in commercial products being 15–22).

Molecules containing furanosyl units undergo acid-catalyzed hydrolysis much more easily than do those containing pyranosyl units. (Remember that sucrose undergoes acid-catalyzed hydrolysis much more readily than do disaccharides such as lactose and maltose.) On the biological side, inulin is a storage polysaccharide, and at any time, molecules in various stages of synthesis and, perhaps, breakdown might be present. As a result, inulin preparations are mixtures of fructooligosaccharides with a glucosyl unit at the reducing end, fructooligosaccharides without a glucosyl unit at the reducing end, and small polysaccharides (DP 15 to about 60). (Those molecules with a glucosyl unit at the reducing end have the glucosyl unit attached in the same type of linkage as found in sucrose [that is, 2 ↔ 1] and can be considered to have a sucrose unit at the reducing end, so those molecules are nonreducing.) The enzymes required to break down plant storage polysaccharides usually occur compartmentalized in vacuoles in the same tissue with the polysaccharide; so when the tissue is macerated, the degrading enzyme(s) is released. Understanding inulin's ease of hydrolysis is not only important for preparing inulin with the highest possible molecular weight but also important in terms of maintaining its molecular weight during processing. Other factors related to inulin's DP are its source, the growing conditions of the source, harvesting conditions, and storage time after harvesting. On the other hand, as will be discussed later, fructooligosaccharides (FOS), which are also food ingredients, are produced by deliberate depolymerization of inulin.

Source and preparation

Inulin occurs naturally as a storage carbohydrate in thousands of plant species, including asparagus, banana, garlic, and onion. The primary commercial source is chicory (*Cichorium intybus*) root where inulin constitutes 16%–18% of the dry matter. Some are also obtained from Jerusalem artichoke (*Helianthus tuberosus* L.) tubers.

Inulin has been consumed by humans since antiquity. People on all five inhabited continents have eaten various roots and tubers, some containing starch and some containing inulin or a similar glucofructan. Before the potato was introduced into Western

Europe in the latter half of the 16th century, the Jerusalem artichoke tuber (14%–19% inulin) was a major source of carbohydrate for the people there. Some groups of people still consume significant amounts of underground storage organs containing inulin or a similar glucofructan.

Chicory roots (and Jerusalem artichoke tubers) must be processed soon after harvesting (preferably on the same day, but no longer than 7 days after harvesting) because they deteriorate rapidly. The roots are washed, sliced, and extracted with hot water. Colloidal materials, colored substances, and other solubles (salts, amino acids, sugars, etc.) are removed from the extract by a purification/clarification process involving treatment with calcium hydroxide (lime) to precipitate protein, followed by treatment with carbon dioxide to neutralize the alkaline mixture and to precipitate any calcium ions in solution as calcium carbonate, centrifugation, filtration through an adsorbent earth and/or activated carbon, and ion exchange, a process very similar to that used in purifying sucrose extracts. The extract is then concentrated by evaporation and spray-dried. Concentration may also be done by ultrafiltration. The inulin may also be recovered by crystallization or precipitation. Even with extensive care, inulin extracted from chicory roots is still largely a mixture of oligosaccharides. A typical commercial preparation contains 2% monosaccharides, 5% disaccharides, and 93% molecules of DP 3–60, with an average DP of 8–10. The maximum average DP of chicory root inulin is about 15, indicating a relatively high percentage of molecules of low DP. The maximum amount in chicory roots is about 18% (dry basis).

Jerusalem artichoke flour is also available.

Properties and uses of inulin

Inulin is water soluble; dissolution occurs most easily by heating an aqueous dispersion to 40–80°C (105–175°F). (pH should be neutral or close to neutrality, especially for hot solutions, as inulin is very sensitive to acid-catalyzed hydrolysis.) Solutions can be made at concentrations as high as 50%.

When hot solutions of inulin of concentrations greater than 25% are cooled, thermoreversible gels are formed. Gel strength increases as the concentration increases. Synergetic[1] interactions that increase gel strength have been reported between inulin and alginates (Chapter 14), κ- and ι-carrageenans (Chapter 13), gellan (Chapter 12), LM-pectin[2] (Chapter 15), starch (Chapter 6), and gelatin.

Inulin gels of all concentrations and textures are particle gels (especially after shearing). Because inulin gels have a creamy, fatlike texture, inulin can be used as a fat mimetic (Chapter 17) in reduced-fat products. Inulin improves the texture and mouthfeel of low-fat ice creams and sauces. It functions as a processing aid in no-fat icings and whipped toppings and provides a creamier mouthfeel to potato products. Inulin is also an ingredient in breakfast, meal replacement, nutrition and sports/energy bars, soy beverages, and vegetable patties.

[1] A synergistic interaction is one in which the observed value is greater than what would be predicted by summing the values for the independent parts.

[2] For definition, see Chapter 15.

More important than its functional characteristics, inulin has nutritional/health benefits. Inulin is not digested by the enzymes in either the stomach or the small intestine. Therefore, it is designated as soluble dietary fiber (Chapter 17), and its consumption gives the positive attributes of soluble fiber (that is, anticonstipation activity, pH reduction of the contents of the large intestine, and cholesterol and triglyceride reduction). It has a glycemic index (Chapter 17) of zero; it raises neither glucose nor insulin levels in the blood. It is a substrate for beneficial colonic bacteria and therefore is a prebiotic.[3] Absorption and catabolism of the short-chain fatty acids produced by these beneficial bacteria via anaerobic fermentation of inulin in the large intestine (colon) contribute about 2 kcal/g of inulin.

Fructooligosaccharides

Inulin is often deliberately depolymerized into fructooligosaccharides (FOS), which may also be called inulodextrins, and which have structures like the microbial FOS produced by fermentation (section on Fructooligosaccharides in Chapter 3), with or without the glucosyl/sucrosyl unit at the reducing end. Inulinase is used for the depolymerization. Such FOS products are sold as syrups or powders. A typical composition is as follows:

	%
Monosaccharides	2 (>95% fructose)
Disaccharides	3 (includes about 10% sucrose)
Trisaccharides	33
Tetrasaccharides	34
Pentasaccharides	17
Hexasaccharides	9
Heptasaccharides	1
Degree of polymerization > 7	0.4

The oligosaccharides may or may not be terminated with a glucosyl/sucrosyl unit. The probability that a glucosyl unit is present increases with increasing DP up to a maximum of about 35% of the molecules being terminated with a glucosyl/sucrosyl unit at DP 5.

FOS have the physical properties of other oligosaccharides of comparable size. As opposed to maltooligosaccharides, but like oligosaccharides from guaran, they are

[3] As defined in Chapter 17, prebiotics are substances that are not digested by the enzymes of the human small intestine, but which provide beneficial physiological effects to the host by selectively stimulating the growth and/or the bioactivity of beneficial microorganisms already resident in the large intestine (colon).

nondigestible in the small intestine and, therefore, components of soluble dietary fiber (Chapter 17). Because they are excellent prebiotics,[3] they are most often used for their health benefits.

Konjac glucomannan

Structure

Konjac glucomannan is also known as konjac, conjac, and konnyaku. Its basic structure is that of a chain of β-D-mannopyranosyl and β-D-glucopyranosyl units linked (1 → 4) (that is, the chains contain D-glucopyranosyl units joined as they are in cellulose (Chapter 8) and D-mannopyranosyl units joined as they are in the backbone chains of the galactomannans (Chapter 9)). (The β-D-mannopyranosyl and β-D-glucopyranosyl units differ from one another only in the configuration of the C2 carbon atom.) The ratio of mannosyl units to glucosyl units is about 1.6:1. The two saccharide (glycosyl) units appear to be randomly distributed along the chain. The molecule is basically linear but is slightly branched with 1.2%−1.6% of the units being branch points. The native polysaccharide is also slightly acetylated at O6. The degree of substitution is about 0.05−0.11. The acetyl groups are randomly located. The neutral polysaccharide has a high molecular weight (about 10^6).

Source and preparation

Commercial konjac glucomannan is a flour obtained from the tubers of *Amorphophallus konjac* K. Koch (known as elephant yam, devil's tongue, and voodoo lily), which is grown in mountainous or hilly areas in Southeast Asia. The tuber is eaten throughout Asia, especially in China, Japan, Indonesia, Thailand, and India. It is added to traditional soups, stews, and similar dishes. It is also cooked by itself in the presence of lime (calcium hydroxide), and the resulting gel is pressed into blocks called konjac tofu.

To produce konjac flour, 2- to 3-year-old tubers are harvested, cleaned, and sliced. The slices are dried and dry-milled. Subsequent air classification separates the sacs containing the glucomannan from the starch granules and other materials present. This raises the glucomannan content from 30%−50% (dry basis) in the tuber to 60%−70% in the flour. The purified flour may then be washed with aqueous alcohol and dried and further ground to produce a purer product that is composed of 80%−95% of the glucomannan. The purified forms consist of oval, 100−500 μm sacs containing the polysaccharide.

Properties and uses

Particles in preparations of konjac glucomannan hydrate rapidly, absorbing up to 200 times their weight in water, depending on purity. The glucomannan forms viscous, pseudoplastic (Chapter 5) dispersions. A typical preparation of konjac glucomannan produces a viscosity of 18,000 mPa•s at a concentration of 1%. As with other hydrocolloids, increasing the temperature and degree of shear during hydration decreases the time required to achieve full viscosity/hydration.

Thermally reversible gels are produced by cooling hot solutions of combinations of konjac glucomannan (KG) and κ-carrageenan (Chapter 13) or xanthan (Chapter 11). The synergistic interaction with κ-carrageenan produces strong, elastic, thermally reversible gels, the texture of which can be varied by varying the ratio of the two hydrocolloids and the total concentration of polysaccharide. Such gels are more heat stable than are pure κ-carrageenan gels. Heat-stable gels are formed when such a mixture is heated to retort temperatures. Konjac glucomannan is roughly twice as effective in its synergism with κ-carrageenan as is locust bean gum (LBG). The optimal ratio of KG to κ-carrageenan is about 2:3; varying the ratio changes the texture of the gel.

A combination of xanthan and KG can result in a solution viscosity which is as much as three times the solution viscosity of either hydrocolloid used alone at the same concentration. Although neither native KG nor xanthan will form a gel when used alone, hot (85°C) solutions of blends of the two hydrocolloids (at the optimum ratio of about 3:2 KG to xanthan) will produce very elastic, strong, thermally reversible gels with good freeze-thaw stability. Varying the ratio will change the gel texture. As with KG—κ-carrageenan gels, heat-stable gels are formed when KG—xanthan mixtures are heated to higher temperatures. KG is roughly twice as effective in its synergism with xanthan than is LBG.

Konjac glucomannan also interacts with starches and modified starches to produce an increase in viscosity. Heat-stable gels are formed by freezing cooked mixtures.

The acetyl groups on native KG molecules prevent them from associating with themselves and forming gels. However, when the native polysaccharide is de-esterified, thermally stable gels can be formed. Traditionally, the acetyl groups are removed by treatment with calcium hydroxide (lime). However, any food-grade weak base, such as potassium carbonate, can be used to raise the pH. Raising the pH of a KG dispersion to pH 9—10 and heating it effects de-esterification. Cooling the deacetylated KG solution produces a gel that will withstand even retort temperatures without melting. The gels are strong and somewhat elastic (but less elastic than those of KG plus κ-carrageenan or xanthan). Neutralization is possible after gelation; gel quality is not affected. The gels are stable to low pH once they are set.

In summary, thermally reversible gels can be formed from KG plus κ-carrageenan and KG plus xanthan. Heat-stable gels can be formed with KG treated with alkali, KG treated with alkali plus pasted starch, KG plus pasted starch and freezing, heated KG and xanthan dispersions, and KG and κ-carrageenan dispersions heated to retort temperatures. These seven different kinds of gels made under seven different gelling conditions provide a great variety of gel types. An even greater variety can be obtained by varying the total polysaccharide concentration, the ratio of KG to the other hydrocolloid or starch used, the alkali concentration, the temperature to which dispersions are heated, and the shear employed.

Even with all this range of gelling conditions and gel properties and the additional attributes of improving cling of starch pastes and KG's ability to form heat-stable, flexible, transparent or opaque, high-strength, protective coatings and films, konjac glucomannan is little used. Because of the stability of its gels to low pH values, it is possible to use KG to make shelf-stable dessert gels and gummy candies, but its use in making certain types of gelled confectioneries is limited or banned in the United States, Canada and the European Union because of the risk of children choking on the nondissolving candies. Because each flour particle can form a gel particle, KG can be used as a fat mimetic in prepared meat products and in vegetable burgers and other soy analogs, in which it also functions as a binder. It has been long used as a binder in surimi. KG has also been used for many years in several traditional oriental foods, primarily in noodles and gels. It will help retain the texture and shape of pasta held on steam tables.

Konjac glucomannan is not allowed as an ingredient in human food in the United States, but it is sold as a dietary supplement,[4] even though several companies have been fined for claims without supporting clinical data.

Additional resources

Inulin

Flamm, G., Glinsman, W., Kritchevsky, D., Prosky, L., Roberfroid, M., 2001. Inulin and oligofructose as dietary fiber: a review of the evidence. Critical Reviews in Food Science and Nutrition 43, 353–362.

Flickinger, E.A., Loo, J.V., Fahey Jr., G.C., 2003. Nutritional responses to the presence of inulin and oligofructose in the diets of domesticated animals: a review. Critical Reviews in Food Science and Nutrition 43, 19–60.

Franck, A., De Leenheer, L., 2002. Inulin. In: Vandamme, E.J., DeBaets, S., Steinbüchel, A. (Eds.), Biopolymers, Polysaccharides II, Polysacchrides from Eukaryotes, Vol. 6. Wiley-VCH, Weinheim, pp. 439–479.

Izzo, M., Niness, K., 2001. Formulating nutrition bars with inulin and oligofructose. Cereal Foods World 46, 102–106.

Meyer, D., Blaauuhoed, J.-P., 2009. Inulin. In: Phillips, G.O., Williams, P.A. (Eds.), Handbook of Hydrocolloids. CRC Press, Boca Raton, pp. 829–849.

Praznik, W., Cieślik, E., Huber, A., 2004. Fructans: occurrence and application in food. In: Tomasik, P. (Ed.), Chemical and Functional Properties of Food Saccharides. CRC Press, Boca Raton, pp. 197–215.

Konjac Glucomannan

Takigami, S., 2009. Konjac mannan. In: Phillips, G.O., Williams, P.A. (Eds.), Handbook of Hydrocolloids. CRC Press, Boca Raton, pp. 889–901.

Thomas, W.R., 1997. Konjac gum. In: Imeson, A. (Ed.), Thickening and Gelling Agents for Food, second ed. Blackie Academic & Professional, London, pp. 169–179.

[4] When sold as a nutritional supplement, konjac glucomannan is sold in "diet pills" with claims by the sellers that it is effective in treating obesity, constipation, and high cholesterol.

Xanthan

11

Chapter Outline

Key information and skills that can be obtained from study of this chapter will enable you to

1. Describe the source and preparation of xanthan.
2. Give a shorthand structure for xanthan.
3. Give the structure of a pyruvylated β-D-mannopyranosyl nonreducing end unit of xanthan.
4. Describe how the structure of xanthan differs from the structures of plant polysaccharides (other than cellulose).
5. Describe the molecular (three-dimensional) structure of the xanthan molecule.
6. Relate the molecular (three-dimensional) structure of xanthan to its high degree of pseudoplasticity.
7. Describe the unusual properties of xanthan solutions with respect to (1) pH and (2) temperature and tell what is unusual about them.
8. Give applications of xanthan that make use of (1) the effect of temperature on, (2) the pseudoplasticity of, and/or (3) the yield stress of its solutions. Explain.
9. Describe how the Newtonian plateau of xanthan solutions is related to its ability to stabilize emulsions and suspensions.

Introduction

Microorganisms produce polysaccharides as structural elements, as protective coatings to ward off invasive organisms, and/or to prevent excessive moisture loss under drying conditions. A polysaccharide coat or capsule usually is produced when nutrients become less available or as the water supply diminishes. Capsular material is composed of high-molecular weight polysaccharides as a gel layer or, under drying conditions, a cohesive film that, on later contact with water, rehydrates to a thick gel coating. Capsules do not develop to significant extents when the organisms have available to them an ample supply of nutrients and are undergoing rapid cell division in the logarithmic growth phase. *Xanthomonas campestris*, a bacterium commonly

Carbohydrate Chemistry for Food Scientists. https://doi.org/10.1016/B978-0-12-812069-9.00011-X

found on leaves of plants of the cabbage family, when grown under the right conditions, produces a polysaccharide termed xanthan that encapsulates the cell and diffuses into the surrounding medium. The polysaccharide is known commercially as xanthan gum.

Xanthan is an extensively and widely used hydrocolloid. It is produced commercially by a pure culture in a submerged, aerobic fermentation. The characteristics of xanthan vary with variations in the strain of the organism, the sources of nitrogen and carbon, the degree of medium oxygenation, the temperature and pH of the fermentation, and the concentrations of various inorganic ions. When the fermentation is finished, the broth is pasteurized to kill the organisms, and the gum is recovered by precipitation with 2-propanol (isopropanol) or ethanol.

Structure

The backbone chain of xanthan molecules has the same structure as that of cellulose molecules (Chapter 8). In the xanthan molecule, every other β-D-glucopyranosyl unit in the backbone chain has attached (at the O3 position) a β-D-mannopyranosyl-(1→4)-β-D-glucuronopyranosyl-(1→2)-6-*O*-acetyl-α-D-mannopyranosyl trisaccharide unit (Fig. 11.1).[1] About half of the terminal β-D-mannopyranosyl units have pyruvic acid attached as a 4,6-cyclic acetal (Fig. 2.3). The trisaccharide side chains in essence wrap around the main chain, making the molecule a rather stiff rod (Fig. 11.2) with extraordinary stability to heat, acid, and alkali. The average molecular weight is probably about 2 million, although much larger values, presumably due to aggregation, have been reported. The structure, molecular weight, and physicochemical properties of xanthan do not vary as long as the same strain of the microorganism is used and it is grown under constant conditions, which is what is done commercially. However, different producers use different strains of *X. campestris*, so xanthan preparations from different producers are not identical. Variations in structure, molecular size, and properties also result from changes in growth conditions (that is, in the types and concentrations of nitrogen and carbon sources, concentrations of certain inorganic salts, the degree of oxygenation, the pH, and the temperature of the growth medium). Xanthans of the highest pyruvic acid content have the highest viscosities and thermal stabilities.

Properties

Xanthan is soluble in hot or cold water. Because of their stiffness, xanthan molecules are extended in solution. As a result, solutions of xanthan have high viscosities in comparison with most other polysaccharide solutions and are highly pseudoplastic

[1] Bacterial heteroglycans usually have regular, repeating-unit structures, distinguishing them from plant heteroglycans, which do not.

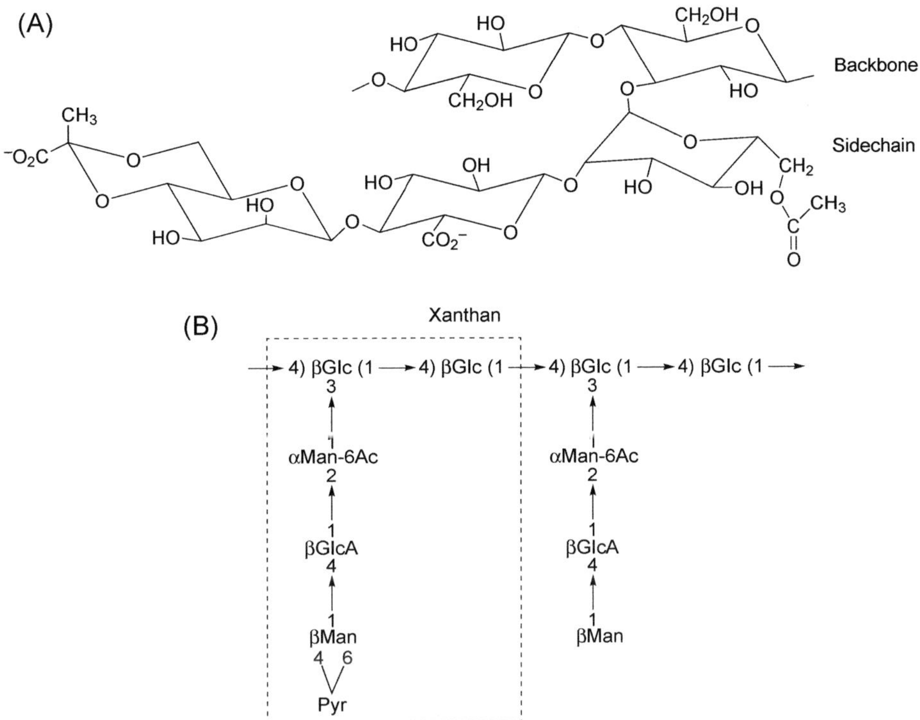

Figure 11.1 Structure of the pentasaccharide repeating unit of xanthan, actual (A) and shorthand designation (B) The upper chain in structure A is the main cellulosic chain (the backbone chain); the lower chain is the trisaccharide side chain. Inside the box in structure B is a pyruvylated pentasaccharide building block unit as in the upper structure; outside the box is a nonpyruvylated unit.

(Chapter 5). These characteristics are exhibited over a wide range of pH values and a temperature range from 0 to 100 °C (that is, from freezing to boiling temperatures), a property that is unique, not only for food hydrocolloids but also for water-soluble polymers in general. The high degree of pseudoplasticity ensures that final products in which it is used have good flavor release and mouthfeel (that is, are nonslimy) (Chapter 5). It also results in greater efficiency and less energy usage during mixing and pumping and good pourability of the final product.

Xanthan solutions have a Newtonian plateau at low shear stress values (Fig. 11.3), which means that xanthan solutions do not thin at very low rates of shear. Its solutions also have a high elastic modulus (G′ value; Chapter 5). Both properties make it excellent for generating and stabilizing emulsions and suspensions (for example, oil-in-water emulsions and suspensions of relishes and herbs in pourable salad dressings).

Xanthan interacts with galactomannans, such as guar gum and locust bean gum (LBG). In the case of guar gum, the result is a synergistic increase in solution viscosity (Fig. 9.3). The interaction with LBG produces a heat-reversible gel (Fig. 9.5). It is

Figure 11.2 Two representations of the relationship of the trisaccharide side chains to the backbone helix of xanthan molecules. The right-hand representation is a view down the helix axis; the *small circles* represent the positions of the trisaccharide side chains and the *large central circle* the backbone helix.

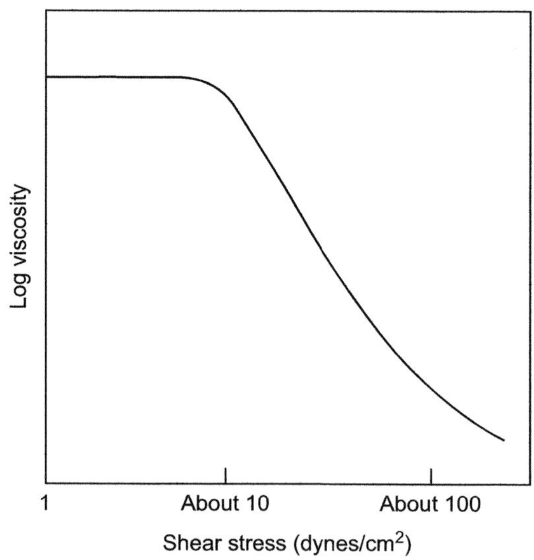

Figure 11.3 An example of the response of a xanthan solution to shear rates showing the Newtonian plateau and the pseudoplastic region.

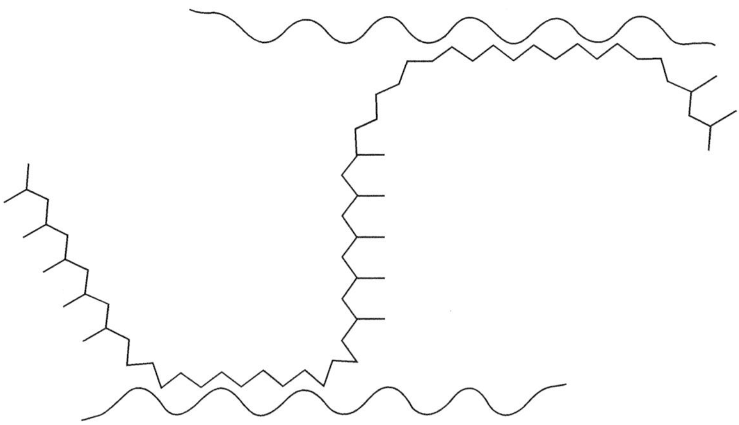

Figure 11.4 A representation of the hypothesized interaction of a locust bean gum (LBG) molecule with single-helical portions of xanthan molecules to form a three-dimensional network and a gel.

possible that the interaction also occurs with single-helical xanthan (Fig. 11.4; compare with Fig. 9.6). LBG molecules interact with bundles of xanthan molecules. Whether the interaction is with single-helical portions, double-helical portions, or bundles of single- or double-helical xanthan molecules, the smaller LBG molecules effectively cross-link the xanthan structures (Fig. 11.5) forming a network. Xanthan also interacts synergistically with κ-carrageenan (Chapter 13) and konjac glucomannan (Chapter 10).

In summary, some properties of xanthan that make it useful in food systems are

- solubility in both hot and unheated water
- high solution viscosity at low concentrations
- solution viscosity stability from 0°C to more than 100°C (unique thermal stability), even at low pH
- rheological properties that make it ideal for stabilizing emulsions and suspensions
- highly pseudoplastic solution rheology
- a Newtonian plateau at low shear
- excellent acid stability
- good solution freeze-thaw stability
- stability in high-salt systems (compatibility with high concentrations of salt)
- resistance to the action of enzymes found in food systems
- synergetic interactions with κ-carrageenan and LBG (to produce unique gels) and to a lesser extent with guar gum (to increase viscosity)

Uses

Xanthan is extensively used in processed food products (Box 11.1) because of its unusual and very useful properties. Many of these properties undoubtedly result from the structural rigidity of its molecules, which in turn results from its linear, cellulosic

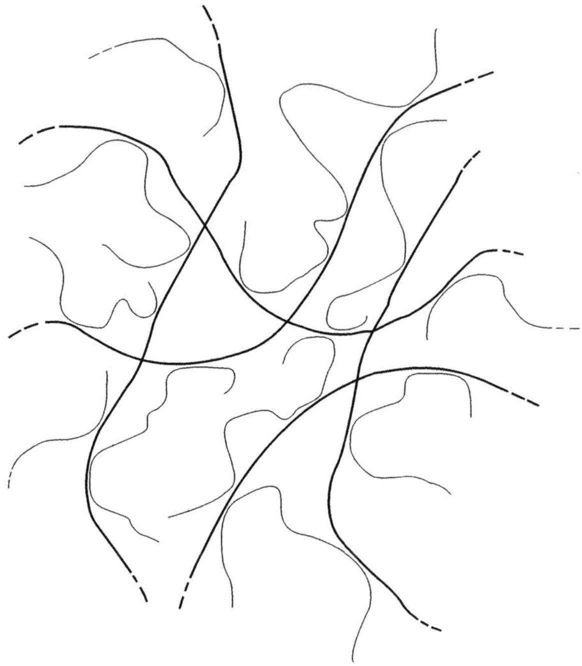

Figure 11.5 A representation of the interaction of locust bean gum (LBG) molecules (*thin lines*) with single- or double-helical xanthan molecules or bundles of xanthan molecules (*thick lines*).

backbone which is stiffened and shielded by the trisaccharide side chains. As already mentioned, the high degree of solution pseudoplasticity imparted by the presence of xanthan makes liquid/fluid systems easier to mix and pump and results in clean organoleptic properties in the mouth (that is, no perception of sliminess).

Xanthan is an excellent hydrocolloid for stabilizing aqueous dispersions, emulsions, and suspensions. The fact that the viscosities of its solutions do not thicken upon cooling makes it irreplaceable for thickening and stabilizing such things as pourable salad dressings and chocolate syrup and other squeezables, which need to flow as easily when taken from the refrigerator as they do at room temperature. In regular salad dressings, it is both a thickener and a stabilizer for the oil-in-water emulsion and prevents any particles, such as those of spices, from settling out. It is an effective stabilizer of both suspensions and emulsions because its solutions have a yield value (Chapter 5) and the Newtonian plateau (that is, its solutions do not thin at the low shear rates that would occur if an oil droplet or a solid particle were to move up or down in the system). If shear thinning was to occur as a result of the movement of the oil droplet (due to the differences in density) or solid particle (due to the force of gravity due to the differences in density) and flow of the solution around it, the droplet or particle would move even faster and cause even more shearing and even more thinning so on and on, but this does not occur in solutions of xanthan; so the oil droplets or solid particles do not move (either up or down). It thickens and suspends seeds and spice particles in no-oil (reduced-calorie) salad dressings. In both full oil and "light" salad dressings,

Box 11.1 Typical Products Containing Xanthan

- Bakery products
  - Cake mixes
  - Crescent rolls
  - Danish fillings
  - Pie crusts
  - Pie filling mixes
  - Pop tarts
  - Refrigerated dough
- Beverages
  - Chocolate drinks
  - Nutritional/diet
  - Protein shake
  - Soy milk
- Cereal bars
- Condiments
  - Pickle relish
  - Salsa
- Dairy products
  - Cheesecakes
  - Cheese spreads and dips
  - Cheese substitutes
  - Chocolate drinks
  - Cottage cheese, including low-fat
  - Cream cheese
  - Frozen cheese
  - Sour cream
  - Whipped toppings
- Egg substitutes
- Frozen foods
  - Frozen cakes
  - Frozen cheese
  - Frozen cheesecake
  - Frozen gravy
  - Frozen guacamole
  - Frozen lasagna
  - Frozen pizza
  - Frozen rice
- Honey-roasted peanuts
- Meat products
- Breakfast slices
  - Chicken strips
  - Poultry breast slices
- Mixes
  - Cake mixes
  - Cocktail mixes
  - Fruit drink mixes
  - Gravy mixes
  - Pancake mixes
  - Pie filling mixes
  - Pudding mixes
- Relishes
- Salad dressings, etc.
  - Low-calorie pourable salad dressings
  - Pourable salad dressings
  - Reduced-calorie mayonnaise
- Sauces
  - Barbecue sauces
  - Cocktail sauces
  - Mushroom sauces
  - Oriental sauces
  - Pasta sauces
  - Pizza sauces
  - Sloppy Joe sauces
  - Taco sauces
  - Tartar sauce
- Spreads
  - Margarine spreads
  - Sandwich spreads
  - Sugar-free jams
- Syrups
  - Chocolate syrup
  - Pancake syrup
- Toppings
  - Butterscotch toppings
  - Fudge toppings
  - Marshmallow creme
  - Whipped toppings
- Vegetable patties

and dressings. Especially in the latter, xanthan may be used in combination with propylene glycol alginate (PGA) (Chapter 14). PGA by itself produces solutions of much lower viscosity and much less pseudoplasticity. So when the two hydrocolloids are used together, they produce a solution of less viscosity than would be realized if only xanthan was used. Also when used in combination, the two hydrocolloids provide both the pourability associated with a pseudoplastic xanthan solution and the creaminess sensation (viscosity at the shear rates encountered in the mouth) associated with a nonpseudoplastic solution. They are generally used in a xanthan–PGA ratio of about 1:2, with 0.3% xanthan and 0.6% PGA allowing the oil content to be reduced partially or all the way to zero. (In a regular salad dressing containing oil, most of the viscosity is produced by the emulsion.) The hydrocolloids change the rheology from that of an emulsion to that of a hydrocolloid solution.

Blends of xanthan and LBG and/or guar gum are employed as stabilizers of ice creams and frozen novelties (Chapters 9 and 13). Carrageenan is also added to the mix to prevent whey separation during freezing (Chapter 13). Other applications of these two- or three hydrocolloid blends are in the preparation of canned frostings, chip dips, cream cheese, cottage cheese, cottage cheese products, processed cheese products, and pizza sauce.

Xanthan is useful in barbecue sauces, spaghetti sauces, soy sauce–based products, and taco sauce, in part because of its stability in low pH and high salt systems. It provides good cling in dipping sauces (for example, for chicken nuggets). In most sauces, xanthan is used together with a modified food starch (Chapter 7) to provide optimum body and texture. It can be used in gravies that do not thicken appreciably as they cool and in which there is little skin formation after hours on a steam table (because of its great ability to hold water and prevent evaporation). Use of a properly modified food starch and xanthan together provides freeze-thaw stability to frozen sauces and gravies.

After dressings and sauces, the largest applications of xanthan are in bakery mixes. In cake and related mixes, xanthan provides recipe tolerance, an increase in cake volume, and moisture retention. However, use of too much xanthan in flour-based systems may result in denseness, gumminess, graininess, or other undesirable changes in texture. Use levels in dry mixes are typically less than 0.1% (finished formulation weight).

No enzymes found in food ingredients or products will degrade xanthan. Ordinary preparations of xanthan have some cellulase activity and, as a result, cannot be used in combination with carboxymethylcellulose (CMC) or methylcellulose products. However, cellulase-free preparations of xanthan are available. A cellulase-free xanthan is at times used together with CMC (Chapter 7) in reduced-calorie pancake syrups, especially ones containing emulsified butter.

Ingredient labels for products containing xanthan may read xanthan or xanthan gum.

Additional resources

Born, K., Langendorff, V., Boulenguer, P., 2002. Xanthan. In: Vandamme, E.J., De Baets, S., Steinbüchel, A. (Eds.), Biopolymers, Polysaccharides I, Polysaccharides From Prokaryotes, vol. 5. Wiley-VCH, Weinheim, pp. 259–297.

Stokke, B.T., Christensen, B.E., Smidsrød, O., 1998. Macromolecular properties of xanthan. In: Dumitriu, S. (Ed.), Polysaccharides. Marcel Dekker, New York, pp. 433–472.

Sworn, G., 2009. Xanthan gum. In: Phillips, G.O., Williams, P.A. (Eds.), Handbook of Hydrocolloids, second ed. CRC Press, Boca Raton, pp. 186–203.

Urlacher, B., Noble, O., 1997. Xanthan gum. In: Imeson, A. (Ed.), Thickening and Gelling Agents for Food, second ed. Blackie Academic & Professional, London, pp. 284–311.

Gellans, Curdlan, Dextrans, Levans, and Pullulan

12

Chapter Outline

Key information and skills that can be obtained from study of this chapter will enable you to

1. Give the complete chemical structure of the repeat unit of native (high-acyl) gellan.
2. Give the shorthand structure of the repeat unit of deacylated (low-acyl) gellan.
3. Compare the polymolecularity of fermentation gums to those of gums from higher plants and marine algae.
4. Describe the molecular basis for formation of gellan gels.
5. Describe and explain the principles involved in dissolving gellan in tap (hard) water at room temperature using a sequestrant and then forming a gel by lowering the pH of the solution.
6. Explain the reasoning for blending high-acyl and low-acyl types of gellan.
7. Give the shorthand and complete chemical structure of curdlan.
8. Describe the unique property of curdlan.
9. Describe the differences between low-set and high-set curdlan gels.
10. Describe dextrans and levans and their origin.

Introduction

In addition to xanthan (Chapter 11), two other bacterial/microbial polysaccharides (fermentation gums), namely, gellan and curdlan, are available to food product developers. However, both their volumes and breadths of use are dwarfed by those of xanthan.

Carbohydrate Chemistry for Food Scientists. https://doi.org/10.1016/B978-0-12-812069-9.00012-1

Gellan

Production and structure

As with xanthan, this polysaccharide is known by two names: gellan is the name of the polysaccharide; gellan gum is its commercial name.

Gellan is an extracellular, anionic polysaccharide produced by the bacterium *Sphingomonas elodea.* The gellan molecule is linear and is composed of β-D-glucopyranosyl, β-D-glucuronopyranosyl, and α-L-rhamnopyranosyl units in molar ratios of 2:1:1. The repeat unit structure of native gellan is given below. Native gellan (also called high-acyl gellan) contains two ester groups (namely, an acetyl group and a glyceryl group), both on the same glucosyl unit. On average, there is about one glycerate ester group (on O2) per tetrasaccharide repeat unit and one acetate ester group (on O6) for every two repeat units.

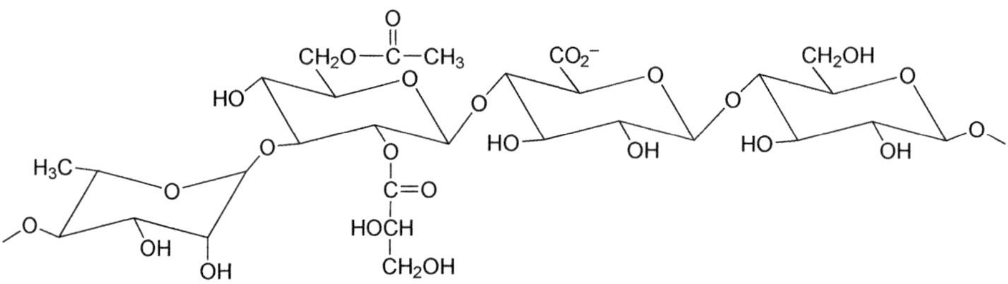

As in the production of xanthan, the starter culture and fermentation process variables are carefully controlled so that the same structure is made each time, which is in contrast to land plant and seaweed hydrocolloids whose structures often vary with species harvested and seasons or crop years. After an aerobic submerged fermentation under controlled conditions, the medium is heated to kill the microorganisms; the pH is adjusted; the medium may be clarified; the polysaccharide is isolated by precipitation with 2-propanol (isopropanol), and the precipitate is dried and milled. The average molecular weight of gellan is $1.5{-}3.0 \times 10^6$.

Some of the gellan for food use is deesterified by treatment with alkali. Deesterification removes the acyl groups, which has a dramatic effect on the properties of gels produced with gellan. The deesterified form is known as low-acyl gellan. Its tetrasaccharide repeat unit structure (that is, the deacylated structure of the unit shown above) is →4)-αLRha*p*-(1→3)-βGlc*p*-(1→4)-βGlc*p*A-(1→4)-βGlc*p*-(1→.

Two basic forms of gellan are available: high-acyl (native) and low-acyl. As described below, blending of high- and low-acyl types is used to prepare products with properties intermediate between the high- and low-acyl types.

Gellan forms a coaxial, three-fold double helix (Fig. 5.23) with the acetate ester groups on the outside. The glycerate ester groups are involved with stabilizing intermolecular associations between the two left-handed helical chains. A cation (any cation, even H^+) is required for double helix formation. Pairs of antiparallel double helices form junction zones (Chapter 5).

Properties

Gellan, at very low concentrations (0.02%–0.1%), is used for thickening. These dilute solutions are excellent at stabilizing suspensions because of their high yield values (Chapter 5).

Gellan is, perhaps, best known for its ability to gel aqueous systems. It is quite effective at concentrations of 0.15%–1.0% in forming gels with both monovalent and divalent cations, with divalent cations (Ca^{2+}) being about 10 times more effective than monovalent cations. Gels can be formed with as little as 0.05% polysaccharide (99.95% water), but at this concentration, the gel is not self-supporting.

Gellan can be dispersed in room temperature tap water without clumping (or dissolving appreciably) because it gels in the presence of any cation, and because hydrocolloids are insoluble in water under conditions at which they form gels (Chapter 5). Cations increase the hydration temperature, but generally suspensions must be heated to 75–85°C, depending on the hardness of the water, to make a solution. (Types of gellan that will hydrate in water without heating are available. These products contain a sequestrant, such as trisodium orthophosphate, tetrasodium pyrophosphate, or most commonly, sodium citrate.) The resulting solutions can then be gelled by addition of a divalent cation (Ca^{2+}) or by addition of an acid to release the Ca^{2+} from the sequestrant.

Gelation is often effected by cooling a hot solution containing the required cation, a process often useful in food systems. An advantage of gellan is that it is quite stable to heating at low pH (that is, in acidic systems).

The low-acyl type of gellan forms firm, brittle, nonelastic gels (with textures similar to those of gels made with agar and κ-carrageenan [Chapter 13]). The high-acyl (native) type forms soft, elastic, nonbrittle gels (with textures similar to those made with mixtures of xanthan and locust bean gum [Chapter 11]). A range of intermediate gel textures can be achieved by mixing the two basic types of gellan. Particularly is the degree of brittleness affected by combinations of the two types of gellan. The majority of gellan used in food products is the low-acyl type.

To make a dessert, gel requires about 0.2% gellan and calcium ions. Different types of gels can be made with different cations, in addition to blends of the two types of the hydrocolloid. In general, sodium salts form gels with lower melting temperatures (as compared to gels made in the presence of potassium or calcium ions). Potassium and calcium salts of gellan have higher hydration/dissolution temperatures. Sucrose (at concentrations of more than 20%) reduces gel strength. Other hydrocolloids used in combination with gellan also modify gel properties. Those most often used are κ-carrageenan, gelatin, and agar. So a variety of gels can be prepared with variations in gel strength, elasticity, brittleness, setting temperatures, melting temperatures, and clarity by controlling the type of gellan, its concentration, the type and concentration of cation, and the types and amounts of other ingredients.

Gellan gels do not melt in the mouth as do gelatin gels,[1] but they do disintegrate easily, giving excellent flavor release. This is probably due to the fact that gellan gels contain a very low concentration of hydrocolloid.

Shearing during cooling of a hot gellan solution prevents the normal gelation mechanism from occurring and produces a smooth, homogeneous, thixotropic fluid (a pourable gel) that stabilizes emulsions and suspensions very effectively. Gentle agitation of a weak gellan gel will also disrupt the gel structure and turn the gel into a smooth, pourable, thixotropic fluid with excellent emulsion and suspension stabilizing properties.

Uses

As already mentioned, gellan is useful in producing water gels (as are several other food hydrocolloids). The ability of its solutions to suspend particles at low hydrocolloid concentrations (without producing high viscosity) make it useful as a stabilizer in nutritional and two-phase beverages. It can be used to make chewy candy because of its gelling ability, outstanding flavor release, and acid stability. Its ability to gel without high soluble solids is applied in preparing fruit toppings, especially those that will be served hot. Gellan can also be used as a stabilizer in sour cream and yogurt products.

Gellan may be incorporated into bakery mixes (such as brownie, cake, and muffin mixes) to take advantage of the fact that, at room temperature, it does not hydrate appreciably and thereby increase the viscosity of the batter. It does, however, hydrate upon heating and holds moisture in the baked product. Its moisture-retaining property can also be exploited in formulating nutritional bars.

Gellan may be identified as gellan or gellan gum on ingredient labels.

Curdlan

Production and structure

Curdlan is a neutral extracellular polysaccharide produced by the microorganism *Agrobacterium biovar.* The curdlan molecule is unbranched and unsubstituted (that is, neither esterified nor etherified nor derivatized with cyclic acetal structures as are xanthan and gellan). It is a (1→3)-linked β-glucan.

$$[\beta\mathrm{Glc}p\text{-}(1\rightarrow 3)]_n\text{-}\beta\mathrm{Glc}p$$

The curdlan in the fermentation medium is dissolved by raising the pH of the medium; bacterial cells are removed; the solution is purified, and the polysaccharide is recovered by precipitation with 2-propanol [isopropanol] or ethanol.

[1] No polysaccharide gel melts in the mouth.

In solution, curdlan is usually present as a mixture of single- and triple-helical structures. (Under certain conditions, random coil structures will also be present.) Under gelling conditions, curdlan is primarily present as a rigid, tightly coiled, triple helix.

Properties

The uniqueness of curdlan is its ability to form thermally irreversible gels upon heating aqueous solutions of it. No ions, pH adjustment, or other solutes are required. Unlike the gels formed by heating solutions of methylcelluloses and hydroxypropylmethylcelluloses (Chapter 8), in general, curdlan gels do not revert to fluids when cooled. This property allows the creation of products shaped and set while hot that do not melt or change shape when either cooled or reheated.

Curdlan is insoluble in room temperature water, in which it can be dispersed without clumping. On raising the temperature, little happens until the temperature reaches about 50°C. Between 50°C (120°F) and 60°C (140°F) curdlan particles hydrate and the viscosity increases rapidly.

Upon heating of aqueous solutions to temperatures above 60°C, curdlan forms two types of gels, distinguished by differences in thermoreversiblity. A thermally reversible gel is formed when dispersions of curdlan are heated to about 65°C (150°F), then cooled. The curdlan will redissolve, forming a solution, when such a gel (termed a low-set gel) is then held at a temperature of about 60°C.

The most common type of curdlan gel used in making food products is termed a high-set gel. High-set gels are formed when dispersions of curdlan are heated to about 80°C (175°F). High-set gels are not thermally reversible. When a low-set gel is disintegrated by shear, it can be reformed by heating and cooling; however, when a high-set gel is disintegrated by shear, it cannot be reformed. The remainder of the discussion of properties refers to high-set gels.

Higher temperatures result in stronger gels. Any temperature above 80°C will produce a completely irreversible gel. However, gel strength continues to increase with increasing temperature up to a temperature of about 130°C (265°F). Gel strength is also a function of concentration (that is, higher curdlan concentrations result in higher gel strengths). Sucrose concentrations above 10% reduce gel strength, while 20% concentrations of glucose and fructose have almost no effect on gel strength. The presence of salt also decreases gel strength. Curdlan gels are reasonably stable to freeze-thaw cycles. The syneresis[2] that occurs can be reduced by addition of a modified starch (Chapter 7).

The elasticity of curdlan gels is intermediate between those of gelatin gels and agar or gellan gels.

Some applications are not based on gelation of the entire system. Rather, curdlan (as a powder) is dispersed throughout the system. Then, during heating, small gel particles are formed. Texture and mouthfeel are affected by this type of process. Increased weight of product (that is, product yield) may also result. All these

[2] Syneresis is the expulsion/separation of liquid from a gel caused by shrinking of the gel (Chapter 5).

attributes are due to curdlan's ability to absorb and hold water (which is not unique to curdlan, but the formation of dispersed small gel particles is somewhat unique).

Uses

Currently, most curdlan is used in Japan. It is claimed to increase juiciness and tenderness in processed meats and to reduce shrinkage when they are frozen. It is also claimed that it will improve appearance, mouthfeel, and texture of frozen dairy products and that it will reduce moisture loss and oil absorption in batter-coated and breaded products. It can be used as a fat mimetic (Chapter 17) to add creaminess to products such as low-fat yogurt, reduced-fat cheeses, and nonfat salad dressings. When used with egg white, it improves the texture and shape retention in noodles for canned soups and canned macaroni products. Tofu noodles containing curdlan retain their shape when boiled or sautéed.

The labeling designation is curdlan.

Dextrans and levans

Dextrans are not used directly as food additives but are related to food and food processing. Dextran is a generic term for a family of glucans made by polymerization of the α-D-glucopyranosyl moiety of sucrose in a reaction catalyzed by the enzyme dextransucrase. The common feature is a preponderance of (1→6)-linked α-D-glucopyranosyl units.

Several microorganisms produce dextrans as exopolysaccharides[3] with ranges of molecular weights and with structures varying from slightly to highly branched. Commercial dextran is biosynthesized by the microorganism *Leuconostoc mesenteroides*.[4] The basic reaction catalyzed by dextransucrase is n sucrose → (α-D-glucopyranosyl unit)$_n$ + n D-fructose. Branches arise from position O3 of the glucosyl units. The degree of branching of commercial dextran is about 5%. About 40% of the side chains are single α-D-glucopyranosyl units; about 45% are two units long, and about 15% contain more than two units. The average molecular weight of commercial dextran can range from 9×10^6 to 5×10^8 (The molecular size can be controlled by varying growth conditions of the organism). Dextran produces relatively low-viscosity solutions, which distinguishes it from other high-molecular weight polysaccharides.

Uses of dextrans in foods have been described and patented, but dextrans are not permitted as food additives in the United States or Europe. (The generally recognized as safe status was removed by the US Food and Drug Administration in 1977 because dextran was not being, and had not been, used as an ingredient.) The principal potential uses of dextran in foods appear to be related to its capacity to prevent crystallization,

[3] Exopolysaccharides are found outside the organism in its medium.

[4] Dextran is produced commercially for medical use.

retain moisture, and provide body, but the fact that it is not being used as an additive indicates that it provides no particularly useful functionality that cannot be provided by another hydrocolloid. Small amounts of dextran, primarily produced by species of *Leuconostoc* and *Lactobacillus,* are likely to be present in fermented products, such as sourdough and yogurt that contained sucrose before fermentation, and can have a noticeable effect on the properties of the product—especially can bacterial growth produce a "ropey" texture in yogurt.

Dextran formation is detrimental to sugar (sucrose) production because dextran inhibits crystallization; increases viscosity and, hence, decreases filterability and heat transfer in crystallizing pans;[5] and results in poor clarification. As a result, steps are taken to minimize dextran formation by contaminating microorganisms. Dextranases may be employed to break down any dextran produced.

Dextran formation is also related to tooth decay (dental caries). In the presence of sucrose, organisms in the mouth (strains of *Streptococcus mutans*) produce dextran which provides adhesion and allows the organism to colonize tooth surfaces. Colonization results in plaque formation.

There are also organisms that use the D-glucosyl unit of sucrose (rather than the D-fructosyl unit) for energy and polymerize the D-fructosyl units to prepare an exopolysaccharide called levan. As with dextrans, the in situ production of levan during sourdough fermentation, for example, will affect dough properties and bread quality. Levans can also be prebiotics (Chapter 17).

Pullulan

Pullulan, another microbial polysaccharide, is not used as a food ingredient, but is of interest as a water-soluble coating and packaging material. Pullulan is a linear, mixed-linkage glucan that can be made into flexible films and fibers much as thermoplastic synthetic polymers are. Water soluble, edible films of pullulan were first proposed as edible pouches for premeasured portions of ingredients or foodstuffs themselves that could be dissolved in hot systems, but the interest shifted to use of pullulan as an edible coating.

Additional resources

Gellan

Giavasis, I., Harvey, L.M., McNeill, B., 2000. Gellan gum. Critical Reviews in Biotechnology 20, 177–211.

Gibson, W., Sanderson, G.R., 2000. Gellan gum. In: Imeson, A. (Ed.), Thickening and Gelling Agents for Food, second ed. Blackie Academic & Professional, London, pp. 119–143.

[5] Industrially, sucrose crystallization takes place in specially designed, heated, and evacuated vessels called pans.

Kremer, F., Lagaly, G. (Eds.), 1999. Physical Chemistry and Industrial Application of Gellan Gum. Progress in Colloid and Polymer Science, vol. 114. C.H.I.P.S, Weimar, Texas.

Sworn, G., 2009. Gellan gum. In: Phillips, G.O., Williams, P.A. (Eds.), Handbook of Hydrocolloids. CRC Press, Boca Raton, pp. 204–227.

Curdlan

Lee, I.-Y., 2002. Curdlan. In: Vandamme, E.J., De Baets, S., Steinbüchel, A. (Eds.), Biopolymers, Polysaccharides I, Polysaccharides From Prokaryotes, vol. 5. Wiley-VCH, Weinheim, Dextrans and Levans, pp. 135–158.

Nishinari, K., Zhang, H., Funami, T., 2009. Curdlan. In: Phillips, G.O., Williams, P.A. (Eds.), Handbook of Hydrocolloids. CRC Press, Boca Raton, pp. 567–591.

Dextrans and Levans

Leathers, T.D., 2002. Dextran. In: Vandamme, E.J., De Baets, S., Steinbüchel, A. (Eds.), Biopolymers, Polysaccharides I, Polysaccharides From Prokaryotes, vol. 5. Wiley-VCH, Weinheim, pp. 299–321.

Park, J.K., Khan, T., 2009. Other microbial polysaccharides: pullulan, scleroglucan, elsinan, levan, alternan, dextran. In: Phillips, G.O., Williams, P.A. (Eds.), Handbook of Hydrocolloids. CRC Press, Boca Raton, pp. 593–614.

Rhee, S.-K., Song, K.-B., Kim, C.-H., Park, B.-S., Jang, E.-K., Jang, K.-H., 2002. Levan. In: Vandamme, E.J., De Baets, S., Steinbüchel, A. (Eds.), Biopolymers, Polysaccharides I, Polysaccharides From Prokaryotes, vol. 5. Wiley-VCH, Weinheim, pp. 351–377.

Carrageenans

13

Chapter Outline

Key information and skills that can be obtained from study of this chapter will enable you to

1. Define and/or identify
   processed Eucheuma seaweed (PES)
   semirefined carrageenan (SRC)
   Philippine natural grade (PNG) carrageenan
   alkali-modified seaweed flour
2. Describe the preparation of carrageenans.
3. Describe and give the physicochemical basis for the preparation of PES/SRC/PNG carrageenan.
4.
   a. List the five members of the family of red seaweed polysaccharides in order of solubility.
   b. List the five members of the family of red seaweed polysaccharides in order of gel strength (at equal concentrations).
   c. Explain the order by discussing the factors involved in determining each list order.
5. Give idealized structures of κ-, ι-, and λ-types of carrageenan.
6. Describe and explain the interaction between κ-carrageenan and locust bean gum (LBG).
7. Explain why there is no synergistic interaction between κ-carrageenan and guar gum.
8. Explain why carrageenans are such effective thickeners of milk products.
9. Explain why κ-carrageenan is effective as a secondary stabilizer of ice cream.
10. Diagram the hypothesized mechanism of gelation of solutions of κ- and ι-types of carrageenan.

Sources and manufacture of carrageenans

More than in the case of any other food carbohydrate, with the probable exception of modified food starch, the term carrageenan on a food label could refer to any of a multitude of products. In other words, the term carrageenan denotes a family of

Carbohydrate Chemistry for Food Scientists. https://doi.org/10.1016/B978-0-12-812069-9.00013-3

polysaccharides, all of which are sulfated polysaccharides composed of 3,6-anhydro-D-galactopyranosyl and/or D-galactopyranosyl units. All carrageenans contain sulfate half-ester groups attached to their monosaccharide units (that is, they are galactan monosulfates). There are three basic types of carrageenan, which may be obtained in relatively pure form or in a mixture with another type(s) (section on Structures in this chapter). All are obtained from red marine algae (red seaweeds of the family Rhodophyceae). Only one of the three basic types of carrageenan is obtained from some species; mixtures of the three general types are obtained from other species. Some plants are harvested from the wild, primarily along the North American coast from Boston to Halifax and Prince Edward Island, the Brittany coast of France, and the coast of the Iberian Peninsula. Other plants are grown in seaweed farms in shallow waters around The Philippines, Indonesia, the east coast of Africa, and Madagascar. Carrageenans constitute 30%–80% of the cell walls of the marine algae in the Rhodophyceae family, the amount depending on species, season, and growing conditions.

Most often, carrageenans are obtained from the appropriate washed seaweed or seaweeds by extraction with solutions of sodium hydroxide or sodium carbonate, which produces the normal commercial product (sodium carrageenate [that is, the extracted polysaccharide(s) in the sodium salt form]). Following filtration and concentration, the product is usually recovered by precipitation with 2-propanol (isopropanol). The resulting precipitate is dried, ground, blended, and often standardized (see below). The product is polymolecular (that is, it is a mixture of molecules with similar but different structures). For labeling purposes, any carrageenan preparation is simply called carrageenan, without regard to source or principal structure.

Also available and used is a product called **PES, SRC, PNG carrageenan,** or **alkali-modified seaweed flour**. To prepare this product, *Eucheuma* species, which yield a κ-type, an ι-type, or a mixture of κ- and ι-type carrageenans (section on Structures in this chapter) are treated with about 2M potassium hydroxide solution. Because the potassium salts of the types of carrageenans found in these seaweeds are insoluble (section on Properties in this chapter), the polysaccharide is neither dissolved nor extracted/removed from the seaweed; rather low molecular weight soluble components are primarily extracted from the plants during the treatment. The remaining seaweed is dried and ground to a powder. The product is, therefore, a composite material that contains the molecules of carrageenan (now as insoluble potassium carrageenate) and other cell wall materials. Calcium salts of these products can be made in the same way.

Two other hydrocolloids, agar and furcellaran (also called Danish agar), are also extracted from red seaweeds and are closely related structurally and in properties to the carrageenans. There are also several genera and species from which agar is extracted, and as with the carrageenans, agar-producing plants are both collected from the wild and grown in seaweed farms. Furcellaran is not obtained pure because the species of seaweed that produce carrageenan grow amongst those that produce furcellaran.

Structures

The term carrageenan denotes a family of sulfated galactans extracted from red seaweeds, rather than a single polysaccharide. The principal structures are termed κ-, ι-, and λ-carrageenans. Carrageenan molecules are linear chains of alternating 3-*O*-substituted β-D-galactopyranosyl units and 4-*O*-substituted α-D-galactopyranosyl units. In κ- and ι-type carrageenans, the latter units contain an internal 3,6 ether bond. These units are termed 3,6-anhydro-α-D-galactopyranosyl units[1] (Fig. 13.1). In all three types of carrageenan molecules, most sugar units have one or two sulfate groups esterified to a hydroxyl group at carbon atoms C2 or C6. Sulfate contents range from 18% to 40% (Table 13.1). Because the sulfate ester groups are esters of a very strong acid, they are always ionized and the carrageenan molecules are always anionic (that is, the pH cannot be lowered enough to remove the charge on the sulfate group).

No commercial carrageenan preparations have the precise chemical structures shown in Fig. 13.1. Because the types of carrageenan are primarily defined by their properties, they should be properly called κ-, ι-, and λ-types but are referred to here as κ-, ι-, and λ-carrageenans as they are in the food industry. The repeating

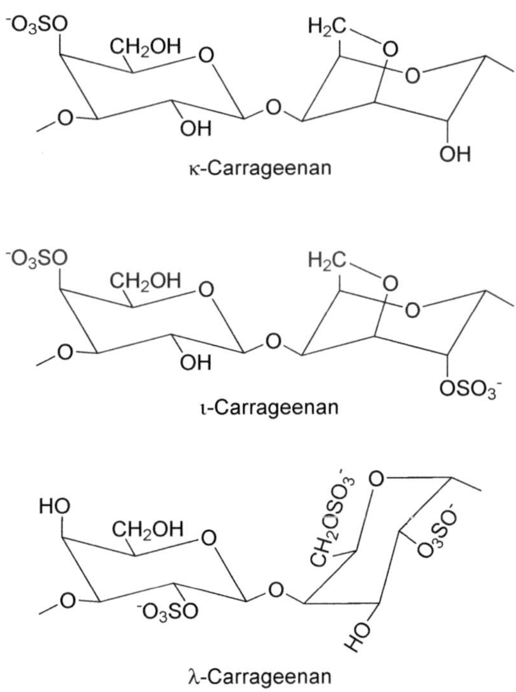

Figure 13.1 Idealized repeating unit structures of κ-, ι-, and λ-type carrageenans.

[1] Anhydro indicates that HOH has been removed between the two hydroxyl groups, forming a cyclic ether (Fig. 13.1).

Table 13.1 Gross structural differences between red seaweed polysaccharides

Hydrocolloid	3,6-AnGal*p*[a] units[b]	Approximate sulfate content (%)
Agarose/agaran	+	0
Furcellaran	+	16–20
κ-Carrageenan	+	18–25
ι-Carrageenan	+	25–34
λ-Carrageenan	−	30–40

[a]3,6-Anhydro-α-D-galactopyranosyl.
[b]λ-Carrageenan contains α-D-galactopyranosyl units in place of 3,6- AnGalp units.

disaccharide units shown in Fig. 13.1 represent the predominate building block of each type. For example, molecules of κ-carrageenan will be composed predominately of the disaccharide unit shown in Fig. 13.1, but there will be microheterogeniety within the polysaccharide chain (that is, some disaccharide repeat units within the chain will differ from those shown). (They may have more or less sulfate half-ester groups or may be missing the 3,6-anhydro ring.) The same is true of ι- and λ-types (that is, all preparations are polymolecular [and polydisperse] [Chapter 5]). So carrageenans, as extracted, are mixtures of polysaccharides with mixtures of these units. (The previous sentence says as extracted because the hot alkaline solutions used for extraction change the structures of the molecules.) Under alkaline extraction conditions, and especially if strong alkaline solutions are used, some units of λ-carrageenan are converted into the type of units in κ- or ι-carrageenans due to nucleophilic displacement of some sulfate groups at the C6 position by alkoxy groups (-RO^-) produced from the hydroxyl groups at C3 positions to create 3,6-anhydro rings (that is, intramolecular ethers between the hydroxyl groups at C3 and C6).

Differences in the presence of 3,6-anhydro-D-galactopyranosyl units and approximate sulfate half-ester contents for the entire family of red algal polysaccharides are given in Table 13.1. In general terms, 3,6-anhydrogalactosyl units are required for gelation. Solubility, gel texture, and protein reactivity are functions of the degree of sulfation (Table 13.2).

Furcellaran is closely related to κ- and ι-carrageenans (the gelling carrageenans). (λ-Carrageenan is classified as a nongelling carrageenan.) The close structural relationships of κ- and ι-carrageenans are shown in Fig. 13.2—this time using the Haworth representation. The gel strengths of furcellaran gels are in between those of agar and those of κ-carrageenan, but closer to the latter.

Properties

Carrageenan products, of which there are very large numbers for different applications, contain different proportions of the three main behavioral types: κ, λ, and ι. The composition and properties of carrageenan preparations depend on the species

Table 13.2 Relation between structure and some properties of red seaweed polysaccharides

Gum	3,6-Anhydro rings	Sulfate[a] half-ester content	Solubility[a,b]	Gelling ability and gel strength[a,c,d]
Agar	+			↑
Furcellaran	+			
κ-Carrageenan	+			
ι-Carrageenan	+			
λ-Carrageenan[e]	−	↓	↓	

[a]Arrows point in the direction of increases in the constituent or property.
[b]Of the sodium salt of the carrageenan in water.
[c]Water gels.
[d]See Table 13.3.
[e]Nongelling.

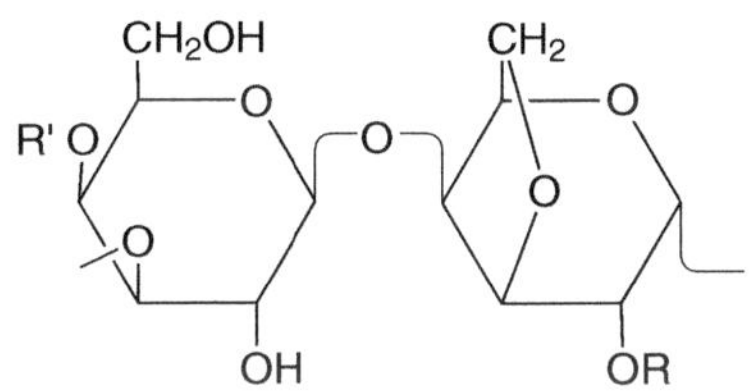

$R = R' = SO_3^-$ ι-Carrageenan

$R = H; R' = SO_3^-$ κ-Carrageenan

$R = H; R' = 50\%\ SO_3^-$ 50% H Furcellaran

Figure 13.2 Idealized repeating unit structures of ι- and κ-carrageenans and furcellaran.

collected (or grown), growth conditions, and treatment during production. Generally, commercial preparations are not pure κ-, ι-, or λ-types because carrageenan isolates are often blended (or the seaweeds may be blended before extraction of the carrageenan). Properties that can be controlled by blending or processing include dispersibility, rate of hydration, solubility, solution viscosity, gel strength, and interaction with protein. Sugar is often added to standardize the product (usually to a specified gel strength at a given concentration) so that the user does not have to adjust the amount added to a formulation for each different lot of the product. Food-grade salts (most often potassium salts with κ-carrageenan) are frequently added, again to standardize gel or viscosity characteristics for specific applications. Properties of the three types of carrageenans are summarized in a generalized form in Table 13.3.

Table 13.3 Summary of general properties of carrageenans

Property	Kappa-types	Iota-types	Lambda-types
Solubility			
In hot water	Soluble above 60°C (140°F)	Soluble above 60°C (140°F)	Soluble
In cold water	Na^+, K^+, and Ca^{2+} salt forms are insoluble. Particles of Na^+ salt form swell.	Na^+ salt form soluble. Ca^{2+} salt form produces thixotropic dispersions	Soluble
In cold milk	Particles of Na^+ salt form swell. All salt forms are insoluble.	Insoluble	Soluble
In hot milk	Soluble	Soluble	Soluble
In concentrated sugar solutions	Soluble in hot solutions	Swell, but difficultly soluble	Soluble in hot solutions
In concentrated salt solutions	Insoluble	Soluble in hot solutions	Soluble in hot solutions
Gelation			
Effect of cations on	Strongest gels with K^+	Strongest gels with Ca^{2+}	Nongelling
Type of gel formed	Firm, brittle	Soft, elastic	Nongelling. Viscous solutions formed
Synergism with locust bean gum	High	High	None
Stability			
At pH 7 and above	Stable	Stable	Stable
In acidic systems	Undergoes hydrolysis when a solution of it is heated. Stable in gels	Undergoes hydrolysis when a solution of it is heated. Stable in gels	Somewhat more stable
Of gels to freeze-thaw conditions	Unstable	Stable	Unstable
Syneresis[2] of gels	High	Low	Nongelling

All carrageenan products can be dissolved in water to form highly viscous solutions with pseudoplastic behavior. The solution viscosity is quite stable over a wide range of pH values because the sulfate half-ester groups are always ionized, giving the molecules a negative charge. Coulombic repulsion maintains the molecular segments at a maximum distance from each other, tending to produce a linear arrangement with an overall configuration of an expanded coil. Coulombic repulsion also prevents interchain segmental collisions. Nevertheless, segments of molecules of κ- and ι-type carrageenans can, and do, exist as double helices of parallel chains formed from right-handed, threefold helical stretches.

Gel formation is the most important attribute of κ- and ι-type carrageenans (the gelling carrageenans). Gels may be formed in either water or milk. Various blends of carrageenans in the presence of various cations (K^+, Ca^{2+}, proteins) will form gels that can be either clear or turbid, either rigid or elastic, either tough or tender, and heat stable[3] or thermally reversible (although never melt-in-the-mouth types). Factors affecting gelation of carrageenan solutions are the type of carrageenan product used, concentration of the carrageenan product used, the type of cation(s) (including protein) present, the concentration of cation(s), the temperature to which the "solution" is heated before cooling, the rate of cooling, and the presence of other ingredients.

Potassium ions are the best gelling agent for κ-carrageenans. When a 4% solution of a κ-type carrageenan is cooled below 45–60°C (115–140°F) in the presence of potassium ions, a rigid, brittle gel forms. Calcium ions are less effective in effecting gelation, but potassium and calcium ions together produce a strong gel. Gels made from κ-type carrageenans are the strongest and stiffest of the carrageenan gels but, like others, tend to synerese[2] as junction zones extend within the structure, shrinking and squeezing water out of the gel. The presence of other hydrocolloids retards syneresis.[2] Particles of the sodium salt of κ-carrageenans (the most soluble form of κ-carrageenan) will swell but not dissolve in cold water.

ι-Carrageenans are a little more soluble than are the κ-carrageenans (in that the sodium salt form is soluble in unheated water) because they are a little more sulfated. ι-Carrageenans gel best with calcium ions. The resulting gels are soft and flexible (elastic or resilient), have good freeze-thaw stability, and do not synerese, presumably both because ι-type carrageenans are more hydrophilic than are κ-types and because they produce fewer junction zones than do κ-carrageenans. Both the freeze-thaw stability and lack of syneresis are unique properties among polysaccharide gels. Both κ- and ι-types exhibit synergistic interactions with starches, but the synergism is more pronounced with ι-carrageenans.

Gelation occurs because, as solutions of κ- or ι-carrageenans are cooled, the linear molecules form double or triple helices of restricted length due to the presence of structural irregularities (such as the absence of a α-D-galactopyranosyl unit containing a 3,6-anhydro ring). The linear helical portions then associate to form a rather firm,

[2] Syneresis is the expulsion/separation of liquid from a gel caused by shrinking of the gel (Chapter 5).

[3] Up to a certain temperature depending on type(s) and amount(s) of cation(s) present.

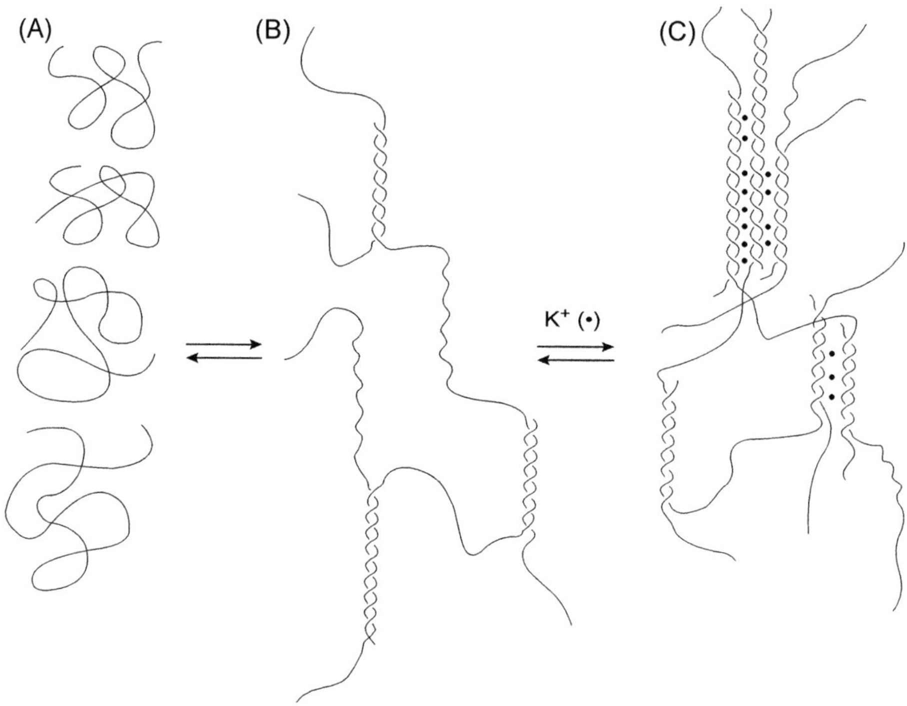

Figure 13.3 A representation of the hypothesized mechanism of gelation of κ- and ι-type carrageenans. The polysaccharide molecules in a hot solution are in a coiled state (A). As the solution cools, they intertwine in double-helical structures (B). On further cooling, the double helices are believed to nest together with the aid of potassium or calcium ions (C).

three-dimensional, stable gel network in the presence of the appropriate cation (Fig. 13.3). All salts of λ-carrageenans are soluble and nongelling. Blending of gelling and nongelling types to create the desired gel strength and texture is commonly done. Also common are preparations containing different proportions of sodium, potassium, and calcium ions to provide products with the end-use properties desired.

Under conditions in which double-helical segments occur, carrageenan molecules, particularly those of the κ-type, form junction zones with the naked segments of LBG to produce gels (Figure. 11.5) that are still rigid and brittle and undergo syneresis, but are stronger, more elastic, and undergo less syneresis[2] than those made with a κ-type carrageenan alone. Only one-third as much of the hydrocolloid mixture is needed to form a gel equivalent to that formed using a κ-carrageenan alone.

Carrageenans are quite susceptible to acid-catalyzed hydrolysis, undergoing rapid and extensive loss of viscosity and gelation potential when solutions below pH 5 are heated. The loss of viscosity and gelation potential is primarily due to cleavage of the (1 → 3) glycosidic linkages (except when an O2 sulfate group is present).

Solutions of PES/SRC/PNG are turbid (because of the presence of insoluble cell wall substances) and have a brown color. PES/SRC/PNG products will produce gels of high gel strength because of their high content of κ- and/or ι-type carrageenans.

Uses

Commercial carrageenans can be viscosity builders, gelling agents, and/or stabilizers. Carrageenan products are most often used because of their ability to form gels with milk and water (Box 13.1). Blending provides a wide range of products that are standardized with various amounts of sucrose, glucose (dextrose), buffer salts, or gelling salts, such as potassium chloride, so that the user always obtains the same performance from the same product. The available commercial products form a variety of gels: gels that are may be either clear or turbid (with water), rigid or elastic, tough or tender, heat-stable or thermally reversible, do or do not undergo syneresis, and are more or less freeze-thaw stable. Carrageenan gels do not require refrigeration because they do not melt at room temperature (Note in Box 13.1 the large number of milk-based products and the similarity of applications for carrageenans and LBG [Chapter 9] because the two are often used together.).

A useful property of carrageenans is their reactivity with proteins, particularly those of milk. κ-Carrageenans link together casein micelles of milk by means of charge–charge interactions, forming a weak,[4] thixotropic (Chapter 5), nonthermally reversible, pourable gel. The thickening effect of κ-carrageenans in milk is 5–10 times greater than it is in water. This property is used in the preparation of chocolate milk where the weak gel structure prevents settling of the cocoa particles, thereby maintaining a uniform distribution, whereas the shear-thinning behavior allows the product to be poured and swallowed. Such stabilization requires a hydrocolloid concentration of only about 0.025%. This property is also applied in the preparation of evaporated milk, ice cream, other milk-based desserts, infant formulas, freeze-thaw stable whipped cream, many other dairy products, and emulsions in which milk fat is replaced with a vegetable oil. Carrageenans are often used as a secondary stabilizer in ice cream to prevent whey separation during freezing. κ- and ι-carrageenans can gel milk at 0.1%–0.2% concentration. Making flan-type milk gels is also the principle use of furcellaran. In evaporated milk, carrageenans stabilize the alpha-casein and prevent creaming at concentrations as low as 0.005%.

Carrageenans of the ι-type are more soluble than are the κ-types, but with both, only the sodium salt forms are soluble in cold water. While water gels of κ-type carrageenans are firm, rigid, and characterized by syneresis,[2] ι-carrageenans form elastic, syneresis-free, thermally reversible gels that are stable to repeated freeze-thaw cycling. They are used to make milk gels/puddings. Blending is common, and blends of ι- and κ-type carrageenans are used in eggless custards and flans, instant whipped desserts, water-dessert gels that do not require refrigeration (as opposed to gelatin gels), and whipped toppings. Gelled water desserts containing fruit concentrates or flavors, usually at a pH of about 3.8, contain 0.5%–1.0% carrageenan. Acid is generally added just before rapid cooling to minimize degradation of the carrageenan molecules. Carrageenan is often used in combination with LBG (Chapter 9) to yield a fast-setting, temperature-stable gel.

[4] A weak gel is one that does not support its own weight (Chapter 5).

Box 13.1 Typical Products Containing a Carrageenan Product

- Bakery products
  - Cheesecake
  - Cookies
  - Fillings
- Bars
  - Cereal
  - Diet
  - Granola
- Beverages
  - Nutritional/diet
  - Soy
- Confections
  - Chewy candy
  - Chocolate bars
  - Fruit gushers
- Coffee creamer, nondairy
- Dairy products
  - Bakery fillings
  - Cheese dressings
  - Cheesecake
  - Chocolate milk
  - Chowder
  - Coffee cream
  - Coffee creamer
  - Cottage cheese, including low-fat
  - Cream cheese
  - Dips
  - Evaporated milk
  - Flan
  - Frozen cheese
  - Frozen cheese products
    - Cheesecake
    - Lasagna
    - Pizza
  - Frozen desserts
  - Frozen novelties
  - Frozen yogurt
  - Ice creams, including fat-free
  - Infant formulas
  - Milk shakes
  - Nutritional/diet beverage
  - Processed cheese
  - Protein shake
  - Puddings
  - Yogurt, fat-free
  - Whipped cream, pressurized
  - Whipped toppings
  - Whipping cream
- Egg substitute
- Frozen meals
- Gelled desserts
- Gel sticks and snacks
- Imitation crab meat, frozen
- Mayonnaise, low-fat
- Meat products
  - Luncheon meats
  - Sausages
  - Turkey breast slices
  - Weiners
- Meatless frankfurters
- Mixes, dry
  - Breakfast drink
  - Brownie
  - Cake
  - Cappuccino, instant
  - Chocolate milk
  - Chai tea latte
  - Diet dairy
  - Dips
  - Hot cocoa
  - Pancake
  - Pie filling
  - Pudding, including low-fat
  - Soup
- MRE,[a] microwaveable
- Pudding snacks
- Salad dressing, low-calorie
- Sauces
  - Fudge
  - Horseradish
  - Stir fry
- Soups, low-fat, low-calorie
- Vegetable patties

[a]Meals ready to eat.

Carrageenans also interact with proteins other than milk proteins, but while interactions with κ-casein occur above the protein's isoelectric pH value (pI), other proteins must be below their isoelectric pH values for complexes to form. Thus, the natures of carrageenan–protein interactions are a function of the pH of the system and the pI of the protein, as the anionic nature of the carrageenan molecules are unaffected by pH. In every case, the protein functions as a polycation in its interaction with carrageenan molecules. Other factors that affect carrageenan–protein interactions are the type of carrageenan product, the type of protein, the rate of cooling, and the presence of other cations.

The synergistic effect between κ-carrageenan and LBG (Chapter 9) produces gels with greater elasticity, greater gel strength, and less syneresis[2] as compared with gels of potassium κ-carrageenate alone. As compared with κ-carrageenan alone, the carrageenan–LBG combination provides greater stabilization and air bubble retention (called overrun) in ice cream but also a little too much chewiness; so guar gum is added to increase mix viscosity and to soften the gel structure. The combination is also used in canned pet foods to provide body, to suspend fruit in yogurt, and to stabilize low-fat yogurt. Tart fillings and glazes can be gelled with a mixture of a κ-type carrageenan (1%–2%) and LBG (0.7%–1%). High gelation and melting temperatures are obtained upon addition of a potassium salt. Freeze-thaw stability is improved by adding carboxymethylcellulose (CMC) (Chapter 8). Konjac glucomannan (Chapter 10) is more synergistic with κ-carrageenan than is LBG. At the optimum ratio (as with LBG about 1:1 w/w), the combination produces about double the gel strength of a κ-carrageenan–LBG gel of the same concentration.

In summary of what has been discussed here and in previous chapters, most ice creams and related milk shake and frozen dessert products contain three to four different hydrocolloids. Carboxymethylcellulose (CMC) (Chapter 8) is the primary stabilizer. κ-carrageenan is added as a secondary stabilizer to prevent whey separation. Addition of LBG (Chapter 9) converts the mix to a weak gel structure via its interaction with the carrageenan. Guar gum (Chapter 9) is added to soften and make the gel structure formed with LBG and κ-carrageenan less chewy, providing a nongrainy (smooth) texture. The hydrocolloids increase mix viscosity, which increases air incorporation (overrun). They also enhance the body of the finished product by improving meltdown.[5] They inhibit shrinkage of the ice cream and improve stability (that is, they protect the ice cream against thermal stress) by reducing the rate of ice crystal formation and growth. Stabilization can also be achieved with blends of xanthan (Chapter 11), LBG, and guar gum, which control the viscosity of the mix and provide

[5] Meltdown refers to the rate of loss of structure and shape when ice cream is warmed to room temperature. When ice cream is warmed, two things happen: (1) the ice crystals melt, and (2) the foam structure collapses. The ice crystals can melt without a complete loss of shape due to stability of the foam provided by the milk proteins and emulsified fat globules, the degree of which stability being what is indicated by the meltdown test. Ideally, an ice cream should melt rather quickly, and the melt should have characteristics similar to those of the original mix before freezing.

tolerance to heat shock.[6] Reduced butterfat products often contain microcrystalline cellulose (Chapter 8). Mixes containing 3–5 of these polysaccharides plus other ingredients (such as emulsifiers [mono- and diglycerides]) are prepared and sold as a single component to be added to milk of various fat contents. Flavors may also be included or may only need to be added by the ice cream maker.

Reactivity with protein is also useful in the preparation of sliceable meat products. Cold hams and poultry rolls take up 20%–80% more brine when the brine contains 1%–2% of a κ-type carrageenan and slicing is improved. In the preparation of luncheon meats, turkey rolls, etc., a brine containing a κ-type carrageenan (insoluble) is injected into or tumbled with the meat. The hydrocolloid goes into solution during cooking. Normal use is generally 0.25%–0.50%. Addition of 0.2%–1.0% of carrageenan alone or in combination with LBG (Chapter 9) improves the texture of gelled broths of canned meat and fish products. Carrageenan coatings on meats such as hams can serve both as a barrier and a carrier for seasonings and flavors. Meat emulsions, such as those used to make wieners and sausages, often contain κ- or ι-type carrageenans to bind water and, thereby, maintain the water content and softness of products during cooking. Meat analogs from casein and vegetable proteins in fiber form may contain 1% carrageenan. Addition of a κ- or ι-type carrageenan in the Na^+ salt form to low-fat ground beef maintains texture and hamburger quality. Normally, fat serves the purpose of maintaining softness, but because of the binding power of carrageenan for protein and water, it can be used to replace, in part, this natural animal fat functionality in lean products. PES/PNG and sodium ι-carrageenate may be used in meat applications.

The nongelling λ-carrageenans will thicken milk systems in the presence of phosphate salts in both heated and nonheated applications. They are used as emulsion stabilizers in products such as nondairy coffee creamers, instant breakfast drinks, dry-mix hot cocoa, milk shakes, and whipped cream. A κ-, λ-carrageenan blend is used in milk shakes.

Labels of products containing κ-, λ-, or ι-type carrageenans may read carrageenan, chondrus extract, or Irish moss extract. The label designation of PES/SRC/PNG carrageenan may be carrageenan.

Agar: structure and uses

Agar is a thermoreversible gelling agent composed of two polysaccharides. Agarose (or agaran) is a linear chain of 3-*O*-substituted β-D-galactopyranosyl units joined by (1 → 4) linkages to 3,6-anhydro-α-L-galactopyranosyl units. (Note that the anhydrosugar in agar has the L configuration.) Agarose is the gel-forming component. Agaropectin is a branched, nongelling component of agar. Small amounts of sulfate half-ester groups and pyruvate 3,6-cyclic acetal groups (Chapter 2) may be present.

[6] The heat shock test is a specific test that measures the stability of the ice cream to repeated freeze-thaw treatments (−20 and 20°C [−4 and 68°F]) for prescribed times.

Considerable amounts of methyl ether groups are also present. These structural characteristics make agar the least hydrophilic and least water soluble of the red seaweed polysaccharides. Usually, agar can only be dissolved in water at 100°C (boiling) or higher temperatures, although preparations that hydrate and dissolve at about 80°C are available (with and without standardizing materials).

The principal use of agar is in bakery icings and frostings because it is compatible with large amounts of sugar and its products neither melt at high storage temperatures (such as might be found in a delivery truck) nor stick to the packaging material. It is also used in light sour cream, no-oil salad dressings, and yogurt. Its use is limited by its high price. Agar may be designated agar or agar—agar on product labels.

Additional resources

Structures and Manufacture

Armisén, R., Galatas, F., 2009. Agar. In: Phillips, G.O., Williams, P.A. (Eds.), Handbook of Hydrocolloids, second ed. CRC Press, Boca Raton, pp. 82—107.

Imeson, A., 2009. Carrageenan and furcellaran. In: Phillips, G.O., Williams, P.A. (Eds.), Handbook of Hydrocolloids, second ed. CRC Press, Boca Raton, pp. 164—185.

van de Velde, F., De Ruiter, G.A., 2002. Carrageenan. In: Vandamme, E.J., De Baets, S., Steinbüchel, A. (Eds.), Biopolymers, Polysaccharides II, Polysaccharides from Eukaryotes, Vol. 6. Wiley-VCH, Weinheim, pp. 245—273.

Distribution of Carrageenans

Craigie, J.S., 1990. Cell walls. In: Cole, K.M., Sheath, R.G. (Eds.), Biology of the Red Algae. Cambridge University Press, Cambridge, pp. 221—257.

Dairy Applications

Hansen, P.M.T., 1993. Food hydrocolloids in the dairy industry. In: Nishinari, K., Doi, E. (Eds.), Food Hydrocolloids. Plenum Press, New York, pp. 211—224.

Analysis of Carrageenans

Roberts, M.A., Quemener, B., 1999. Measurement of carrageenans in food: challenges, progress, and trends in analysis. Trends in Food Science & Technology 10, 169—181.

Algins/Alginates

14

Chapter Outline

Key information and skills that can be obtained from study of this chapter will enable you to

1. Define or describe
   M/G ratio
   M block
   G block
   propylene glycol alginate (PGA)
   egg-box model
2. Give the source of algins/alginates.
3. Explain the difference between alginic acid and an alginate.
4. (a) Describe the impact of the D-mannuronic acid-L-guluronic acid ratio on the gelling ability and gel strength of alginates, and (b) give the molecular explanation for the phenomenon.
5. (a) Describe the effect of lowering the pH of sodium alginate solutions from 6.5 to 3.5, and (b) give the molecular explanation for the phenomenon.
6. Explain the behavior of solutions of sodium alginate on addition of calcium ions. Explain why this phenomenon occurs.
7. Describe the differences between diffusion setting, internal setting, and setting by cooling for the preparation of calcium alginate gels.
8. Describe the preparation of propylene glycol alginates (PGAs).
9. Describe in molecular terms why PGA (a), in contrast to unmodified alginates, can be used in low pH systems, (b) can be used in milk systems, and (c) has the ability to stabilize emulsions and foams.

Sources and manufacture

Commercial algin is a salt, usually the sodium salt, of alginic acid, a linear poly(glycuronic acid) that constitutes 18%–40% of the dry weight of brown marine algae (brown seaweeds) (family Phaeophyceae). The major source of

Carbohydrate Chemistry for Food Scientists. https://doi.org/10.1016/B978-0-12-812069-9.00014-5

algin/alginates in the United States has been the giant kelp, *Macrocystis pyrifera*, growing in the Pacific Ocean, near the shore, from just north of San Diego south into costal areas of Mexico. This seaweed, which attaches to the rocky seafloor, can reach a length of more than 40 m (125 ft). It has carbon dioxide-filled bladders of about 2.5 cm (1 in.) in diameter along the strand, which float it in an upright position so that it can receive sunlight. Under optimal conditions, fronds[1] will grow up to 60 cm (2 ft) per day and even under less than optimal conditions will grow 30–35 cm (12–14 in.) per day. Harvesting is done by a large ship that moves through the floating kelp cutting the plants about 1 m below the surface and continuously lifting the cut kelp by means of a conveyor into the ship. *Macrocystis* species also grow along the coasts of Chile, Peru, and Argentina mixed with species of *Lessonia*, another brown algae containing algin, and in cooler waters along the coasts of South Africa, Australia, New Zealand, and China, mixed with *Eklonia* species. Two other species of algin-producing brown algae (namely *Ascophyllum* species and *Laminaria* species) grow in cool water along the coasts of Northern Europe, the British Isles, and Eastern North America. Seaweeds of the genus *Ascophyllum* are attached to rocks near a shore. They grow to a length of 1-1.5 m (3-5 ft) and are harvested by hand or by using a small mechanical harvester. Algins obtained from different species of brown algae in different locations differ from each other in structures and properties.

Harvested seaweed is first washed. Then, the alginate is extracted with a sodium carbonate solution, which converts the polysaccharide into its soluble sodium salt (sodium alginate). The polysaccharide is then precipitated from the extract by addition of a solution of calcium chloride, which converts it into an insoluble calcium salt. The precipitate is washed free of solubles. Addition of hydrochloric acid converts the preparation into alginic acid, which is also insoluble. Finally, the alginic acid is neutralized and converted into a monovalent ion salt—most often into sodium alginate. Lesser amounts of potassium and ammonium alginates are produced. Salts with monovalent cations are soluble in water.

Structures

Alginic acid is composed of β-D-mannopyranosyluronic acid (M) and α-L-gulopyranosyluronic acid (G) units. The relative amount of each monomer unit is expressed as the **M/G ratio** of the preparation. M/G ratios vary from less than 0.5 to almost 2 in commercial sources of alginates. The two monomeric units occur in homogeneous regions composed of only one of the two units and in regions containing both units. Segments containing only D-mannuronopyranosyl units are referred to as **M blocks** and those containing only L-guluronopyranosyl units as **G blocks**. MG blocks contain both units in a mixed arrangement.

[1] Fronds are the leaflike branches.

D-Mannuronopyranosyl units are in the $^4C_1$ conformation (Chapter 1), whereas L-guluronopyranosyl units are in the $^1C_4$ conformation, which gives the different blocks quite different chain configurations (Figs. 14.1–14.3). Different percentages of the different block segments cause alginates from different species of brown algae to have different properties.

At least seven different brown seaweed species from different parts of the world are used to obtain alginates. The more commonly used ones vary in L-guluronic acid content from about 40 to about 70% and in G block content from about 18 to about 60%. Alginates with high G block contents produce gels of high strength with calcium ions. Blending of alginates from different species is done to create preparations with desired properties. Average degrees of polymerization from 80 to 1000 (MW 16,000–2,00,000) have been reported.

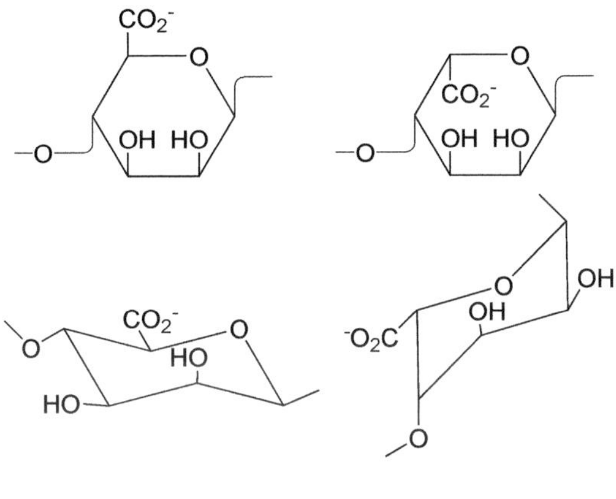

βMan*p*A unit αLGul*p*A unit

Figure 14.1 Ionized units of β-D-mannopyranosyluronic acid (βMan*p*A) in both the Haworth projection and the $^4C_1$ conformation and of α-L-gulopyranosyluronic acid (αLGul*p*A) in both the Haworth projection and the $^1C_4$ conformation.

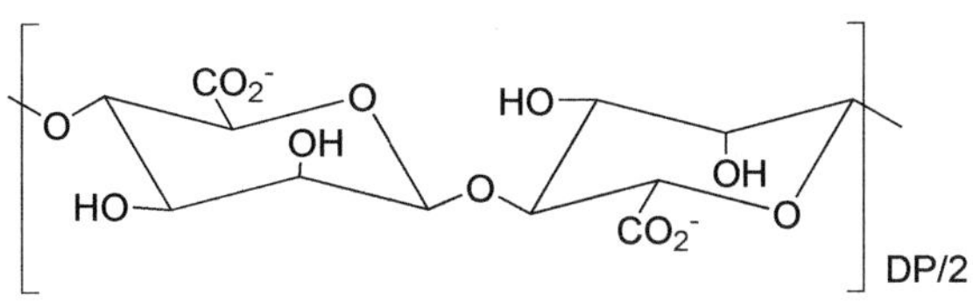

Figure 14.2 A segment of an M-block region in the ionized form. Note the similarity of this structure to that of cellulose (Chapter 8) and of the mannan backbone of the galactomannans (Fig. 14.1) and the flat ribbon-like structure formed by the equatorial→equatorial glycosidic bonds.

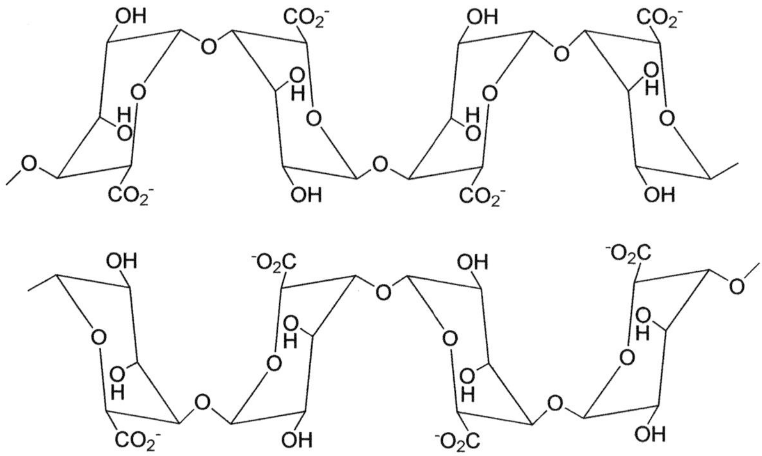

Figure 14.3 Segments of G-block regions. Note the pleated conformations of the chains formed by the axial→axial glycosidic bonds and the "holes" that complex calcium ions. Compare with Fig. 15.3.

Propylene glycol alginates (PGAs) are made by reacting moist alginic acid with propylene oxide until 50%–90% of the carboxyl groups are esterified. PGA products can vary in degree of esterification and molecular weight.

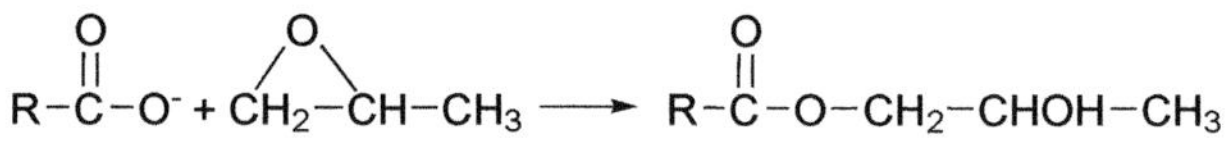

Where R–COO$^-$ = a unit of D-mannopyranosyluronic acid or L-guluronopyranosyluronic acid

Properties

Alginate solutions are slightly pseudoplastic (Chapter 5). Addition of very small amounts of calcium ions will increase viscosity. Slightly more will make alginate solutions thixotropic (Chapter 5). And even more will convert alginate solutions into permanent gels.[2]

Sodium alginate solutions gel on addition of calcium ions which coordinate with hydroxyl and carboxylate groups on adjacent chains (Fig. 14.4) and partially fill the cavities formed between parallel G-block chain regions. These cavities contain two carboxylate and two hydroxyl groups, one each from each chain (Fig. 14.5). In fact, calcium ions connect four poly(L-glucuronic acid) chains and four water molecules in an octahedral coordination. The result is a junction zone that has been called an

[2] Presented is the classic explanation for gelation of alginates. It has been concluded from the results of several research projects that this model (the egg-box model) is not exactly correct. However, the egg-box model effectively explains the behaviors of alginates from different sources.

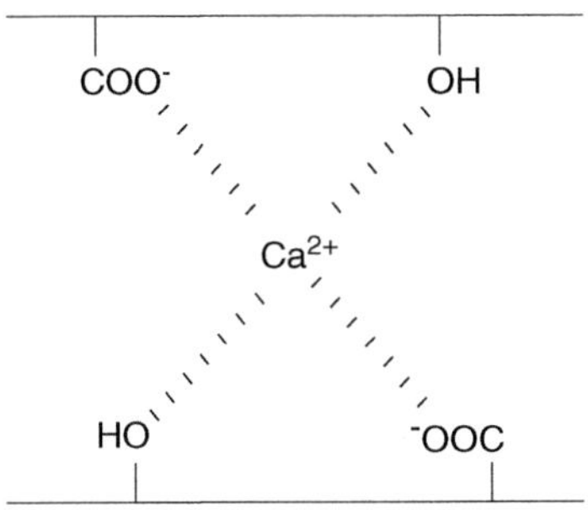

Figure 14.4 The coordination of a calcium ion with hydroxyl and carboxylate groups on adjacent G-block chains.

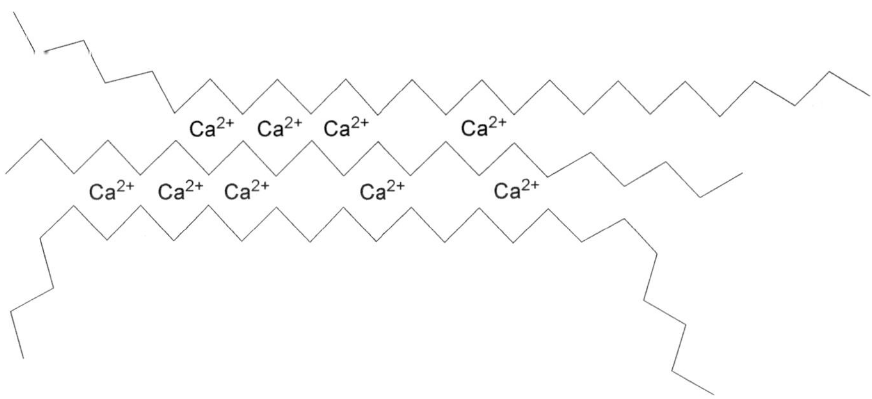

Figure 14.5 A representation of the proposed formation of a junction between two alginate molecules promoted by calcium ions.

egg-box arrangement with the calcium ions being likened to eggs in the pockets of an egg carton. Strength of the gel depends on the content and size of the G blocks in the alginate used and the concentration of calcium ions. Because of the effect of cross-linking of alginate molecules by polyvalent ions, such as calcium ions, the hardness of the water used to dissolve algin has an important effect on solution viscosity.

Because alginate molecules are salts of weak carboxylic acids with pK_a values of 3.38 (ManA) and 3.65 (LGulA), the extent of carboxyl group ionization decreases as the pH is decreased and is greatly reduced at pH values of 3 or less. As the pH approaches and drops below the pK_a values of the carboxylate groups (-COO^-), more and more of them are converted into uncharged carboxylic acid (-COOH) groups. This loss of ionization at low pH values means that the alginate chains lose their negative charges, giving the molecules more and more of the characteristics of long neutral molecules as the pH drops, which allows intermolecular associations to develop. As intermolecular associations increase, the viscosity of the solution increases. Finally, when the pH is low enough, alginic acid precipitates from solution because of extensive intermolecular associations.

Solutions of propylene glycol alginate (PGA) are thixotropic (Chapter 5) and much more tolerant of low pH values and polyvalent cations, including calcium ions and proteins, than are solutions of nonesterified alginates because esterified carboxyl groups cannot ionize. Hence, propylene glycol alginates have much less negative charge than do alginate salts. The propylene glycol group also introduces a bulge on the chain that blocks interchain associations, giving stability to PGA solutions. Because propylene glycol alginates can tolerate calcium ions, they can be used in dairy products. The hydrophobic propylene glycol groups also give the molecules interfacial activity (that is, emulsifying, emulsion-stabilizing, and foaming properties). For example, PGA will stabilize beer foam and, as described in Chapter 11, it is often used in combination with xanthan in oil-in-water, pourable salad dressings.

Physical properties that can vary between specific commercial preparations are powder particle size (a characteristic of all hydrocolloids), particle form (granular vs. fibrous), color, odor, load of microorganisms (a characteristic of most hydrocolloids), and texture profiles (stress–strain relationships) of solutions and gels. Molecular characteristics that can vary are M/G ratio, G and M block contents, molecular weight (viscosity type), salt type (cation present), calcium content, and degree of esterification (of PGA). The presence or absence of other substances, such as sequestrants[3] of calcium ions, is another variable. Sequestrants that may be added to tie up calcium ions are sodium hexametaphosphate, sodium tripolyphosphate, sodium hydrogen pyrophosphate, and sodium citrate.

Uses

Alginates can be used to provide viscosity at low concentrations. Higher viscosity (under low shear conditions) can be achieved by adding small amounts of calcium ions to the alginate solution. Thickening action, and particularly thixotropic behavior brought about by low concentrations of calcium ions, tends to hold insoluble particles in suspension.

When PGA is used, the calcium ion cross-linking of chains that occurs through the remaining carboxylate groups produces thickening without producing a weak gel. Propylene glycol alginates with less than 60% esterification produce thixotropic solutions. Their ability to produce thixotropic solutions decreases with increasing esterification. The mild interfacial activity of PGA, which is imparted by the somewhat hydrophobic ester groups, stabilizes oil-in-water emulsions.

The most extensive use of alginates is to produce gels. Calcium alginate gels are obtained by diffusion setting, by internal setting, and by setting by cooling. All three processes are used to prepare structured foods. In diffusion setting, calcium ions first contact the surface of a system containing sodium alginate and diffuse from the outside in. Two examples of products of diffusion setting are structured pimento strips and structured onion rings. In the production of pimento strips for stuffing green olives, pimento puree is mixed with water containing a small amount of guar gum (Chapter 9), which thickens the diluted puree and reduces shrinkage of the final strip. Then, sodium alginate is added. The mixture is applied to a conveyor belt as a sheet.

[3] Syneresis is the expulsion/separation of liquid from a gel caused by shrinking of the gel.

Spraying the sheet with calcium chloride solution sets the surface. The moving belt is then passed through a calcium chloride setting bath, where more calcium ions diffuse into the sheet, gelling it even more. The final sheet is tough, flexible, and resilient. Equilibration for several days in a solution of sodium and calcium chlorides allows the alginate gel to undergo syneresis[4] and shrink to an acceptable level of consistency. The sheet is then cut into strips, which are bent double and stuffed into pitted green olives by a machine.

Structured, frozen onion rings are made in much the same way from minced onions or rehydrated onion flakes, sodium alginate, salt, and flavor. The resulting paste is formed into rings and gelled by immersion in a setting bath of 3%–5% calcium chloride. After setting, the rings are battered and fried.

Products made by diffusion setting may have liquid centers. Whether or not they do is controlled by the time they are in the calcium chloride bath because the calcium chloride solution will diffuse relatively slowly through the outer gel layer that is formed on initial contact with calcium ions. Liquid centers are useful in such products as structured blueberries.

Internal setting is used to produce structured fruit (such as apple, peach, and pear) pieces. Internal setting involves slow release of calcium ions within the mixture. In the preparation of fruit pieces, a two-mix system is employed. One mix contains sodium alginate and an insoluble calcium ion source such as dicalcium phosphate or calcium sulfate dihydrate. (No calcium alginate is formed because the calcium salt is insoluble.) The second mixture contains minced fruit, sequestrant,[3] and acid. The two mixtures are then mixed and allowed to stand under shear-free conditions in their final container or shaped form on a conveyer belt. Setting occurs due to the controlled release of calcium ions by the acid, resulting in partial cross-linking of alginate chains. The purpose of the sequestrant is to tie up the calcium ion released by the acid so that the product does not set too fast (that is, it delays setting until the mixture is transferred to the mold). Use of dicalcium phosphate results in a faster release of calcium ions and more rapid gel development, which may be undesirable. Slow release of acid, for example, via use of D-glucono-δ-lactone (GDL; Chapter 2), is also effective. The specific ingredients to be used are a function of the product being made and the equipment used.

Setting by cooling involves dissolving a calcium salt, a slightly soluble acid (such as adipic acid), and a sequestrant in hot water and allowing the mixture to set on cooling. The calcium salt may be calcium citrate, calcium gluconate, calcium lactate, or dicalcium phosphate. Which one is used is often a function of the pH of the mixture. It may also be a matter of regulation. GDL may be used in place of a slowly dissolving acid to slowly lower the pH. In this process, calcium ions are released at a moderate rate when the mixture is heated, bringing about a low degree of cross-linking. Sufficient thermal energy is provided by the hot water that the alginate chains form only weak, temporary junction zones. Gelation does not occur until cooling slows polymer

[4] Sequestrants are substances that form chelate complexes with divalent or trivalent cations. In this context, sequestrants are substances that chelate with calcium(II) ions and prevent them from combining with an alginate, which would otherwise form a gel in the presence of the calcium ions.

chain movement. Gels produced in this way exhibit little or no syneresis,[4] allowing the process to be used to make structured fruit pieces. The gels are also heat stable, so those containing fruit can be used for pie fillings that remain stable through pasteurization and cooking.

Unlike gelatin gels, alginate gels are not thermoreversible and can be used as dessert gels in warm climates, but also unlike gelatin gels, they do not melt in the mouth. In fact, no polysaccharide gels have this property.

An alginate solution whose pH has been lowered so that some of the carboxylate groups (-COO$^-$) are converted into carboxylic acid (-COOH) groups, with or without addition of some calcium ions, is employed in the preparation of soft, thixotropic, nonmelting gels.

Strong complexes and precipitates are often formed when the pH of a mixture containing proteins and alginate is lowered from 7 to 5. Interaction between protein and alginate molecules occurs at pH values below the protein's isoelectric pH value (pI) where the protein molecules have a net positive charge. (Alginate molecules remain

Box 14.1 Typical Products Containing Algins

Sodium alginate[a]	**Propylene glycol alginate (PGA)**
Alfredo sauce	Buttered pancake syrups
Barbecue sauce	Cheesecake
Breakfast bars	Cheese spreads
Caramels	Chocolate milk mixes
Dry soup mixes	Cocktail mixes
Frozen gravy	Frozen desserts and novelties
Frozen guacamole	Fruit punch
Fruit purees	Fruit syrups
Macaroni and cheese	Icings and frostings
Nonrefrigerated dessert gels	Instant cocoa mix
Pimento pieces for Spanish olives	Low pH syrups
Sauce mixes	Milk shakes
Structured bacon substitute pieces and structured fruit pieces	Pourable salad dressings (regular and reduced-calorie)
Structured nut pieces	Sandwich spreads
Structured onion rings	Seasonings
Vegetable patties	Tarter sauce
	Whipped toppings

Sodium alginate + acid
Fruit-containing, filled breakfast cereal products and cereal bars
Jelly-type bakery fillings
Nonrefrigerated dessert gels
Tomato aspic

[a]Many of these applications require natural or added calcium ions.

negatively charged at pH 5.) Alginate-minced fish, alginate-comminuted meat, and alginate–whey complexes are formed in this way.

Ingredient labels for products containing an algin may read algin, alginic acid, sodium alginate, potassium alginate, or ammonium alginate depending on the product used.

Propylene glycol alginate (PGA) is used when stability to acid, nonreactivity with calcium ions (for example, in milk products), or its surface-active property is desired (Box 14.1). It is a common ingredient in salad dressings, including low-calorie salad dressings. In both regular and low-calorie pourable dressings, it is often used together with xanthan (Chapter 11). The PGA plus xanthan combination produces more uniform flow than does xanthan alone. PGA is also used as a thickener for fruit syrups. Propylene glycol alginate and algin derivative are acceptable label designations.

Alginic acid, although insoluble, swells in water and is used commercially as a tablet disintegrant. It is sensitive to heat, a property that is characteristic of all polysaccharides with free carboxylic acid groups, except xanthan (Chapter 11).

Additional resources

Donati, I., Holtan, S., Mørch, Y.A., Borgogna, M., Dentini, M., Skjåk-Bræk, G., 2005. New hypothesis on the role of alternating sequences in calcium-alginate gels. Biomacromolecules 6, 1031–1040.

Draget, K.I., Smidsrød, O., Skjåk-Braek, G., 2002. Alginates from algae. In: Vandamme, E.J., De Baets, S., Steinbüchel, A. (Eds.), Biopolymers, Polysaccharides II, Polysaccharides from Eukaryotes, Vol. 6. Wiley-VCH, Weinheim, pp. 215–224.

Draget, K.L., 2009. Alginates. In: Phillips, G.O., Williams, P.A. (Eds.), Handbook of Hydrocolloids, second ed. CRC Press, Boca Raton, pp. 807–828.

Helgerud, T., Gåserød, O., Fjæreide, T., Anderson, P.O., Larsen, C.K., 2009. Alginates. In: Imeson, A. (Ed.), Food Stabilisers, Thickeners and Gelling Agents. Wiley-Blackwell, Oxford, pp. 50–72.

Siew, C.K., Williams, P.A., Young, N.W.G., 2005. New insights into the mechanism of gelation of alginate and pectin: charge annihilation and reversal mechanism. Biomacromolecules 6, 963–969.

Pectins

15

Chapter Outline

Key information and skills that can be obtained from study of this chapter will enable you to

1. Define, describe, and/or identify:

galacturonoglycan
galacturonan
poly(α-D-galactopyranosyluronic acid)
pectin (as the term is defined in the food industry)
degree of methylation/methoxylation (DM)
degree of esterification (DE)
HM pectins
LM pectins
amidated pectins
degree of amidation (DA)
gelling temperature

2. List the two principal sources of pectin in order of importance
3. Describe the structure of pectins with respect to L-rhamnopyranosyl units
4. Describe the structural differences between HM pectin, conventional LM pectin, and amidated pectin
5. Explain the gelling behavior differences of the three types of pectin listed in objective 4
6. Describe the two ways that junction zones can be formed from pectin molecules

Carbohydrate Chemistry for Food Scientists. https://doi.org/10.1016/B978-0-12-812069-9.00015-7

7. Describe the molecular differences between gel formation by HM and LM pectins; and
8. Relate DM with the ability of a pectin preparation to gel with (a) acids and high solids and (b) calcium ions.

Introduction

The principal component of commercial pectins is D-galacturonic acid in an α-D configuration of the pyranose ring form. As polymers of D-galacturonic acid, they are properly named **galacturonoglycans [galacturonans, poly(α-D-galactopyranosyluronic acids)].** Commercial pectins contain various contents of methyl ester groups (methyl esters of the D-galacturonic acid carboxylate groups) (Fig. 15.1). Native pectins (that is, the parent molecules of the commercial preparations as they occur in plant tissues) are more complex molecules that are more properly termed rhamnogalacturonans. They are found in cell walls and intercellular layers of all higher land plants.

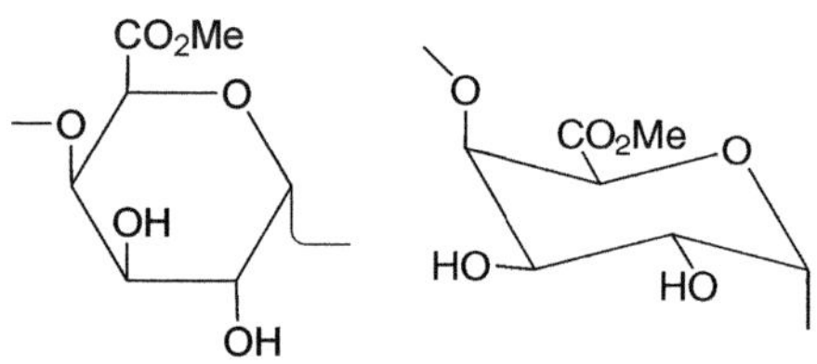

Figure 15.1 The most prevalent monomeric unit of a high-methoxyl pectin in Haworth and conformational representations. (Me = CH_3).

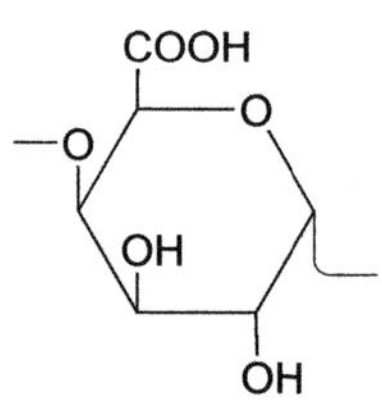

An α-D-galacturonic acid unit

Commercial pectin is obtained by acid extraction of citrus peel (20%—30% pectin, dry basis) and apple pomace (10%—15% pectin, dry basis), primarily the former. Both are by-products of juice manufacturing. That which is obtained from lemon and lime peel is the highest quality pectin. For pectin production, citrus peel is extracted with dilute acid (pH 1.5—3.0) at 60—100°C (140—212°F). The extract is filtered, and pectin is precipitated by the addition of 2-propanol (isopropanol). Lesser amounts of pectin are produced from orange peel, which pectin is of lower quality for most applications. Sunflower heads and sugar beet chips are minor sources.

Structures

Compositions and properties of pectin preparations vary with the source material, the processes used for handling and drying of the peel, the type of extraction, and subsequent treatments. During extraction with dilute acid, some hydrolytic depolymerization and hydrolysis of methyl ester groups occur. Therefore, the term pectin actually denotes a family of compounds (as with several other hydrocolloids), which is part of a larger family of pectic substances that includes polysaccharides often associated with pectins, such as arabinans (arabans). In the food and food ingredient industries, the term **pectin** is used in a generic sense to designate water-soluble galacturonoglycan [poly(galacturonic acid)] preparations of varying methyl ester contents and degrees of neutralization that are capable of forming gels. Some of the carboxyl groups of commercial pectins are in the methyl ester form, some in the free-acid form, and some in a sodium, potassium, or ammonium salt form, most often the sodium salt form. Ratios of the various forms of the carboxyl groups of the D-galacturonic acid units are determined by the source material and the time, the temperature, and the concentration of reagents used in the extraction process.

Pectin products are subdivided according to their **degree of methylation (DM),** which is the percentage of carboxyl groups esterified with methanol. (This value is also referred to as the **degree of esterification (DE),** not to be confused with dextrose equivalency [Chapter 6]. It is also commonly, but erroneously, referred to as the **degree of methoxylaton.**) Pectins with DM values greater than 50% (that is, with more than one-half of the carboxyl groups in the methyl ester form [-$COOCH_3$]) are called **high-methoxyl pectins (HM pectins)**. (They would more properly be called high-methyl-esterified pectins.) Those with DM values less than 50% are **low-methoxyl pectins (LM pectins)** (more properly low-methyl-esterified pectins). In both cases, the remainder of the carboxyl groups are present as a mixture of free acid (-COOH) and salt (-COO^-Na^+) forms (Fig. 15.2).

Treatment of a pectin preparation with ammonia dissolved in methanol converts some of the methyl ester groups into carboxamide (-$CONH_2$) groups. Because in the process, methyl ester groups are lost as they are converted into carboxamide groups, a low-methoxyl (LM) pectin (by definition) is formed. **Amidated LM pectins** may have 15%–25% of the carboxyl groups converted into carboxamide groups. The **degree of amidation (DA)** indicates the percentage of carboxyl groups in the amide form. The DM and DA values strongly influence the solubility, gel-forming ability,

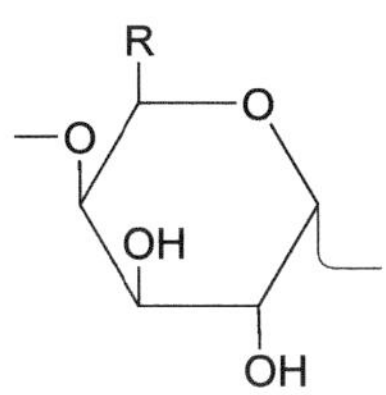

R = $-CO_2CH_3$, $-CO_2^-$, or $-CO_2H$

Figure 15.2 A representative segment of a nonacetylated, nonamidated pectin molecule.

Table 15.1 Types of pectins

High-methoxyl (HM)	Low-methoxyl (LMC)	Amidated
-$COOCH_3$ (>50%)	-$COOCH_3$ (<50%)	-$COOCH_3$
-COOH	-COOH	-COOH
-COO^- Na^+	-COO^- Na^+	-COO^- Na^+
		-$CONH_2$ (<25%)

conditions required for gelation, gelling temperature (below), and gel properties of the pectin (Table 15.1).

Acetyl ester groups are sometimes found at the O2 or O3 positions. Acetyl groups inhibit gelation and are especially numerous in sugar beet and sunflower head pectins.

The key feature of all pectin molecules is a linear chain of (1→4)-linked α-D-galactopyranosyluronic acid units. Pectin is, therefore, an α-D-galacturonan [a poly(α-D-galactopyranosyluronic acid) or an α-D-galacturonoglycan] (Fig. 15.2). Neutral sugars, primarily L-rhamnose, are also present. In at least citrus peel, apple pomace, and sunflower head pectins, α-L-rhamnopyranosyl units are inserted into the backbone chain at rather regular intervals. Some pectins contain highly branched arabinogalactan chains and/or short side chains composed of D-xylosyl units attached to the rhamnogalacturonoglycan backbone. The junction zones, which are critical to gel formation, are formed between regular, unbranched pectin chains when the negative charges on the carboxylate groups are removed by the addition of acid, hydration of the molecules is reduced by the addition of other solubles to a solution of HM-pectin, and/or when portions of chains where the carboxyl groups are ionized (that is, in the carboxylate form) are bridged by multivalent cations such as calcium ions. The inserted L-rhamnopyranosyl units may provide the irregularities in structure that limit the size of the junction zones and effect gelation. The presence of side chains may sterically limit the extent of chain association.

Sodium and calcium pectates and pectic acid occur in the solid state as right-handed helices. Junction zones in gels of HM pectin plus sucrose are believed to be formed by a columnar stacking of methyl ester groups to form cylindrical hydrophobic areas parallel to the helix axes. The same general "egg-box" model used to describe the formation of calcium alginate gels (Chapter 14) is used to explain the gelation of solutions of LM and amidated LM pectins upon addition of calcium ions. This model is used because of the close similarity in molecular structures of the (1→4)-linked poly(α-D-galactopyranosyluronate) segments of pectin molecules with carboxyl groups in the carboxylate form (Fig. 15.3) and the (1→4)-linked poly(α-L-gulopyranosyluronate) segments of alginates, as these two segments are mirror images of each other, except for the configuration at C3.[1]

[1] However, just as the "egg-box model" for alginate gelation may not be exactly correct, using the model for LM pectin gelation may not be exactly correct.

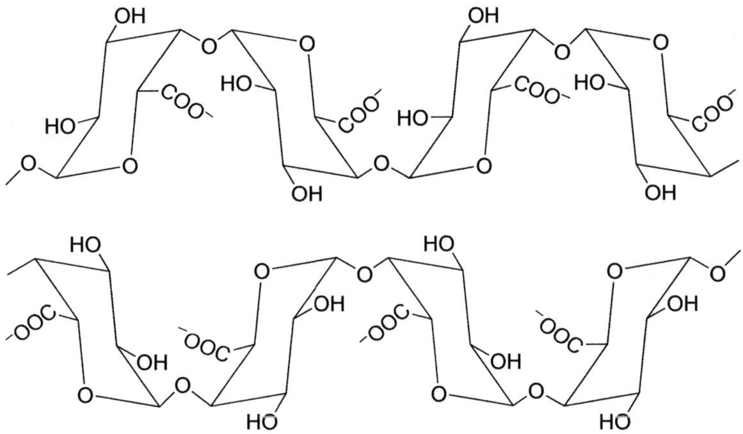

Figure 15.3 Pectate-type segments of nonamidated LM pectin showing the cavities where calcium ions are bound. Compare with Figure 14.3.

Properties and uses

Pectins are used primarily because of their unique ability to form spreadable gels (jams, jellies, marmalades, preserves). Pectins can be used in acidic foods because of their excellent stability at low pH values. The temperature and rate of gel formation of HM pectins depend upon the type of pectin employed, the pH, and the concentration of water-binding components (primarily sugar) present (Box 15.1).

HM pectins

A solution of HM pectin of the required concentration, which also contains the required concentration of sugar (sucrose) and is at a proper pH value (prepared at a

Box 15.1 Products Containing Pectin

Acidified milk drinks
Caramel topping
Cheese spreads
Chewy candies
Confections
Cupcake mixes
Fillings for chocolates
Fruit bakery fillings
Fruit drinks
Fruit juices
Fruit-milk drinks
Fruit preserves in fruit yogurts
Fruit roll-ups
Fruit smoothies
Fruit syrups
Frozen desserts and novelties
Fat-free ice cream
Icings and frostings
Jams, including sugar-free and low-sugar
Jellies, including sugar-free and low-sugar
Low-calorie salad dressings
Marmalades
Prepared mustard
Preserves
Nonfat sour cream
Nonfat yogurt

temperature above the gelling temperature and then cooled to decrease the thermal energy of the polymer molecules), will form junction zones when the molecules collide and associate with each other. The temperature at which gelation occurs is the **gelling temperature.** (It is also called the setting temperature, but more properly, it is the temperature of gelation.)

Solutions of HM pectin molecules gel when sufficient sugar (sucrose) and acid is present to reduce the hydration of the pectin. Sugar (a hydrophilic, water-binding substance) competes for water in the solution and reduces the amount of water molecules available for hydrogen bonding to the pectin. This lowered solvation of the pectin allows more intermolecular bonding, or junction zone formation, and promotes gel formation. As the pH of a pectin solution is lowered (generally to pH 2.5−3.8), the highly hydrophilic and hydrated carboxylate (-CO_2^-) groups are converted into nonionic, less hydrated carboxylic acid (CO_2H) groups. Upon losing some of their charge and hydration, segments of the polymer molecules associate, forming junctions and a network of polymer chains that entraps portions of the aqueous solution. Dehydration and junction zone formation is assisted by the presence of a high concentration of a solute (sugar; 55%−85%, generally about 65%), which competes with the hydrocolloid molecules for water. The gelling temperature of an HM pectin solution (sol to gel transition) can be varied from about 35°C (95°F) to about 90°C (195°F). HM pectin gels are generally not thermoreversible.

The ability of a solution of a pectin preparation to gel upon the addition of acid and soluble solids (sugar) and upon the addition of calcium ions varies with the degree of esterification/methylation in an inverse relationship. As the degree of methylation of an HM pectin decreases, the ability of its solutions to gel upon the addition of acid and soluble solids (sugar) decreases (Fig. 15.4).

Pectin preparations are evaluated on the basis of gel strength (that is, the breaking strength as measured by detecting the force that is needed on a plunger to break the

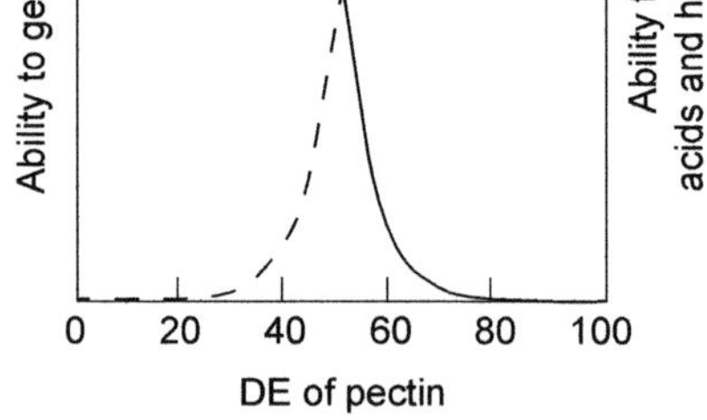

Figure 15.4 A representation of the relationship of methyl ester content to conditions required for gelation.

surface of a gel made with prescribed standard concentrations of the pectin and sugar [sucrose] and pH). HM pectins are standardized, usually to 150-grade USA-SAG. The SAG value (a measure of quality) is the decrease in height of a gel (the amount of sag), again made with prescribed concentrations of the pectin and sugar and pH during a specific period after the gel is removed from a container of defined size. 150 Grade refers to the number of parts of sugar that, under the standard conditions, must be added to produce a gel of standard gel strength by one part of the HM pectin. The gel strength of LM pectins can be determined by similar procedures, but no standards for LM pectin (below) have been adopted because the conditions under which LM pectins are used (concentrations of soluble solids, concentration of Ca^{2+}, pH) vary too much to allow a single standard assay.

For the standardization of HM pectins, unstandardized HM pectins are diluted to 150-grade USA-SAG by the addition of sucrose, glucose, or other sugars. Gelling temperatures and/or gelling times can be adjusted by blending different lots of pectins with different degrees of esterification.

The various gelling grades of HM pectins are a function of their methyl ester content. Rapid-set pectins have a methyl ester content of 72 ± 4% (DM 72 ± 4) and form a self-supporting gel (that is, "set") within 20–70 s at pH 3.0–3.1. Gelation occurs with 0.3% pectin and about 65% soluble solids (sugar). Slow-set pectins have DM values of 62 ± 2 and set in 180–250 s. Extra–slow-set pectins have DM values of 60 ± 2 and set in 200–240 s. Medium-set pectins have DM values of 69 ± 3. All require an acidic medium for gelation; slow-set pectins require the lowest pH. Rapid-set pectins are used when rapid gelation is needed to prevent settling or flotation of fruit pieces before solidification (for example, when making jams), while slow-set pectins are used in the preparation of jellies to allow time for air bubbles to rise and escape before gelation.

About half of the HM pectin produced is used in the manufacture of jams and jellies. Concentrations in these products generally range from 0.1% to 0.4%. Gelation does not occur at pH values above 3.5[2] or at sugar concentrations of less than 55%. The gelling temperature can be varied from about 35°C (95°F) to about 90°C (195°F). HM pectin gels are rigid and will hold a cut surface. Syneresis[3] occurs. The presence or absence of calcium ions is not a factor in setting of HM pectins.

Conventional LM pectins

Conventional LM pectin (LMC pectin) solutions gel only in the presence of divalent cations, which provide cross-bridges. Increasing the concentration of divalent cations (only calcium ion is used in food applications) increases the gelling temperature and gel strength. Its ability to gel with calcium ions decreases as the degree of esterification/methylation increases (Fig. 15.4). Because its solutions do not require sugar for gelation, LM pectin is used to make dietetic jams, jellies, and marmalades.

[2] At pH values greater than 6.5, both HM and LM pectins undergo depolymerization by the β-elimination mechanism described in Chapter 4 (Fig. 4.12, where OR = OMe), resulting in a loss of gel-forming ability.

[3] Syneresis is the expulsion/separation of liquid from a gel caused by shrinking of the gel (Chapter 5).

LM pectin gels can be repeatedly remelted and reformed. Therefore, LMC pectins are more suitable for industrial bakery jellies. The remelt temperature can be as high as 200°C (390°F). These gels are pumpable, spreadable, thixotropic (Chapter 5), and generally shear reversible. LM pectins will set at pH values of up to about 6.5 (but the usual pH range is 1—5)[2]. LM pectins require no sugar for gelation because gel formation is dependent only on the presence of Ca^{2+}, which ions are always required for gelation of LM pectin solutions. Calcium ions effect gel formation by cross-linking in the same way that they produce gels with alginate solutions (Figs. 14.4 and 14.5). Gel strengths and gelling/setting temperatures increase as the calcium ion concentration increases. Gelling temperatures can range from 40 to 100°C [105—212°F]). However, an increase in calcium ion concentration above the concentration required to saturate the polymer (that is, to cross-link all carboxylate groups) lowers both the gel strength and the gelling temperature and may produce an undesirable apple sauce—like structure. Gelation will occur in the presence of from 0% to 80% soluble solids. The amount of sugar present determines the concentration of Ca^{2+} required. The pH of a gel does not influence its texture. LM pectins are used in barbecue sauces because of their ability to provide cling. An unusual, unique feature of an LMC pectin gel is that it is shear reversible.

LM pectin is a reasonably good stabilizer for milk products below pH 4.0—4.5. Pasteurized, acidified milk drinks, and milk-fruit drinks can be stabilized by an LM pectin because the negatively charged pectin molecules combine with milk protein molecules, preventing agglomeration and settling. A similar behavior is seen with certain alginates (Chapter 14) and carboxymethylcelluloses (CMC) (Chapter 8).

Amidated LM pectins

Amidation is the introduction of carboxamide (-$CONH_2$) groups by reaction of some of the methyl ester (-CO_2Me) groups with ammonia—concurrently reducing the methyl ester content and making the pectin an LM pectin. Amidated pectins are more sensitive to calcium ions (Fig. 15.5) than are LMC pectins, so less Ca^{2+} is required, and the amount normally present in fruit suffices for strong gel production. In fact, amidated pectins (LMA pectins) are so sensitive to Ca^{2+} that the normal calcium ion content of tap water can be a problem because neither any pectin nor any other gelling hydrocolloid will dissolve in an aqueous system at a temperature at which the system would otherwise be gelled by it. At pH values less than 3.5, LMA pectin gels are similar to those made with HM pectin, except that they are more rubbery. Above pH 3.4—3.5, LMA pectin gels are spreadable, thixotropic, and shear reversible. The gelling temperature of LMA pectin gels is generally in the range 30—70°C (85—160°F). LMA pectin gels are thermoreversible, with the remelt temperature generally being below 75°C (165°F). They can be used to make low-calorie jams and jellies because sugar is not required for gelation as it is with HM pectins.

A comparison of LMC and LMA pectins reveals that some LMC pectins require added calcium ions for gelation, while LMA pectins usually do not. Solutions of LMC pectins generally have a higher temperature of gelation than do solutions of LMA pectins. (The gelling temperatures of LMC solutions are generally between

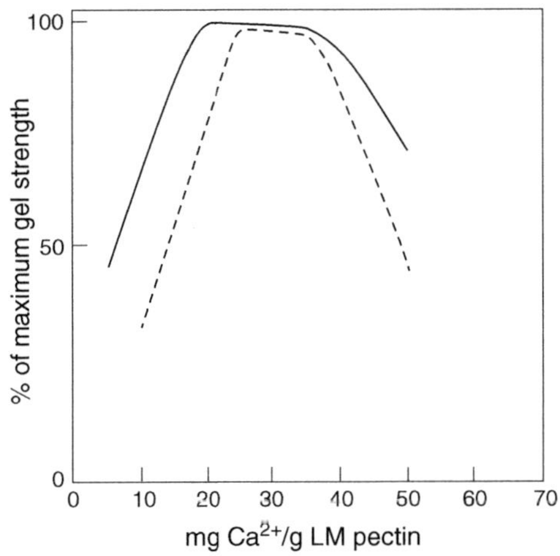

Figure 15.5 The effect of Ca^{2+} ion concentration on the maximum gel strength of amidated (—) and nonamidated (—) LM pectins.

40°C and more than 100°C [105°F and >212°F], while those of LMA pectin solutions generally range from 30 to 70°C [85–160°F].) LMC pectin gels are spreadable and thixotropic, while LMA pectins form firm, rigid gels that will maintain a cut surface. LMC pectin gels are shear reversible (an unusual, unique property), while LMA pectin gels generally are not. The shear-reversible property is important in cases where the product must be pumpable below the gelation temperature, such as in the preparation of fruit-yogurt products. Both types of gels are thermoreversible.

Small gel particles made by shearing an LM pectin gel produce some of the mouthfeel of fat (that is, they act as a fat mimetic) (Chapter 17). Likewise, a pectin specially prepared by means of an acid extraction and processed into a fine powder swells in water to form small gel particles that give sensations similar to those given by the presence of fat globules. Calcium-reactive pectin technology has also been used to reduce oil uptake by as much as 50% in French-fried potatoes and battered and breaded food products.

It needs to be reemphasized that all statements concerning properties of HM pectin, LMC pectin, and LMA pectin gels are general statements. Their actual properties are a function of the pH and the type and amounts of solutes, such as sugar and calcium ions, present.

Both HM pectins and LM pectins (including both LMC and LMA pectins) may be labeled pectin.

Additional resources

Braccini, I., Rodriguez-Carvajal, M.A., Pérez, S., 2005. Chain-chain interactions for methyl galacturonate: models for high methyl-esterified pectin junction zones. Biomacromolecules 6, 1322–1328.

Brejnholt, S.M., 2009. Pectin. In: Imeson, A. (Ed.), Food Stabilisers, Thickeners and Gelling Agents. Wiley-Blackwell, Oxford, pp. 237–265.

Endress, H.-U., Christensen, S.H., 2009. Pectins. In: Phillips, G.O., Williams, P.A. (Eds.), Handbook of Hydrocolloids, second ed. CRC Press, Boca Raton, pp. 274–297.
MacDougall, A.J., Ring, S.G., 2004. Pectin polysaccharides. In: Tomasik, P. (Ed.), Chemical and Functional Properties of Food Saccharides. CRC Press, Boca Raton, pp. 181–195.
Ralet, M.-C., Bonnin, E., Thibault, J.-F., 2002. Pectins. In: Vandame, E.J., DeBaets, S., Steinbüchel, A. (Eds.), Biopolymers, Polysaccharides II, Polysaccharides From Eukaryotes, vol. 6. Wiley-VCH, Weinheim, pp. 345–380.
Rolin, C., Nielsen, B.U., Glahn, P.-E., 1998. Pectin. In: Dumitriu, S. (Ed.), Polysaccharides. Marcel Dekker, New York, pp. 377–431.
Schols, H.A., Visser, R.G.F., Voragen, A.G.J. (Eds.), 2009. Pectins and Pectinases. Wageningen Academic Publishers, Wageningen.

Gum Arabic and Other Exudate Gums

16

Chapter Outline

Key information and skills that can be obtained from study of this chapter will enable you to

1. Define and/or identify
   exudate
   tear (as applied to an exudate gum)
   protein–polysaccharide (vs. glycoprotein)
2. List the four exudate gums.
3. Give the sources of gum arabic and gum tragacanth.
4. Explain in molecular terms why gum arabic has the ability to both emulsify and stabilize emulsions.
5. Describe the three fractions of gum arabic.
6. Describe the (a) chemical and (b) molecular structures of the fraction of gum arabic responsible for its emulsifying properties.
7. List the sources of heterogeneity of gum arabic.
8. Give the two (probably) most significant variations in gum arabic structures.
9. List three unique properties of gum arabic.
10. List the seven properties of gum arabic that make it important in the preparation of spray-dried flavors.

Introduction

When the bark of certain trees and shrubs is injured by insects or by manual tapping, the plants (commonly those that grow in semiarid conditions) exude a sticky substance that hardens to seal the wound and give protection from infection and desiccation.

Carbohydrate Chemistry for Food Scientists. https://doi.org/10.1016/B978-0-12-812069-9.00016-9

The hardened **exudate** attached to the plant is called a **tear**. Tears were articles of commerce long before recorded history. They could be easily gathered, broken up, powdered, and dissolved in water to produce a thick and generally adhesive paste that was useful as a food ingredient, for medicinal applications, for sizing fabrics, for making early types of paper, and for printing. Gum arabic (gum acacia), gum karaya, and gum ghatti are tree exudates; gum tragacanth is a shrub exudate. Only gum arabic is still used in significant amounts, mostly in food and beverage applications.

All exudate gums require cleaning and pasteurization before use as a beverage, confectionary, or food ingredient because when freshly exuded, they are sticky and trap dust particles, insects, and/or bacteria. Pieces of bark that may remain attached to the tear when it is removed may provide slight color.

Gum arabic

Source

Gum arabic (also known as arabic gum, gum acacia, and acacia gum) is officially defined as "a dried gummy exudation obtained from the stems and branches of *Acacia senegal* (L.) Willd. or of related species of *Acacia* (Fam. Leguminosae)." It has had a long and fascinating history and is the principal exudate hydrocolloid used in food products.[1] It is known to have been used as an adhesive for mineral paints in Egypt in the 17th century BCE and to have been an article of commerce even before then. It is the material exuded by wounds on acacia trees of which there are more than 1000 species distributed over tropical and subtropical areas of Africa, India, Australia, Central America, and southwest North America. The most important growing area for the species that produce gum arabic of the highest quality is the Sudan (where *A. senegal* grows). Several other species of *Acacia* grow in other regions of Africa. Gum arabic from *A. senegal* (about 80%, the highest quality material) and *Acacia seyal* (about 10%, lesser quality material) account for more than 90% of the commercial hydrocolloid. Adulteration of higher quality preparations with lower quality preparations is a problem.

In central Sudan, which is the most important gum arabic—yielding area and the source of the highest quality material, gum arabic is obtained from trees both growing in the wild and planted in groves. In either case, the acacia trees are tapped in the dry season by stripping an approximately 60 cm by 5 cm (2 feet by 2 in.) section of bark and hand collecting dried tears weighing 20—200 g over a 4—8 week period. The Sudanese government controls production and price. Acacia trees are legumes and beneficial for the soil and are often planted to prevent desert encroachment.

After harvesting, tears are sorted into grades according to color (the lightest color being the best) and content of extraneous matter. Tears deteriorate on storage

[1] The fact that these ancient polysaccharidic materials were called gums is probably the reason other water-soluble, nonstarch polysaccharides (hydrocolloids) have been called food gums or just gums.

(probably because they contain enzymes), the rate being a function of the temperature and the time of storage. For food use, the better grades are usually dissolved in water, and the resulting solution is filtered, centrifuged, and then microfiltered to remove pieces of bark, insects, and dust that adhered to the sticky exudates. The final filtrate is sterilized (pasteurized) and (usually) spray-dried to yield a high-quality, cold water–soluble powder.

Structure

Gum arabic has unique properties due to a unique structure. It is a neutral or slightly acidic (creates a solution pH of 4.5–5.0) hydrocolloid containing calcium, magnesium, potassium, and sometimes other cations. As described below, preparations of gum arabic contain at least three kinds of polymers.

Compositions of the gum arabics vary as the climate varies and with the age of the tree. High rainfall during the wet season and high temperatures during the dry season result in best yields. Selection of trees for higher yields and gum quality is practiced where trees are planted in groves. There are also basic differences in composition of different lots of *A. senegal* gum (highest quality) and *A. seyal* gum (lesser quality) (Table 16.1).

Gum arabic is a highly heterogeneous material. It consists primarily of a spectrum of polysaccharide species that have highly branched structures. Gum arabic can be divided into three main fractions. One, which accounts for about 70%–90% of the gum, is composed of polysaccharide molecules. These polysaccharide molecules are acidic arabinogalactans (glucuronoarabinogalactans). This fraction contributes to gum arabic's anionic nature, but seems to have little to do with its important functional properties. Another fraction contains molecules of much higher molecular weight that have a hydrophobic protein as an integral part of their structures. This fraction makes up less than 10% of the total hydrocolloid, but it is the fraction that provides the most useful functionality of gum arabic. The polysaccharide structures in this fraction are covalently attached to a hydrophobic protein by linkage to hydroxyproline and,

Table 16.1 Typical compositions of gum arabic preparations

	From *Acacia senegal*		From *Acacia seyal*	
Monosaccharide	**Average (%)**	**Range (%)**	**Average (%)**	**Range (%)**
D-Galactose	45	37–53	36	34–38
L-Arabinose	25	20–30	49	46–52
D-Glucuronic acid	15.5	12–19	10	6–14
L-Rhamnose	13	10–16	2.6	1–4
4-*O*-Methyl-D-glucuronic acid	1.5			

Figure 16.1 Proposed model for the structure of the functionally active component of gum arabic as it accumulates at an oil–water interface during emulsion formation.
Adapted from Fincher, Stone, Clark, 1983. Annual Reviews of Plant Physiology 34, 47–70.

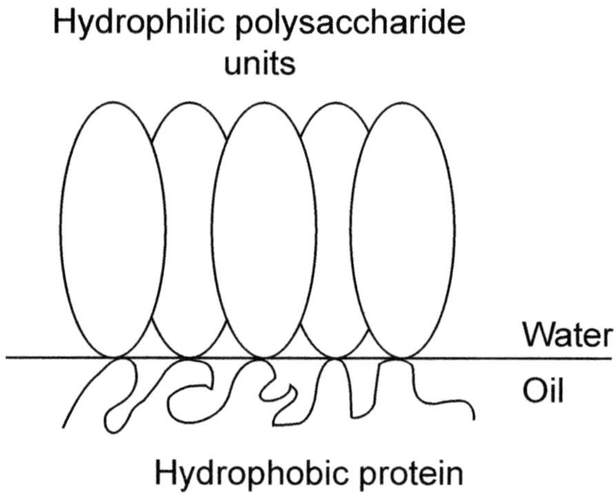

perhaps, serine and/or threonine units. The overall protein content is about 2 wt%, but specific fractions may contain as much as 25 wt% protein. These components of gum arabic are classified as **protein–polysaccharides**[2] and are responsible for the gum's excellent emulsifying properties. (The effectiveness of a given preparation of gum arabic as an emulsifier and emulsion stabilizer is strongly correlated with its nitrogen/protein content.) A molecular model for these molecules which accounts for their ability to emulsify has been proposed. In it, five polysaccharide units are attached to one side of the polypeptide unit (Fig. 16.1). Overall, the structure is assumed to be very compact (that is, to occupy a very small volume) because of the extremely low viscosity of its solutions (see below). The third fraction, a minor fraction representing only about 1% of the total gum, is a glycoprotein[3] (protein content 20%–50%).

Gum arabic molecules also contain, as part of their structures, glycosylphosphatidylinositol (GPI)[4] lipid moieties. The fatty acids of this structure are long (C_{22}–C_{25}) saturated molecules, some of which are hydroxylated at C2. The glycosyl parts are oligosaccharides. It is believed that these components of the protein part of gum arabic molecules contribute to the surface activity of the hydrocolloid by providing amphiphilic units.[5]

[2] **Protein–polysaccharides** (or sometimes polysaccharide–proteins) are naturally occurring plant molecules in which one or more polysaccharide chains are covalently attached to a polypeptide. (The same term is now used in immunology for conjugates in which polysaccharide molecules are joined to protein molecules with a cross-linking reagent. It is also used for complexes of proteins and polysaccharides formed by electrostatic interactions.) The term proteoglycan is used to describe molecules in animal tissues in which polysaccharide chains are attached to protein chains. In proteoglycans, the polysaccharide parts belong to a class of polysaccharides called glycosaminoglycans. Both proteoglycans and plant protein–polysaccharides are predominately carbohydrate/polysaccharide in composition.

[3] Glycoproteins are proteins containing saccharide moieties (often, but not always, oligosaccharide chains) covalently to a polypeptide. Glycoproteins are predominately protein in composition.

[4] See Chapter 2 for the structures of inositols.

[5] Molecules or units of molecules with both hydrophilic and hydrophobic parts.

The polysaccharide structures, both those attached to protein and those that are not, contain main chains of (1 → 3)-linked β-D-galactopyranosyl units having two- to four-unit side chains consisting of (1 → 3)-linked β-D-galactopyranosyl units joined to it by (1 → 6)-linkages. Both the main chain and the numerous side chains have attached α-L-arabinofuranosyl, α-L-rhamnopyranosyl, β-D-glucuronopyranosyl, and 4-*O*-methyl-β-D-glucuronopyranosyl (Chapter 2) units. The uronic acid units occur most often as nonreducing end units of branches. A repeating unit structure with a molecular weight of about 8000 (DP about 50) has been suggested.

The molecules of any gum arabic preparation can differ significantly in molecular weight, protein content, and composition of the polysaccharide component as a function of species, variety, part of the tree from which the tear is obtained, geographic origin, and climate before harvesting.

Properties

Powdered gum arabic dissolves easily when stirred in water. It is unique among the hydrocolloids because of its high solubility, the low viscosity of its solutions, and the Newtonian flow of its solutions at concentrations below about 40% concentration. Solutions of up to 55% concentration can be made. At this concentration, the dispersion is somewhat gel like. Highest (but still quite low) viscosities are produced at its normal pH of 6. As with other polysaccharides containing carboxylic acid groups, its solubility is minimum when ionization of the carboxyl groups is repressed (below pH about 3). In general, concentrations 20–40 times greater, compared with other hydrocolloids, are required to produce equivalent viscosities, which is an advantage in its primary applications.

Gum arabic has surface/interfacial activity and forms a thick, sterically stabilizing, macromolecular layer around oil droplets. It is a good emulsifying agent for citrus and other flavor oils and a very good stabilizer for these emulsions. For a hydrocolloid to have both an emulsifying power and an emulsion-stabilizing effect, it must have an anchoring group with a strong affinity for the surface of the oil droplet and a molecular size large enough to cover a significant area of droplet surfaces. The anchoring group has been assumed to be the hydrophobic protein moiety, but the fatty acid-containing GPI groups may also be involved. To increase the amounts of the hydrophobic anchoring groups, octenylsuccinate esters (Chapter 7) of gum arabic is manufactured.

Stabilized citrus and other flavor oil emulsions prepared with gum arabic can be spray-dried to produce an encapsulated flavor (Chapter 7). In addition to its encapsulating properties, high-solids, low-viscosity solutions can be made with gum arabic. Its heat stability permits spray-drying at high temperatures, which results in a more rapid removal of water, thereby exposing the flavor to heat for a shorter time. Because gum arabic is nonhygroscopic, it is not necessary to dry the air to which the powder is exposed as it is cooled and conveyed out of the drier. Properties that make gum arabic

Table 16.2 Properties that make gum arabic effective in the preparation of spray-dried flavor powders

Lowering of interfacial tension (emulsifying property)
Protective colloid action (emulsion-stabilizing property)
Ability to form high-solids solutions with low viscosity (so that highly concentrated emulsions can be made, resulting in less energy-intensive expensive removal of water via spray-drying)
Newtonian rheology over a wide concentration range
Ability to encapsulate essential oils
Tolerance to spray-drying at higher temperatures
No special conditioning of air during cooling required
Colorless
Odorless
Tasteless

so effective in the preparation of spray-dried flavor powders are given in Table 16.2. These properties are mimicked rather well by certain starch octenylsuccinate products (Chapter 7). As already mentioned, the properties of gum arabic itself can be improved by addition of octenylsuccinate ester groups. Reaction of either starch or gum arabic with an alkenylsuccinic anhydride (octenylsuccinic anhydride is primarily used) adds lipophilic ester groups to hydrophilic polysaccharide molecules. In the case of gum arabic, the overall effect is a lowering of the required usage levels over those of ordinary gum arabic preparations.

Another important characteristic of gum arabic is its compatibility with high concentrations of sugar, which property accounts for its use in confections in which the sugar content is high and the water content is low (below).

Uses

More than half the world's supply of gum arabic is used in confections such as caramels, jujubes, pastilles, and toffees. In confections, gum arabic both retards or prevents sucrose crystallization and emulsifies and distributes fatty components so as to retard their migration to, and accumulation on, the surface of the product. Avoidance of lipid migration to the surface is important because such migration results in a greasy whitening called bloom. Another use of gum arabic is as a component of the glaze or coating of pan-coated candies (Box 16.1).

Box 16.1 Typical Products Containing Gum Arabic

Apple chips
Baked crackers
Caramels
Chewy candies
Citrus flavors for soft drinks
Coated peanuts
Cocktail mixes
Corn chips
Cottage cheese
Dry drink mixes
Dry salad dressing mixes
Dry soup mixes
Fillings for chocolates
Frozen yogurt
Fruit juices
Grated and shredded cheese
Gravy mixes
Gum drops
Instant cappuccino mix
Instant hot cereals
Jujubes
Liquid diet products
Low-fat puddings
Lozenges
Mints
Nutritional/diet beverages
Nutrition bars
Pan-coated candies
Pasta sauce
Pastilles
Pickle relish
Soft drinks
Toffees

Another primary use of gum arabic is in emulsification of flavor oils. It and starch octenylsuccinate products (Chapter 7) are the hydrocolloids of choice for emulsification of citrus oils, other essential oils, and imitation flavors used as bakers' emulsions and concentrates for soft drinks. Their ability to stabilize flavor oil-in-water emulsions, both as a concentrate (10%–20% oil by weight) and when highly diluted (about 100:1) in beverages, without a need for an emulsifier or secondary stabilizer, is a unique property. Because solutions of gum arabic (and those of the specially prepared octenylated starch products) at high concentrations (relative to those that can be used with other hydrocolloids) have low viscosities, rather concentrated flavor oil emulsions can be prepared; and the hydrocolloid maintains the emulsion when the concentrated form is diluted (for example, in making a lemon-lime soft drink)—a very important attribute. Gum arabic–stabilized emulsions are also used as cloud agents. In the United States, the soft drink industry consumes about 30% of the gum arabic.

Emulsions made with flavor oils and gum arabic can be spray-dried to produce dry, nonhygroscopic flavor powders. By encapsulating in this way, flavor oils are protected from oxidation and volatization, and the liquid flavor is converted into an easily handled powder form, from which there is rapid release of flavor when the powder is dispersed in an aqueous system (without affecting product viscosity). These stable flavor powders are used in dry package products such as beverage, cake, dessert, pudding, and soup mixes. Free-flowing powders made by spray-drying, an emulsion of gum arabic and an oil, are used as cloud agents in beverages and dry drink mixes.

Gum arabic once had a wider range of applications than it does now. It has been replaced in some versions of products in which the functionalities imparted by it can be matched by pectins (Chapter 15) and various low-viscosity starch products, especially octenylsuccinate esters of starch (Chapter 7), because of their greater stability of supply and price. It is still used to some extent in the preparation of gum drops, cough drops, and other lozenges. It provides gloss and flexibility when used as a bun glaze, and its emulsifying action can be employed in frozen desserts, icings, spreads, toppings, and whipped or aerated, reduced-calorie margarines.

Gum arabic may be labeled gum arabic, arabic gum, gum acacia, or acacia gum.

Gum karaya

Gum karaya is an acetylated polysaccharide exudate of *Sterculia urens* trees which grow in the dry, rocky hills and plateaus of central and northern India. Gum tears are collected in April to June and again in September after the monsoons. Trees are tapped by removing bark and later collecting the dried exudate.

Gum karaya is a highly charged (ionic) hydrocolloid. It contains about 37% galacturonic acid. It is partially acetylated (about 8% acetyl groups by weight). It is the least soluble of the exudate gums, usually only forming a dispersion of swollen particles. It is relatively acid stable.

Gum karaya is used in only a few specific food products. Because of its acid stability, it is used in a few pourable salad dressings as a stabilizer, in a few ice pops and sherbets, and in a few icings and frostings. Gum karaya and karaya gum are acceptable label designations.

Gum ghatti

Gum ghatti is exuded from *Anogeissus latifolia* trees, which are found mainly in the dry deciduous forests of India and to a lesser extent in Sri Lanka. It occurs most often in large masses of worm-shaped exudates, but sometimes in tears (about 5−13 cm [2−5 in.] in diameter). It is seldom used in foods in the US today. Gum ghatti and ghatti gum may be used on labels.

Gum tragacanth

Gum tragacanth is another ancient gum of commerce, being described by Theophrastus in the third century BC. It is the exudate of the *Astragalus* bush which grows close to the ground and which has a long root that is tapped. The exudate is in the form of a long curving string or ribbon of rapidly hardening gum. The name tragacanth is derived from the Greek words *tragos* meaning goat and *akantha* meaning horn, which describes the appearance of the exudate. The leguminous *Astragalus* plants are common in Iran, Syria, and Turkey. Collections of gum tragacanth are small.

Gum tragacanth is a mixture of substances. The active component seems to be tragacanthin, a soluble polysaccharide. The main fraction, bassorin (60%–70%), contains methyl ether groups and only swells in water, forming gel particles. The exudate also contains cellulose fibers and protein.

Gum tragacanth produces high-viscosity solutions (the highest of the exudate gums), but it takes up to 24 h of hydration to achieve maximum viscosity. Solutions of it are shear thinning. Like gum arabic, gum tragacanth is both an emulsifier and an emulsion stabilizer. It is rather acid stable.

It is used to only a minor extent in foods in the United States. Gum tragacanth may still be found in some bakery products, barbeque sauces, pickle relishes, pourable salad dressings, and products such as frozen cheese lasagna and frozen gravy. Gum tragacanth and tragacanth gum are allowed label declarations.

Additional resources

Gum Arabic

Nothnagel, E.A., Bacic, A., Clarke, A.E. (Eds.), 2000. Cell and Developmental Biology of Arabinogalactan-proteins. Kluwer Academic/Plenum Publishers, New York.

Phillips, G.O., Williams, P.A., 1993. The specification of the gum Arabic of commerce. In: Nishinari, K., Doi, E. (Eds.), Food Hydrocolloids. Plenum Press, New York, pp. 21–34.

Thevenet, F., 2009. Acacia gum (gum Arabic). In: Imeson, A. (Ed.), Food Stabilisers, Thickeners and Gelling Agents. Wiley-Blackwell, Oxford, pp. 11–30.

Williams, P.A., Phillips, G.O., 2009. Gum Arabic. In: Phillips, G.O., Williams, P.A. (Eds.), Handbook of Hydrocolloids, second ed. CRC Press, Boca Raton, pp. 155–168.

Gum Tragacanth and Gum Karaya

López-Franco, Y., Higuera-Clapara, I., Goycoolea, F.M., Wang, W., 2009. Other exudates: tragacanth, karaya, mesquite gum and larch arabinogalactan. In: Phillips, G.O., Williams, P.A. (Eds.), Handbook of Hydrocolloids, second ed. CRC Press, Boca Raton, pp. 495–534.

Mayes, J.M., 2009. Gum tragacanth and karaya. In: Imeson, A. (Ed.), Food Stabilisers, Thickeners and Gelling Agents. Wiley-Blackwell, Oxford, pp. 167–179.

Gum Ghatti

Al-Assaf, S., Phillips, G.O., Amar, V., 2009. Gum ghatti. In: Phillips, G.O., Williams, P.A. (Eds.), Handbook of Hydrocolloids, second ed. CRC Press, Boca Raton, pp. 477–494.

Carbohydrate Nutrition, Dietary Fiber, Bulking Agents, and Fat Mimetics

17

Chapter Outline

Carbohydrate Chemistry for Food Scientists. https://doi.org/10.1016/B978-0-12-812069-9.00017-0

Key information and skills that can be obtained from study of this chapter will enable you to

1. Define, describe, or identify

glycemic response	psyllium gum
glycemic carbohydrate	resistant starch (RS)
glycemic index (GI)	RS1
glycemic load (GL)	RS2
glycemic impact	RS3
hyperglycemia	RS4
hyperinsulinemia	RS5
dietary fiber (DF)	slowly digesting starch (SDS)
insoluble dietary fiber (IDF)	hydrogenated starch hydrolyzates
soluble dietary fiber (SDF)	nondigestible maltodextrin
total dietary fiber (TDF)	polydextrose
prebiotic	hydrolyzed guar gum
nonstarch polysaccharides (NSP)	chitin
hemicelluloses	chitosan
arabinoxylans	bulking agent
pentosans	bodying agent
wheat flour arabinogalactans	reduced-calorie bulking agent
β-glucans	fat mimetic
mixed-linkage β-glucans	fat sparer
larch arabinogalactan	fat replacer

2. List the only carbohydrates that provide the 4 Kcal/g of energy attributed to carbohydrates and explain why this is so.
3. List the only carbohydrates that are either present in the diet or are produced by digestion in the small intestine that are essentially completely absorbed from the small intestine.
4. Describe the common feature of prebiotics.
5. Describe the common feature of components of dietary fiber.
7. List the six factors that control the rate and degree of starch digestion.
8. Explain why it is so difficult to develop an ideal reduced-calorie bulking agent.
9. Explain the physical concept of a carbohydrate-based fat mimetic.

Introduction

The connection between the topics listed in the title of this chapter is the relationship of carbohydrates to human health. For good reasons, no other aspect of carbohydrate science has received as much attention in recent years as carbohydrate nutrition. This chapter will focus on connections between health-related aspects of carbohydrates and their structures.

Since before recorded history, carbohydrates have been a principal source of metabolic energy for humans, the means for maintaining a healthy gastrointestinal tract, and the principal provider of bulk and texture in plant-based food products. Different carbohydrates fulfill these different roles differently.

Carbohydrates may be caloric, partially caloric, or essentially noncaloric. Oligo- and polysaccharides cannot be absorbed as is. To serve as a source of energy, they must first be converted to monosaccharides via digestion in the small intestine. Oligo- and polysaccharides may be digestible, partially digestible, or nondigestible. When digestion occurs, the monosaccharide products of digestion are absorbed and catabolized. Saccharides of any size escaping the small intestine will likely be metabolized to some degree by microorganisms in the large intestine (colon), producing substances that are absorbed from the colon and catabolized, thereby providing some energy; but even the colonic microorganisms cannot utilize higher-molecular-weight saccharides that enter the colon effectively during the time they are passing through it (that is, fast enough to be completely utilized).

Carbohydrate nutrition

Carbohydrate nutrition remains an area of some debate, controversy, and confusion. Literature data on the relative effects of ingested carbohydrates, proteins, and fats on weight gain and loss are often inconclusive and at times contradictory. A detailed discussion of carbohydrate nutrition and its physiological effects are outside the scope of this book on food carbohydrate chemistry. However, the general concepts of carbohydrate nutrition are relatively simple and are presented here.

Carbohydrates are both the principal source of energy and the means for maintaining health of the gastrointestinal tract for the great majority of humans. The digestible and absorbable carbohydrates (that is, those that can be converted into monosaccharides, which are the only carbohydrates that can be absorbed) in an average human diet in the United States include the following:

Digestible and absorbable carbohydrates in the human diet	Percentage
Starch (polysaccharide, digestible)	ca. 60%
Sucrose (disaccharide, digestible)	ca. 30%
Lactose (disaccharide, digestible)	ca. 7%
Glucose and other monosaccharides (absorbable)	ca. 5%

Only monosaccharides can be absorbed from the small intestine, which is where essentially all digestion and absorption of carbohydrates occurs. So only monosaccharides (D-glucose and D-fructose being the only significant ones in the human diet) need not undergo hydrolysis before absorption. All other carbohydrates must be digested/hydrolyzed before their monomeric units can be absorbed and catabolized to provide energy.

In the small intestine, a sucrose-isomaltase enzyme complex catalyzes the hydrolysis of sucrose into D-glucose and D-fructose, and lactase converts lactose into D-galactose and D-glucose.[1] The other carbohydrases acting in the small intestine, where essentially all digestion and absorption of carbohydrates takes place, are pancreatic α-amylase (Chapter 7) and a maltase-glucoamylase enzyme complex. The enzymes in the two enzyme complexes act together to convert the maltooligosaccharides produced by the action of the α-amylase on soluble starch into glucose. The D-glucose produced by this digestion is absorbed by microvilli of the small intestine to supply the principal metabolic energy source of humans.

As already mentioned, polysaccharides may be digestible (most cooked starch and starch-based products, such as dextrins and maltodextrins), partially digestible (the so-called slowly digesting fraction of starch [later]), or nondigestible (all other polysaccharides). When digestive hydrolysis to monosaccharides (almost exclusively D-glucose from hydrolysis of starch, but with some D-fructose from hydrolysis of sucrose and D-galactose from hydrolysis of lactose, especially in infants) occurs, the products of digestion are absorbed, circulate in the blood stream, and are catabolized for energy. Therefore, carbohydrates may be caloric, partially caloric, or essentially noncaloric. Polysaccharides (other than starch) consumed as natural components of edible vegetables, fruits, and other plant materials and those hydrocolloids added to processed food products are not digested in the upper digestive tract of humans, but pass into the large intestine with little or no change, because no enzymes exist in the human gastrointestinal system for hydrolyzing polysaccharides other than starch. (You will see later that they, therefore, are components of dietary fiber.) (The acidity of the stomach is not strong enough; neither is the residence time of polysaccharides in the stomach long enough, nor is body temperature high enough to effect significant acid-catalyzed hydrolysis.)

When undigested oligo- and polysaccharides reach the large intestine/colon, they come into contact with intestinal microorganisms, some of which produce enzymes that catalyze hydrolysis of certain polysaccharides or certain parts of polysaccharide molecules (see Section Dietary fiber). As a result, polysaccharide molecules not cleaved in the small intestine may undergo some cleavage by microbial enzymes within the large intestine. When this occurs, the result is a reduction in the average molecular weight of the polysaccharides acted upon by enzymes. Most oligosaccharides produced in this way are utilized as carbon and energy sources to some degree by colonic microorganisms if their DPs are less than 10 (see Section Prebiotics).

[1] As explained in Chapter 3 in the discussion on lactase deficiency, not all persons can digest lactose completely.

Monosaccharides released from oligo- and polysaccharides are used by the microorganisms of the large intestine as carbon and energy sources via anaerobic fermentation pathways that produce lactic acid and volatile fatty acids, such as propionic, butyric, and valeric acids. These short-chain fatty acids can be absorbed through the intestinal wall and metabolized (primarily in the liver). In addition, a small, although in some cases significant, fraction of the released sugars can be absorbed through the intestinal wall and transported via the portal blood system to the liver and metabolized. It is calculated that, on average, about 7% of human energy is derived from sugars split from oligo- and polysaccharides by microorganisms in the large intestine and from the short-chain acids produced from them via anaerobic fermentation, so even dietary fiber provides some calories (see Section Dietary fiber). The extent of oligo- and polysaccharide hydrolysis depends on the abundance of particular organisms producing the specific enzymes required. When changes occur in the types of oligo- or polysaccharides consumed in a person's diet, their utilization may be temporarily low until there is an increase in the population of colonic microorganisms that produce the enzymes required for splitting the new saccharide(s) and/or the microorganisms adapt by producing the required enzymes.

There is a legal definition for total carbohydrate in a food product. Currently, it is an imprecise number calculated by difference. To determine total carbohydrate, determined weights of moisture, crude protein, total fat, and ash are added together and the sum is subtracted from the total weight of the food sample. While the remainder is assumed to be carbohydrate, it actually includes compounds such as lignin, lignans, waxes, organic acids, phytate, sterols, and anything else other than moisture, protein, lipid, and ash. All values for the noncarbohydrate components depend on the accuracy of the methods of analysis and the skill of the analysts. So values for total carbohydrate depend on somewhat imprecise numbers and assume that everything other than the substances that must be determined is carbohydrate.

Definitions related to responses to digestible carbohydrates

In any discussion of carbohydrate nutrition, it is necessary to establish definitions. Glycemic is an adjective referring to the amount of glucose in the blood. (Hence, hyperglycemic refers to a higher than normal level of glucose in the blood, and hypoglycemic refers to a lower than normal level.) Three definitions that have been adopted by the AACC International as their official definitions are the following: "Available carbohydrate is carbohydrate that is released from a food in digestion and that is absorbed as monosaccharides and metabolized by the body." "**Glycemic response** is the change in blood glucose concentration induced by ingested food." "**Glycemic carbohydrate** is a carbohydrate in a food that elicits a measurable glycemic response after ingestion." Other terms are also used when discussing carbohydrate nutrition. **Glycemic index (GI)** is a value based on the blood glucose–raising effect of a standard amount (a standard amount being defined as that which contains 50 g of available carbohydrate) compared to that of a control food (dry basis) (usually white bread or D-glucose). To obtain the GI, the area under the blood glucose concentration curve for a 2-h period is compared to that produced by the control substance, whose value is set at 100. (The concentration of glucose in blood serum is measured using the glucose oxidase method described in Chapter 2.) There is a rather high degree of variability in measured GI values between persons and even with the same person at different times. Some factors that affect GI values are the following:

- the rate of gastric emptying, which is impacted by the presence of fat, protein, and alcohol
- the viscosity of the intestinal contents (often a function of the presence of soluble fiber), which impacts the absorption of glucose
- the physical form and chemical nature of the food being consumed, which are functions of factors such as species and/or variety, harvest time/ripeness, pH, the amylose:amylopectin ratio, and whether the food is a solid, gel, or liquid
- the nature of processing and preparation techniques for the food being consumed
- the time of consumption (morning, noon, or night)
- meal frequency
- other foods consumed with the food in question
- what was consumed before (the lente carbohydrate effect)
- the person consuming the food or meal (There are major physiological differences between individuals.).

Primarily, because of these factors, but also because of factors in the analytical methods used, the determination of GI is imprecise and subject to laboratory-to-laboratory variations. Obviously, foods containing larger amounts of gelatinized starch and/or sugar will have higher GI values than those containing larger amounts of non-digestible carbohydrates (dietary fiber). **Glycemic load (GL)** is the product obtained by multiplying the GI by the grams of carbohydrate in a serving divided by 100. Therefore, it takes into account not only the ability of a standard amount of the food to contribute to increases in blood glucose concentration, but also the amount of carbohydrate actually consumed in a serving. The abilities of GI and GL to reflect the actual glycemic response to a food in a complete diet has been questioned, so other indices of the glycemic effect of foods and food ingredients have been proposed. One is **glycemic impact,** which is the response in blood glucose concentration of a serving of a food relative to that of an equivalent weight of glucose; it takes into account the quantity of carbohydrate consumed in a serving, the GI, and the composition of the food.

Increasing evidence suggests that there is an important negative correlation between glycemic load and human health. Postprandial[2] **hyperglycemia** (higher than normal levels of blood glucose after a meal) has significant physiological consequences. For persons without diabetes, hyperglycemia results in the release of large amounts of insulin leading to **hyperinsulinemia** [higher than normal levels of insulin in the blood]), which brings about a series of anabolic reactions, resulting in an increase in liver and muscle glycogen and synthesis and storage of fat, among other things. For persons with diabetes, high blood glucose levels present a risk for complications associated with the disease. Consumption of low-GI foods, resulting in a low GL, by diabetic patients is reported to result in the reduction in serum lipids and improved insulin sensitivity, both of which are desirable. The same type of diet consumed by persons

[2] After a meal.

without diabetes is reported to result in weight loss, reduced body fat, increased serum levels of high-density lipoprotein (HDL) ("good cholesterol"), and a reduced risk for developing diabetes and cardiovascular disease, again desirable outcomes.

Dietary fiber

Definition

Natural food (other than meat and dairy products) gets most of its bulk from remnants of plant cells. These remnants of plant cell wall are resistant to hydrolysis by enzymes of the human small intestine. Cellulose (Chapter 8), hemicelluloses (later in this chapter), pectic polysaccharides (galacturonans, rhamnogalacturonans), and lignin are the major components of these remnants. This nondigestible material is known as **dietary fiber (DF).** (Dietary fiber is a nutritional term that has nothing to do with its physical nature [that is, it is not necessarily fibrous in nature].) In terms of chemical structure, there is no structural similarity between the components of DF other than they are substances that, because of their chemical or physical[3] nature, are not digested by enzymes present in the human stomach or small intestine. Dietary fiber can be subdivided into two components: **insoluble dietary fiber (IDF)** (that which is either insoluble in water or precipitated by 78% ethanol) and **soluble dietary fiber (SDF)** (that which is soluble in 78% ethanol). The sum of insoluble and soluble dietary fiber is **total dietary fiber (TDF).**

Although a simple definition of dietary fiber was given in the preceding paragraph, there actually is no agreement on what constitutes DF or an universally accepted definition of DF. Different definitions have been developed and adopted by different governmental agencies and nongovernmental organizations around the world. (Developing a definition acceptable to everyone is difficult because DF from different sources is usually composed of different mixtures of nondigestible and nonabsorbable carbohydrates and other substances with different effects on human health.) The reason that a definition is necessary is that various agencies of various governments and the European Union require that a content of DF be stated on food product labels, and to measure the content of DF, there must be an official statement of what constitutes DF. (This creates another difficulty in developing a definition, and that is, there must be a method(s) for measuring whatever is included in the definition.) And there must be two components to any definition of DF. What is important to nutritionists are the physiological effects that any and all substances that are to be included in the category of DF must produce. Clearly, if it creates the beneficial physiological effects ascribed to dietary fiber when ingested, it is dietary fiber. What is important to a chemist is that all substances that produce the desired physiological effects be identified so that methods for their determination can be developed (that is, any definition DF must allow for a quantitative measurement of it for labeling purposes). The only common features known are (1) that DF consists of substances that are neither digested

[3] Their physical nature refers to their occurrence as highly associated molecules that make them insoluble and inaccessible to enzymes, as is the case with retrograded starch molecules (see Section Resistant starch).

in the stomach or small intestine nor absorbed from the small intestine and metabolized and (2) that DF has a positive effect on human health. In short, developing a precise definition of DF that is acceptable to everyone is difficult because DF is a mixture of diverse and complex substances with diverse physical properties, and DF from different sources is usually composed of different mixtures of nondigestible and nonabsorbable carbohydrates and other substances with different effects on human physiology/well-being/health.

In terms of carbohydrates, no polysaccharide (the principle component of DF in most foods), other than starch (but not resistant starch [later in this chapter]) is digested in the small intestine, so all hydrocolloids, resistant starch, and naturally occurring polysaccharides, such as cellulose, hemicelluloses (later in this chapter), and pectin, are DF components. Of the oligosaccharides, all those that humans cannot digest (that is, those other than lactose, sucrose, and those oligosaccharides derived from starch [that is, maltodextrins]) are components of DF.

The good and easily understood definition of DF adopted by AACC International is presented here. "Dietary fiber is the edible parts of plants and analogous carbohydrates that are resistant to digestion and absorption in the human small intestine with complete or partial fermentation in the large intestine. Dietary fiber includes polysaccharides, oligosaccharides, lignin and associated plant substances. Dietary fibers promote beneficial physiological effects including laxation, and/or blood cholesterol attenuation, and/or blood glucose attenuation." Both insoluble materials of plant cell walls, primarily cellulose and lignin, and all other nonstarch polysaccharides, whether water soluble or insoluble, are components of DF. Analogous carbohydrates are defined in one of the footnotes to this definition as those carbohydrate-based ingredients that are not digested and/or absorbed in the small intestine but which are not natural (Maltitol [Chapters 3 and 19], for example, is an analogous carbohydrate according to this definition.). The only common characteristic of the components of DF is that they are not digested in the human small intestine. Therefore, not only do most natural components of foods derived from fruits and vegetables contribute to total dietary fiber (TDF), but also do some low-molecular-weight carbohydrates and all hydrocolloids that are added during the preparation of processed food to modify rheological properties and/or to provide other functionalities. Each source or component of TDF produces different physiological effects—quantitatively, if not qualitatively.

Quantitative measurement of dietary fiber provides the closest link of DF to carbohydrate chemistry. There are two reasons why a quantitative determination is necessary. One is that the amount of DF in a serving of a food is required for nutritional labeling. The other is that an allowed method for calculating the caloric content of a serving of a food involves subtracting the amount of insoluble DF from the value for total carbohydrate before summing the calories provided by carbohydrate, fat, and protein. There is debate over which method best measures the content of DF in a food. An ideal method would be based on the physiological responses imparted by the food (next section), but a chemical method is needed for nutritional labeling purposes.

About 20 methods for DF have been developed, tested, and approved as official methods by the Association of Official Analytical Chemists and/or AACC International. (Many of the methods of the two organizations are identical, having been approved by both.) Modern analytical methods for DF follow the general outline: (1) Remove lipid by extraction with an organic solvent and dry to remove moisture. (2) Heat sample in water (such as at 95°C for 35 min) to gelatinize starch granules. (3) Remove starch and protein by treating the sample sequentially with a thermostable α-amylase, a protease, and glucoamylase (Chapter 7). (4) Recover and wash the insoluble residue (IDF). (5) Add ethanol to the soluble portion to precipitate soluble polysaccharides (SDF). (6) Dry, weigh, and ash the IDF and SDF fractions. (7) The amounts of IDF and SDF are summed to obtain TDF. Alternatively, ethanol is added after digestion with the three enzymes (step 2), and the IDF and SDF fractions are dried, weighed, and ashed together to obtain TDF. The only real flaw in this procedure is that it does not include substances that are not digested and the products of digestion absorbed but are not precipitated by the addition of four volumes of 95% ethanol (for example, low-molecular-weight components of fructooligosaccharides [FOS]) [Chapters 3 and 10]. High-performance liquid chromatography (HPLC) can be performed on the soluble fraction to determine such soluble components.

Physiological effects

Overview

Consumption of dietary fiber provides at least five general benefits to humans.

1. Regulates the function of the gastrointestinal tract, especially the colon
   a. Increases fecal bulk (attributed to insoluble DF)
   b. Provides laxation
   c. Decreases transit time
2. Controls body weight/reduces obesity
3. Prevents and controls diabetes
   a. Reduces gastric emptying via an increase in viscosity (attributed to soluble DF)
   b. Attenuates postprandial[2] hyperglycemia and accompanying insulinemia
   c. Enhances satiety
4. Reduces cardiovascular disease
   a. Reduces serum cholesterol (attributed to soluble DF)
   b. Anti-inflammatory effect
5. Prevention of colon and rectal cancer
   a. Promotes the growth of beneficial organisms in the colon microbiota
   b. Promotes the production of short-chain fatty acids in the fermentation process.

Prebiotics

Prebiotics are dietary fiber substances that are not digested by the enzymes of the human small intestine and that provide beneficial physiological effects to the host by selectively stimulating the growth and/or the bioactivity of beneficial microorganisms resident in the colon (large intestine). Prebiotics are principally nondigestible oligosaccharides. They include raffinose and stachyose (oligosaccharides found in soy and other beans [Chapter 3]), fructooligosaccharides (Chapters 3 and 10), lactitol and lactulose (Chapter 3), and all other oligosaccharides (generally of DP less than 7) other than maltooligosaccharides, lactose, and sucrose. Because by definition prebiotics are nondigestible oligosaccharides, they are also by definition components of dietary fiber. Oligosaccharides that have beneficial prebiotic activity (the extent of which varies from structure to structure and which has not been determined in every case) include the following:

- fructooligosaccharides (FOS) (Chapters 3 and 10) (In FOS produced by hydrolysis of inulin, DP 3–5 oligosaccharides usually total about 45%.)
- gentiooligosaccharides (mixtures of oligosaccharides containing primarily (1→6)-linked β-D-glucopyranosyl units)
- inulin (Chapter 10) (In inulin preparations, DP 3–5 oligosaccharides usually total about 10%)
- isomaltooligosaccharides (mixtures of oligosaccharides containing primarily, but not exclusively, (1→6)-linked α-D-glucopyranosyl units)
- lactitol (Chapter 3)
- lactosucrose (Chapter 3) (a trisaccharide in which a β-D-galactopyranosyl unit has been transferred to O4 of the α-D-glucopyranosyl unit of sucrose forming the same type of linkage found in lactose)
- lactulose (Chapter 3) (To produce lactulose, the D-glucosyl unit at the reducing end of lactose is isomerized [Chapter 1] to a D-fructosyl unit)
- mannooligosaccharides (obtained from yeast cell-wall mannans)
- transgalactooligosaccharides (Chapter 3) (mixtures of oligosaccharides consisting of D-glucose and D-galactose obtained by the action of β-galactosidase on lactose)
- specially prepared starch pyrodextrins (see Section Resistant starch and related saccharides)
- soy oligosaccharides (primarily raffinose and stachyose [Chapter 3])
- xylooligosaccharides (composed of (1→4)-linked β-D-xylopyranosyl units obtained by the action of a xylanase on a xylan).

Likewise, polyols, including monosaccharide polyols, the primary one of which is sorbitol, are only partially digested (if a disaccharide or larger) and/or absorbed, so at least some of the consumed polyol will be fermented in the large intestine.

Prebiotics are fermented anaerobically by beneficial microorganisms in the colon (large intestine) that use them as sources of carbon and energy. By stimulating the growth of beneficial microorganisms, prebiotics inhibit the growth of pathogenic bacteria, such as those that cause diarrhea. Benefits to human health also accrue from the short-chain fatty acid products of fermentation, particularly propionic and butyric acids (below).

Effects of dietary fiber on the gastrointestinal tract and general health

The major components of DF are polysaccharides. All nonstarch polysaccharide molecules remain intact or almost intact during their transit through the gastrointestinal tract. These hydrophilic molecules give bulk to the intestinal contents and reduce transit time by maintaining water content in the colonic contents. The result is soft stools and consequent easier passage through the large intestine.

The polysaccharide components of DF often undergo limited hydrolysis to oligosaccharides by bacterial enzymes in the colon. The products of hydrolysis may then act as prebiotics (that is, undergo fermentation), producing lactic and short-chain fatty acids, which have beneficial health effects. Especially, it is indicated that the presence of butyric acid in the colon reduces the incidence of colorectal cancer and inflammation. In addition, indications are that butyric acid is absorbed and provides energy to colonocytes, which increases their ability for self-defense, increases mucosal integrity, and protects against DNA strand breaks.

A problem with determining the physiological effects of dietary fiber is that it is not a single substance. Rather, it can be any mixture of any of a large number of substances that not only differ in their chemical structures and physical properties but also differ in their physiological effects. Most often DF ingredients are materials such as wheat bran,[4] which is a mixture of several quite different substances, each of which meets the definition of dietary fiber. Wheat bran contains substances like phytate (Chapter 2), a variety of antioxidants, and other phytochemicals that are neither digested nor absorbed. However, in general, DF holds water, is an adsorbent, provides a surface for bacterial growth and thereby increases bacterial/fecal mass, promotes peristalsis, maintains normal functioning of the gastrointestinal tract (prevents constipation), produces absorbable metabolites (mostly in the form of short-chain fatty acids) (above), lowers blood cholesterol levels to different degrees (reducing the chance of coronary heart disease to different degrees) without lowering HDL, and reduces the incidence of colonic and rectal cancer. Exact mechanisms behind the physiological effects of DF are not known in every case and are outside the scope of this book. Two general mechanisms behind the reductions in blood cholesterol have been proposed: (1) enhancement of sterol excretion via reduction in resorption of bile acids and (2) inhibition of cholesterol biosynthesis (by short-chain fatty acids such as butyric and propionic acids). Also, in general, soluble non-starch polysaccharides (see below) increase viscosity and thereby delay gastric emptying, decrease mixing, and reduce diffusion within the contents of the gastrointestinal tract and thereby slow absorption of nutrients such as glucose and change transit time. The change in transit time can be in either direction. Small amounts of a nondigestible polysaccharide often reduce transit time (that is, accelerate movement [produce laxation]), the more common effect; large amounts may extend the transit time through the stomach and small intestine by forming a gel-like mass.

The primary known health benefits of whole grains as a source of DF are reductions in body weight/obesity, diverticular disease (that is, improved health of the colon), insulin resistance (that is, improved glucose tolerance), and coronary disease.

[4] Bran is the term used to designate the outer layers of cereal grains. There is a lack consensus on the specific definitions of brans for regulatory purposes.

Some carbohydrate dietary fiber components and sources

Natural sources of IDF include such foods as vegetables, fruits, beans, wheat bran,[5] brown rice, and whole-grain breakfast cereals, breads, and pastas. Natural sources of SDF include foods such as fruits, vegetables, beans, oatmeal, and products made from barley and rye.

Plant cell walls/whole grains

Substances that have a high content of plant cell walls (whole grains, vegetables, fruits) are primary sources of dietary fiber. All annual plant cell walls contain lignin and the same polysaccharide categories, namely, cellulose, hemicelluloses (arabinoxylans, arabinogalactans, xylans, xyloglucans, etc.) (below), and pectic polysaccharides (galacturonans, rhamnogalacturonans, etc. [Chapter 15]); some cell walls contain an additional one or two polysaccharides. Plant polysaccharides other than starch, cellulose, and pectic polysaccharides are designated **nonstarch polysaccharides (NSP)** by food scientists and nutritionists. Chief among these are the hemicelluloses (see below). Added hydrocolloids are also NSP.

Hemicelluloses

Hemicelluloses are plant cell-wall polysaccharides. They vary in amount and in structure depending on the plant type and the location of the cells within the plant. Generally, they constitute 20%–30% of cell walls; they may be present in greater amounts in annual plants. **Hemicelluloses** may be roughly defined as land-plant cell-wall polysaccharides other than cellulose and pectin. They constitute a family of structurally unrelated polysaccharides. They are components of both the soluble and insoluble fractions of the dietary fiber of all plant materials, especially brans,[4] vegetables, and fruits. In addition, they are part of the structures that give fruits and vegetables the crisp, chewy, fibrous characteristics that, in part, remain after cooking.

Arabinoxylans

The hemicelluloses (NSP) of flours, such as wheat flour, are composed primarily of the pentoses D-xylose and L-arabinose. These **arabinoxylans** are often called **pentosans** (more properly, they are pentoglycans). These arabinoxylans have structures that are moderately long polysaccharide chains with a very high incidence of short branches consisting of one to three sugar units. The most common hemicelluloses in annual plants, including cereal grains, consist of a linear main chain composed of $(1\rightarrow4)$-linked β-D-xylopyranosyl units (that is, it is a xylan). The short side chains are from

[5] Wort is the liquid mixture before yeast is added to start the fermentation process during beer making. It is basically a solution of a little sucrose (about 5% of the total carbohydrate) and products of starch hydrolysis (that is, the malted grain). These products are composed of glucose (about 12% of the total carbohydrate), maltose (about 47%), maltotriose (about 15%), and higher saccharides (that is, maltodextrins) (about 25%). In other words, it is primarily a solution of fermentable sugars obtained by the action of amylases on cereal grain starch.

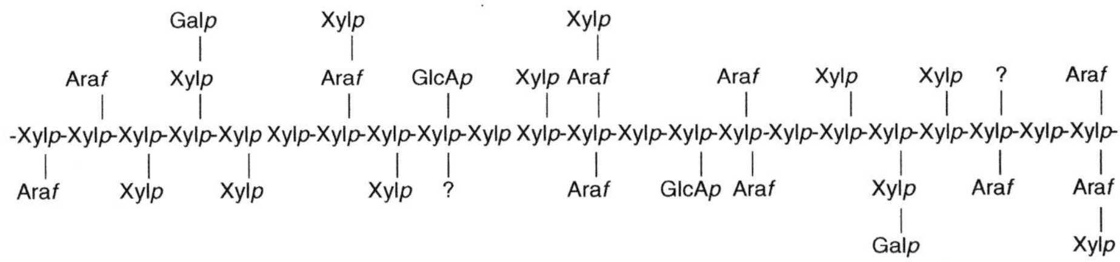

Figure 17.1 Proposed structure for a portion of the water-soluble arabinoxylan of wheat bran. Ara*f* = L-arabinofuranosyl unit; Gal*p* = D-galactopyranosyl unit; GlcA = D-glucuronopyranosyl unit; ? = unidentified sugar unit. Placement of the side units above or below the main xylan chain does not imply any particular linkage position. The depicted structure only indicates the highly branched nature of the molecules and their complexity. It is easily seen why this arabinoxylan is water soluble, for although the backbone chain is very similar to that of cellulose, the highly branched and irregularly substituted xylan chain will not allow intermolecular associations. Corn fiber gum (the arabinoxylan from corn bran) has an even more complex structure with essentially every main chain unit being branched at least once and at least a third of the arabinofuranosyl units attached to the main chain being branched via attachment of two arabinofuranosyl units to it.
Adapted from Rumpagaporn, P., Reuhs, B.L., Kaur, A., Patterson, J.A., Keshavarzian, A., Hamaker, B.R., 2015. Carbohydrate Polymers 130, 191–197.

one to three units long and are usually composed of L-arabinofuranosyl, D-galactopyranosyl, D-glucuronopyranosyl, and/or 4-*O*-methyl-D-glucuronopyranosyl units (Chapter 2). Arabinoxylans from brans are usually esterified with ferulic acid. A possible structure of a segment of soluble wheat bran arabinoxylan (without ferulic acid ester units) is shown in Fig. 17.1.

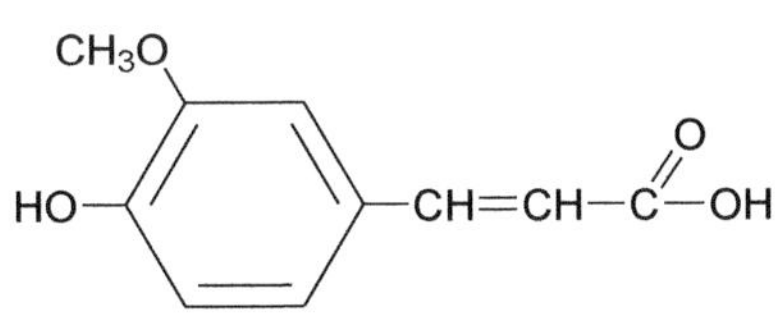

Ferulic acid

Most hemicelluloses can be isolated from plant tissue by extraction with alkaline solutions. During such extractions, some de-esterification occurs. Corn arabinoxylan (extracted from the pericarp of corn kernels, which is separated from other components during isolation of the starch by the wet-milling process [Chapter 6]) is called corn fiber gum. Corn fiber gum is produced commercially and used as an ingredient to increase the DF contents of food products.

Wheat flour arabinoxylans are divided into two classes: water soluble and water insoluble. Both classes have been studied with respect to their structures and their influence on bread making and bread quality. Arabinoxylans constitute 60%–70% of wheat endosperm (and bran) cell walls but less than 2% of wheat flour (water-soluble types being about 0.7%, water-insoluble types being about 1.3%, although reported percentages vary). Arabinoxylans make up about 8% of rye flour.

Wheat flour arabinoxylans are of considerable interest primarily because of their water-absorbing and water-binding capacities. (There is not enough in the flour [unless it is a whole wheat flour] to be a significant source of DF.) Their great propensity to be hydrated is a major factor in the ability of wheat flours to sorb water, even though they are present in small amounts. The high water-holding capacity of cross-linked arabinoxylans (via oxidative coupling of the ferulic acid ester groups present in their structures, which requires an oxidant that generates radicals) affects the distribution of moisture among bread dough constituents, the rheological properties of the dough, gas retention/loaf volume, and the texture and rheological properties of the baked product. Xylanases/hemicellulases may be added to flours to decrease the water-absorbing and water-binding capacities of the arabinoxylans. The result of treatment with a xylanase preparation is the release of water for gluten formation, resulting in a gluten network that is more homogeneous and flexible and producing a more elastic dough.

Wheat flour arabinogalactans

Wheat flour arabinogalactans are also components of wheat flour. Arabinogalactans in wheat flour are attached to protein molecules and thus are part of the structures of arabinogalactan-protein molecules. Like the arabinoxylans, there are variations in both amounts and structures of water-extractable arabinogalactan-protein molecules in wheat flours. Arabinoxylans are believed to play a more important role in the preparation, quality, and shelf life of baked goods than are arabinogalactans.

Beta-glucans

β-Glucan, a water-soluble polysaccharide present in cereal brans, is a commercial product because of its effectiveness in reducing serum cholesterol. β-Glucan is the major NSP in oat and barley. Oat and barley brans are incorporated into a variety of foods and are the source of commercial β-glucan. The principal materials used for β-glucan extraction are the crude oat or barley bran fractions that are particularly rich in β-glucan. These fractions contain some starchy endosperm and have β-glucan contents

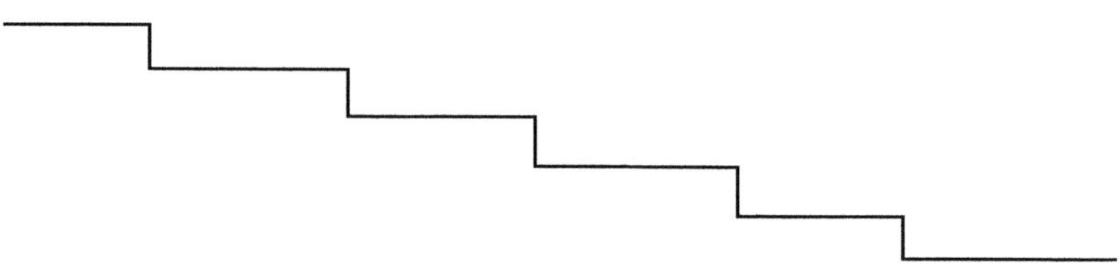

Figure 17.2 A possible segment of a mixed-linkage β-glucan molecule: horizontal lines = (1 → 4) linkages, vertical lines = (1 → 3) linkages. In an entire population of molecules, DP 5–9 segments occur infrequently (<5%) and DP 3 (cellotriosyl) segments make up at least 67% of the total segments.

$$\rightarrow 3)\text{—}\beta\text{Glc}p\text{—}(1 \left[\rightarrow 4)\text{—}\beta\text{Glc}p\text{—}(1\right]_n \rightarrow$$

Figure 17.3 Representative structure of a segment of oat and barley β-glucans where n usually is 2 or 3 but occasionally may be larger.

of 3%–7%. β-Glucan is extracted with water at 65–100°C (150–212°F), and the extract is dried on heated rolls or by spray drying. Lower temperatures can be used if extraction is done with a solution of sodium carbonate at pH 10.

Oat **β-glucan** is a linear chain of β-D-glucopyranosyl units. About 70% of the units are linked (1→4) and about 30% (1→3). The (1→3) linkages occur singly and are separated by sequences of, for the most part, blocks of two or three β-D-glucopyranosyl units joined by (1→4) linkages. Thus, the molecule is composed of (1→3)-linked β-cellotriosyl and cellotetraosyl units. Such (1→4,1→3)-β-glucans are known as **mixed linkage β-glucans** (Fig. 17.2).

The fine structure of cereal β-glucans varies with the source (type of grain). In general, however, 50%–60% of the (1→4)-linked segments are cellotriosyl (DP 3) units and less than 10% of the units are of DP 5–9; the remainder of the (1→4)-linked segments (25%–30%) are cellotetraosyl (DP 4) units. In oat β-glucan, the ratio of cellotriosyl to cellotetraosyl units is about 1.5:1; in barley and rye β-glucans, the ratio is about 2:1, and in wheat β-glucans, about 3:1 (Fig. 17.3).

In solution, β-glucans adopt a typical random-coil conformation (Chapter 5). β-Glucan products can be produced in a range of viscosity grades (Chapter 5), with lower-molecular-weight products having a greater tendency to form gels. Nutritionally, β-glucans are classified as soluble dietary fiber (SDF). When consumed in sufficient amounts, they reduce postprandial[2] serum glucose levels and the accompanying insulin response in both normal and diabetic human subjects, perhaps due to increased viscosity of the intestinal contents. They also are effective in reducing serum cholesterol concentrations. These physiological effects are typical of those of SDF, which is more effective in protecting against hypertension, stroke, cardiovascular disease, and type 2 diabetes than is insoluble DF. The US FDA allows a health claim for oats and soluble fiber from oats. Other soluble nonstarch polysaccharides have similar effects to differing degrees.

A bacterial β-glucanase is used in the preparation of wort[5] for brewing. The enzyme reduces the viscosity of barley, rice, rye, and other cereal grain malt and flour dispersions by hydrolyzing the β-glucans into tri- and tetrasaccharides via cleavage of the (1→3) linkages, thereby reducing wort viscosity, increasing the rate of wort filtration, and improving beer clarity and stability.

Available as food ingredients are several products made from cereal grains and related sources of dietary fiber. One product simply contains a concentrated amount of oat bran β-glucan (20%–50%). Another is made by treating oat flour or bran with just enough α-amylase to convert the oat starch to dextrins (Chapter 7). This product is a gel former. (Junction zones in the gel are formed between consecutive sequences of cellotriosyl units.) The gel made from it is heat stable and can be used as a fat mimetic (see Section Fat mimetics). It can be used in the preparation of extralean ground beef, hot dogs, and other processed meat products; in fat-free cheeses; and in low-fat cookies, granola bars, and muffins. Another product, a zero-calorie product, is produced by disintegration of dietary fiber sources in a two-stage process, each of which involves alkaline pH and high shear. The fiber source my include hulls and brans from corn, oat, pea, rice, soybean, or wheat.

A β-glucan with a structure different from that of the cereal β-glucans is a component of the cell walls of common yeasts and is available as a dietary supplement.

Larch arabinogalactan

A low-viscosity, water-soluble polysaccharide can be extracted with hot water from a number of softwoods. It is obtained commercially from chips of western larch (*Larix occidentalis*) wood. **Larch arabinogalactan** consists of a backbone chain of (1→3)-linked β-D-galactopyranosyl units, each of which bears a side-chain substituent at the O6 position. Most of the side-chain substituents are β-galactobiosyl units. Other common side units are single α-L-arabinofuranosyl and 3-*O*-(β-L-arabinopyranosyl)-α-L-arabinofuranosyl units (Fig.. 17.4). The DP of larch arabinogalactan ranges from about 100 to about 600 (MW range about 16,000 − about 100,000). It is very soluble and produces low-viscosity solutions. Fluid aqueous solutions can be prepared at 60% concentration. Larch arabinogalactan is primarily marketed as a dietary supplement that has the qualities of being a prebiotic and a stimulant of the immune system.

Psyllium gum

Psyllium seed hulls (husks) are the source of **psyllium gum**, which can also be labeled gum psyllium. A product based on psyllium seed hulls has high water-binding properties, leading to rapid transit time in the gastrointestinal tract. It is sold as a dietary supplement to prevent constipation. (A product based on methylcellulose [Chapter 7] is sold for the same purpose.)

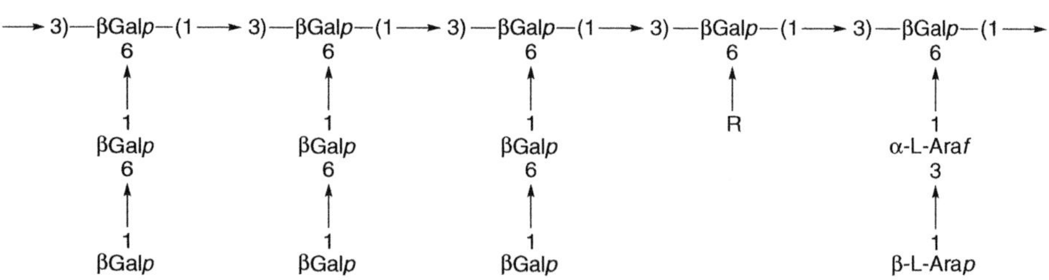

Figure 17.4 Representative structure of larch arabinogalactan. R = a β-D-galactopyranosyl unit or, less frequently, an α-L-arabinofuranosyl or a β-D-glucopyranosyluronic acid unit. The location of the side chains is not intended to indicate order or frequency, only the type of structure that could be expected in a segment of the gum.

The US Food and Drug Administration (FDA) has authorized the placement of health claims on the association between psyllium husk fiber and reduced risk of coronary heart disease on food labels. However, because of its high water-absorption and strong gelling properties, it is a challenge to incorporate the amount of psyllium in one serving of a food product that meets the criteria required by the FDA for a cholesterol-lowering claim. With regards to food products, psyllium husks have primarily, if not exclusively, been added to breakfast cereal products.

Resistant starch, slowly digesting starch, and starch-derived saccharides

Not all starch is equally digestible. In a food product containing starch, there will undoubtedly be rapidly digestible starch (RDS), slowly digestible starch (SDS), and starch that escapes digestion in the small intestine of healthy persons **(resistant starch, RS).** Resistant starch (granular or otherwise) does not raise blood glucose levels (that is, is nonglycemic). It passes intact into the large intestine/colon where it may be fermented by colonic microorganisms (that is, it functions as dietary fiber). There are five categories of RS: **RS1** (starch that is physically inaccessible to digestive enzymes, usually because it is encased inside intact plant cell walls), **RS2** (naive, nongelatinized starch, that is, uncooked, starch granules), **RS3** (retrograded starch in which starch polymer molecules are highly associated and, therefore, attacked by α-amylase only very slowly), **RS4** (starch granules or portions of starch granules that have been chemically modified to such an extent that they cannot be hydrolyzed by α-amylase), and **RS5** (amylose-lipid complexes) (Chapter 6).

Freshly cooked/pasted starch is usually easily and readily digested (that is, it is RDS). Factors that impact starch digestion include the degree of gelatinization, granule size, amylose content, starch-protein interactions, starch-lipid complexes and, perhaps most importantly, the degree of crystallinity, including that formed by retrogradation during processing. Starches have no glycemic index until they are cooked (that is, are resistant to digestion). The GI of cooked starch with a high level of crystallinity (resulting from a high amylose content, from treatments with heat and moisture [Chapter 7], from freeze-thaw cycling, or from other processes) is lower than that of the native starch (that is, it is more resistance to digestion). Cooked high-amylose starch has a low glycemic index. After normal cooking and cooling, this starch becomes even more resistant to digestion. There is no food product in which the starch is entirely resistant to digestion because of retrogradation, but there are products in which some of the starch is resistant to digestion.

There is also a fraction of starch called **slowly digesting starch (SDS). SDS** is converted into D-glucose (which is then absorbed) in the small intestine at a slower rate than is ordinary gelatinized starch and is incompletely converted into glucose because of its slow rate of digestion. Such material has a low glycemic index and is of benefit to diabetic patients. Maltooligosaccharides (maltodextrins) (Chapter 7) in which the reducing-end D-glucose unit has been reduced to sorbitol (Chapter 2), products known as **hydrogenated starch hydrolyzates (HSH)**, are more slowly digested than are ordinary maltooligosaccharides.

Products that are **nondigestible maltodextrins** are made from starch by the following general procedure. Starch (corn, although other sources could undoubtedly be used) is dextrinized (Chapter 7). The process is a pyroconversion (that is, treatment of low-moisture, acidified starch with heat), so three types of reactions occur during the dextrinization: (1) hydrolysis of glycosidic bonds, (2) transglycosidation, (3) formation of new glycosidic bonds via reversion (Chapters 2 and 7). The relative amounts of these three reactions, the first of which depolymerizes the starch and the second and third of which make new glycosidic bonds, depends on the moisture content of the starch, the type and amount of the acid catalyst, and the temperature. As acidified starch (dry, but with some moisture present, often about 13% in the case of corn starch) is heated, glycosidic bonds undergo hydrolysis, some water is consumed, and a rise in temperature occurs. After the temperature surpasses 100°C, dehydration occurs, and transglycosidation and repolymerization begins. New linkages are introduced, and the resulting dextrin is more highly branched than the native starch (Chapter 7). The products, in addition to containing the normal linkages found in starch, contain β-1,6, β-1,2, and β-1,3 linkages and often a levoglucosan unit (Chapter 2) at the reducing end. The dextrins are then treated with amylases to remove most of the digestible structures. The nondigested part is then recovered and spray dried. The average molecular weights of these products is about 2000 (average DP about 12), but there is a wide range of molecular weights (180 [D-glucose] to about 20,000). The products analyze as at least 85% DF. (The linkages that are different than those found in the native starch are what makes the products resistant to hydrolysis, both by the enzymes used in the process and by human digestive enzymes.). The products produce low viscosity. Like polydextrose (next section), they are more stable to acidic conditions than the sucrose it is designed to replace. Their properties, caloric values, and applications are similar to those of polydextrose. While polydextrose is acidic because it contains esterified citric acid, these products are neutral. They also have low degrees of hygroscopicity.

Polydextrose

Polydextrose, whose preparation and structure were described in Chapter 3, analyzes as about 90% dietary fiber (that is, only about 10% of it is digestible). It, therefore, has a very low glycemic index. Its caloric value is about 1 Kcal/g (as compared to about 4 Kcal/g for D-glucose and sucrose).[6] Polydextrose has both dietary fiber and prebiotic properties, although because of its high solubility, it and the nondigestible dextrin described in the previous section are not measured as dietary fiber by most methods. It is employed primarily as a sucrose replacer (that is, as a nonsweet bulking agent) (provides solids to maintain texture in low-sugar products, including sugar-free jam) (see Section Bulking agents), as a nonsweet bodying agent (provides organoleptic/mouthfeel properties), as a humectant (to maintain freshness and tenderness of bakery products and nutrition bars), and to lower the freezing point in frozen dairy desserts. It

[6] Used here is chemical (SI) terminology. In nutritional terms, these values would be about 1 and about 4 calories/gram.

is used in bakery fillings because of its stability and water-holding capacity (that is, to prevent or reduce water migration). It may also be used in low-sugar hard and chewy candies, frostings, toppings, and other products in which amounts of modified food starch, flour, sugar, and/or syrups are reduced (that is, in which the glycemic load is reduced). Polydextrose has fat-sparing properties and is used in fat-free ice cream and low-fat cookies (see Section Fat mimetics).

Cellulose

Cellulose (Chapter 8) and its derivatives are, of course, nondigestible and classified as dietary fiber. Cellulose itself is available in several forms, primary among which are powdered cellulose and microcrystalline cellulose.

Inulin

Inulin, oligosaccharides derived from inulin (fructooligosaccharides, FOS), and fructooligosaccharides made by fermentation (Chapter 3) function as both dietary fiber and prebiotics (Chapter 10).

Hydrolyzed guar gum

All polysaccharides other than the starch polysaccharides are components of dietary fiber. However, they differ in providing health benefits. β-Glucan is very effective physiologically. Another polysaccharide that is quite effective is guaran/guar gum. However, the high viscosity of its solutions prevents its application in amounts that would provide health benefits. A product is available in which the molecular weight of guaran has been reduced by hydrolysis to the extent that it provides high-solid solutions of low viscosity without impairing its effectiveness as dietary fiber. Hydrolyzed guar gum/guaran may also provide functionality to certain food products.

Pectin

Pectin (Chapter 15) is another hydrocolloid that deserves special mention as a source of dietary fiber with proven health benefits. Pectin and combinations of pectin and agar (Chapter 13) have been used to treat diarrhea.

Chitosan

Chitin has a chemical structure identical to that of cellulose with the exception of having an acetamido group (-NH-CO-CH_3) on C2 rather than a hydroxyl group. It is, therefore, a polymer of (1→4)-linked 2-acetamido-2-deoxy-β-D-glucopyranosyl (*N*-acetyl-β-D-glucosaminyl) units, a structure very similar to that of cellulose. (Disaccharide repeating unit structures of both chitin and cellulose are given in Chapter 4.) (The 2-deoxy descriptor means that the glucosyl unit does not have a hydroxyl group/oxygen atom on C2.) Chitin is believed to be the second most abundant biopolymer (after cellulose). It is a component of the exoskeletons and supporting tissues of crustaceans and insects. It is also found in the cell walls of some fungi and in some other organisms.

Chitosan is produced by mostly deacetylating chitin by treatment with a hot, concentrated solution of sodium hydroxide (50%−52% NaOH at 130°C [265°F]). This process converts some of the acetamido groups at C2 into amino (-NH_2) groups (at the sites of deacetylation). Chitosan is only soluble in acidic solutions in which the amino group is in the ammonium ion (−NH_3^+) form, making it the only commercial cationic polysaccharide produced without etherification with a cationic reagent. Acetic, ascorbic, citric, lactic, other organic acids, and mineral acids may be used for the dissolution of chitosan. Neutralization of acidic chitosan solutions results in its precipitation or gelation of the solutions.

Chitosan has been investigated as a nondigestible food ingredient, but is used sparingly, if at all. Rather, it is promoted and sold as a dietary supplement that "prevents fat absorption" and is a "fat blocker". It has also been called a "fat magnet." However, there are no human clinical studies that show that it is effective in this way. In fact, the studies that have been done have led to conclusions that it does not block fat absorption. It may have some efficacy in lowering serum cholesterol levels, but claims that it will reduce the risk of coronary heart disease (heart attacks), high blood pressure, cancer, or obesity are not allowed (in the United States).

Other polysaccharides

Once again, all polysaccharides other than the starch polysaccharides contribute to either the soluble or insoluble dietary fiber content of a food product. Some have received special attention in this section, including methylcellulose under the discussion of psyllium gum. Gum arabic (Chapter 16) also provides the physiological effects of dietary fiber. Another is soluble soy polysaccharide (soy fiber), which is obtained by extracting the water-soluble polysaccharides from soy beans. This product analyzes as up to 70% DF. It has emulsifying and filmmaking properties and has the ability to stabilize proteins in acidic systems, especially at pH values below 4.

Low-molecular-weight carbohydrates

Monosaccharides and products derived from them that are not absorbed as they pass through the small intestine, and oligosaccharides and products derived from them that are not broken down into monosaccharides that can be absorbed as they pass through the small intestine to the colon are by definition components of dietary fiber. (As has been pointed out several times, all oligosaccharides other than the disaccharides sucrose and lactose and the oligosaccharides derived from starch [maltooligosaccharides/maltodextrins] survive digestion in the small intestine.) These components are used by the microbiota of the colon via anaerobic fermentation with the production of short-chain fatty acids and, therefore, function as prebiotics. Absorption of some of the short-chain fatty acids provides some calories to the host.

Polyols such as sorbitol are not absorbed well, undergo fermentation in the colon, and provide, at least, laxation. However, because of a concern that polyols could be consumed in substantial amounts and could produce flatulence and diarrhea and because sugar alcohols and other low-molecular-weight saccharides (DP 1−3) are not measured as DF by most analytical methods (because of their solubility and nonprecipitability) sugar alcohols are often placed in a separate category. So both high- and low-molecular-weight carbohydrates may provide some or all of the physiological effects of dietary fiber, but some of the low-molecular-weight ones (especially the sugar alcohols) may not be defined as components of DF by some.

Dietary fiber as an ingredient

Selecting the proper fiber ingredient from the wide variety of fiber ingredients available, incorporating it into a formulation, and choosing the proper processing equipment to make a desirable product is a challenge to the food product developer. Determining the efficacy of the available ingredients in providing the different health benefits attributed to dietary fiber and the daily amounts needed to effect the benefits is a challenge to nutrition and medical researchers. A special challenge to the food product developer is to incorporate enough of the DF ingredient to have a beneficial physiological effect. (Remember that hydrocolloids are normally added to food products in only small amounts.) Some functionalities that could cause problems can be summarized as follows:

- hydration and water-binding properties (Remember DF is primarily carbohydrate in nature.)
- viscosity building
- incompatibility with hydrocolloids added for another purpose (Phase separation may result.)
- gel formation
- changes in product quality/sensory attributes (such as color, flavor, and texture)
- changes in shelf life.

Adding ingredients containing dietary fiber to processed foods to make the food more healthful usually changes the texture and/or results in changes in processing characteristics. As has already been pointed out, all hydrocolloids, other polysaccharides (especially the polysaccharides associated with plant cell walls, and even some molecules of starch that comprise one of the five categories of resistant starch that are components of DF). Nondigestible and/or nonabsorbable mono- and oligosaccharides may also be included as DF components, depending on the definition used. It has also been pointed out that there is no uniformity of chemical structure or physical or physiological properties of substances that comprise DF. As simple examples, DF products may be soluble or insoluble and differ considerably in their water-binding/holding capacities. In addition, the type of processing equipment that may be required to make high-fiber foods or ingredients may affect the nature of the product.

The recommendation (in the United States) for daily fiber intake is 25−30 g/day. Hydrocolloids are sources of soluble fiber, but such small concentrations are used in food products to produce the desired functionality that they are not a substantial source of fiber. To get the recommended amount, persons need to consume high-fiber foods, such as fruits, legumes, and other vegetables, but most persons do not consume these in the quantities required. A food industry solution is to add a fiber source(s) to a processed food. However, some high-fiber ingredients introduce nonacceptable flavors that are difficult to mask and many alter the texture of the product. A common change is an increase in water binding/holding, making the product "gummy". Increases in viscosity, changes in gelation behavior or gel properties, increases in oil binding, and changes in color may also occur.

Bulking agents

It has already been mentioned that an important function of carbohydrates in foods is to provide bulk and/or body. A **bulking agent** is a substance that contributes solids to provide texture/palatability. A **bodying agent** is a substance that provides or improves mouthfeel (organoleptic attributes), which is generally a function of the specific rheology perceived during chewing and swallowing. Modified food starch products are particularly effective in imparting bulk and body to foods. Sucrose is employed for this purpose in some foods.

In yeast-leavened bakery products, sugar is metabolized by yeast, producing carbon dioxide and providing the leavening. Also, the textural properties imparted to foods by sucrose are often as important as, or more important than, the sweetness it imparts (Chapters 3 and 19). The same is true of syrups (either glucose or high-fructose syrups), although the functionalities imparted by them may be quite different from those imparted by sucrose (Chapter 19). Therefore, to replace either sugar or a syrup with a high-intensity sweetener, another ingredient that will provide the other functionalities provided by sugar or a syrup must also be added, but it must (in the quantity required to provide the functionalities) provide no or fewer calories than the sugar or syrup it replaces. The general term for such an ingredient is **reduced-calorie bulking agent.** In general, reduced-calorie bulking agents are of relatively low molecular weight. Functionalities of sugar or a syrup that may have to be provided by a reduced-calorie bulking agent ingredient or a combination of ingredients might include the following (depending on the specific product). The agent may need to have the ability to

- increase product volume (provide bulk)
- provide humectancy (Ability to prevent or slow undesirable migration of water molecules. Prevention of increases or decreases in the moisture content of the product. Any anticaking function is included.)
- reduce water activity and retard of growth of bacteria
- provide sufficient shelf life
- make an appropriate contribution to viscosity or gel characteristics (texture/consistency)
- provide proper crumb texture (bakery products)
- delay gelatinization of starch (bakery products)
- provide appropriate spread (cookies/biscuits)
- provide appropriate surface cracking (cookies/biscuits)
- form a surface glaze
- form appropriate product color
- have an appropriate effect on flavors/taste
- undergo required crystallization
- depress the freezing point appropriately.

An ideal bulking agent (that is, one that imparts all the required functionalities of sucrose or a syrup without providing close to the same number of calories) is not available. Even combinations of ingredients that impart the necessary functionalities (without providing close to the same number of calories) have yet to be developed. Ingredients that have been investigated as reduced-calorie bulking agents (or just something that imparts the same functionality as sucrose or a syrup) include polydextrose; polyols (Chapters 2 and 19); various hydrocolloids, especially those of low molecular weight (low-viscosity grades); various sources of dietary fiber, including various fruit concentrates; proteins; high-protein flours; and various oligosaccharides. Polydextrose and fructooligosaccharides (FOS) (both earlier in this chapter) have been the extensively investigated as reduced-calorie bulking agents. However, any low-molecular-weight carbohydrate that has a low caloric value (that is, is not digested to absorbable monosaccharide units in the small intestine) will contribute to fermentation in the large intestine and produce gastrointestinal discomfort (diarrhea and flatulence) to different degrees in different individuals if consumed in sufficient amounts.

Fat mimetics

It is recognized by governmental and other health organizations that obesity increases the risk of cardiovascular problems and heart disease. The average American receives about 37% of his/her calories from fats and oils, compared to recommendations of health experts that fats/lipids provide not more than 30% of calories. Dietary lipids may contribute to obesity because they contribute about 9 Kcal/gram, compared to contributions of digestible carbohydrates of about 4 Kcal/g.[6] Carbohydrate consumption that results in hyperglycemia (and, consequently, hyperinsulinemia) also contributes to obesity.

Because most persons are reluctant to give up the sensation of fattiness in their food, fat mimetics have been sought. (A **fat mimetic** [some products in this family being referred to as **fat sparers** or **fat replacers**] is a substance that imitates [mimics] the sensory characteristics of a fat or oil.) An ideal fat replacer would have all functional characteristics of lipids, but with significantly fewer, or zero, calories. The functional properties of lipids that must be mimicked include organoleptic properties (largely rheology and lubricity), the ability to dissolve lipid-soluble flavors and other substances, aeration, aroma, emulsification, flavor, heat stability, and spreadability. Thc ideal replacer does not exist; rather, a variety of products that produce the perception of fats and/or oils are available as fat-sparing ingredients. These products are based on carbohydrates, proteins, or nondigestible lipids, with the largest number in use being carbohydrate based, which are the only ones covered in this book.

The fatty sensation is primarily dependent on particle or droplet size and rheology, much less on the chemical nature of the ingredient. Fat mimetics that produce a fatty sensation are made from carbohydrates. In every case, the mimetic contains microcrystals or microparticles less than 2 μm in diameter. These microcrystals or microparticles may be derivatized, perhaps only on their surface, to make them fat like. One such carbohydrate that contains microcrystals of a size that gives a fat-like mouthfeel is microcrystalline cellulose (Chapter 8). Fat mimetics can also be made from a variety of starches. One mimetic, a microcrystalline maltodextrin made by partial enzyme-catalyzed hydrolysis of potato starch, is claimed to be able to replace up to 50% of the fat in certain food products. Another is a maltodextrin made from tapioca starch that has been octenylsuccinylated to low DS levels. (It is claimed that when 0.5% of a 25% suspension of the latter is added to ice cream containing the required minimum of 10% butterfat, the ice cream has the eating quality of one containing 14% butterfat.). Another fat mimetic is made by treating granular corn starch with dilute acid or with α-amylase to a dextrose equivalencey (DE) of less than 5, preferably about 2 (Chapter 7). In this treatment, the amorphous regions of the granule are preferentially removed by hydrolysis, giving a more crystalline product. Other fat mimetic materials are the products made from cereal grain flours and brans[4] high in β-glucan (especially oat flour and bran, discussed earlier in the section on β-glucans), and small micrometer (micron)-size pectin gel particles made from a special type of pectin (Chapter 15). Slurries of small-granule starches, such as those from amaranth, which are about 1 μm in diameter, produce some of the mouthfeel characteristics of a fat (particularly after the granules are given a mild treatment with glucoamylase) but are not used, at least not to any appreciable extent. These microparticulate materials, in addition to having particle sizes similar to those of fat globules and oil droplets, bind water and provide lubrication (slipperiness), another sensory attribute of fat. Mentioned in Chapter 5 was the fact that solutions of hydrocolloids that are not highly pseudoplastic, because they do not shear thin, coat the mouth and are difficult to swallow, producing a sensation of sliminess or slipperiness. Persistent coating of the mouth is also a characteristic of fats and oils, so a slight amount of a food gum that is classified as one producing slimy solutions (Fig. 5.12) can also give some of the sensation of fat. As yet, no single fat mimetic provides all the sensory characteristics of fat.

Fatty acid esters of sucrose (Chapter 3) provide the physical characteristics of fats and oils while being noncaloric. Members of this family of products melt and can be used for frying (including deep frying), can carry fat-soluble aromas and flavors, and otherwise will provide the organoleptic qualities of triacylglycerols. However, daily intake amounts are limited because of the negative physiological effects associated with the products being nondigestible fats and oils that pass through the digestive tract unchanged, and their use is likewise limited.

Additional resources

Glycemic Index and Glycemic Load

Brand-Miller, J., 2007. The glycemic index as a measure of health and nutritional quality; an Australian perspective. Cereal Foods World 52, 41–44.

Brand-Miller, J., Buyken, A.E., 2012. The glycemic index issue. Current Opinion in Lipidology 23, 62–67.

DeVries, J.W., 2007. Glycemic index: the analytical perspective. Cereal Foods World 52, 45–49.

Monro, J.A., Shaw, M., 2008. Glycemic impact, glycemic glucose equivalents, glycemic index, and glycemic load: definitions, distinctions, and implications. American Journal of Clinical Nutrition 87, 237S–243S.

Raben, A., Macdonald, I.A., Fogelholm, M., 2016. Health aspects of mono- and disaccharides. Chap. 3. In: Eliasson, A.-C. (Ed.), Carbohydrates in Food, third ed. CRC Press, Boca Raton.

Sugars and Health

Benton, D., 2008. Sucrose and behavioral problems. Critical Reviews in Food Science and Nutrition 48, 385–401.

Buxton, C.H.S., Gardner, E.J., McNulty, H.M., 2010. Is sugar consumption detrimental to health? A review of the evidence 1995–2006. Critical Reviews in Food Science and Nutrition 50, 1–19.

Dolan, L.C., Potter, S.M., Burdock, G.A., 2010. Evidence-based review on the effect of normal dietary consumption of fructose on development of hyperlipidemia and obesity in healthy, normal weight individuals. Critical Reviews in Food Science and Nutrition 50, 53–84.

Dietary Fiber – General

Anderson, R., Westerlund, E., Åman, P., 2016. Cell-wall polysaccharides: structural, chemical, and analytical aspects. Chap. 4. In: Eliasson, A.-C. (Ed.), Carbohydrates in Food, third ed. CRC Press, Boca Raton.

BeMiller, J.N., 2017. Carbohydrate analysis. In: Nielsen, S.S. (Ed.), Food Analysis, fifth ed. Springer, New York, pp. 333–360.

Brownlee, I.A., 2010. The physiological roles of dietary fibre. Food Hydrocolloids 25, 238–250.

Chawla, R., Patil, G.R., 2010. Soluble dietary fiber. Critical Reviews in Food Science and Food Safety 9, 178–196.

Cho, S.S., Samuel, P. (Eds.), 2009. Fiber Ingredients: Food Applications and Health Benefits. CRC Press, Boca Raton.

Cui, S.W., Wu, Y., Ding, H., 2013. The range of dietary fiber ingredients and a comparison of their technical functionality. In: Delcour, J.A., Poutanen, K. (Eds.), Fiber-rich and Wholegrain Foods. Woodhead Publishing, Cambridge, pp. 96–119.

Dikeman, C.L., Fahey Jr., G.C., 2006. Viscosity as related to dietary fiber: a review. Critical Reviews in Food Science and Nutrition 46, 649–663.

Dreher, M.L., 2018. Dietary Fiber in Health and Disease. Humana Press, New York.

Grabitske, H.A., Slavin, J.L., 2009. Gastrointestinal effects of low-digestible carbohydrates. Critical Reviews in Food Science and Nutrition 49, 327–360.

van der Kamp, J.W., Jones, J.M., McCleary, B.V., Topping, D.L. (Eds.), 2010. Dietary Fibre: New Frontiers for Food and Health. Wageningen Academic Publishers, Wageningen.

Kritchevsky, D., 2001. Dietary fibre in health and disease. In: McCleary, B.V., Prosky, L. (Eds.), Advanced Dietary Fibre Technology. Blackwell Science, London, pp. 149–161.

McRorie Jr., J.W., Fahey Jr., G.C., 2015. Fiber supplements and clinically meaningful health benefits. Identifying the physicochemical characteristics of fiber that drive specific physiologic effects. In: Wallace, T.C. (Ed.), Dietary Supplements in Health Promotion. CRC Press, Boca Raton, pp. 161–205.

Parrett, A., Harris, H., Edwards, C.A., 2016. Nondigestible carbohydrates: nutritional aspects. Chap. 6. In: Eliasson, A.-C. (Ed.), Carbohydrates in Food, third ed. CRC Press, Boca Raton.

Redgwell, R.J., Fischer, M., 2005. Dietary fiber as a versatile food component: an industrial perspective. Molecular Nutrition & Food Research 49, 521–535.

Rose, D.J., Hamaker, B.R., 2011. Overview of dietary fiber and its influence on gastrointestinal health. In: Paeschke, T.M., Aimutis, W.R. (Eds.), Nondigestible Carbohydrates and Digestive Health. Wiley-Blackwell, Boston, pp. 185–221.

Rose, D.J., DeMeo, M.T., Keshavarzian, A., Hamaker, B.R., 2007. Influence of dietary fiber on inflammatory bowel disease and colon cancer: importance of fermentation pattern. Nutrition Reviews 65, 51–62.

Zhang, G., Hamaker, B.R., 2010. Cereal carbohydrates and colon health. Cereal Chemistry 87, 331–341.

Arabinoxylans

Courtin, C.M., Delcour, J.A., 2002. Arabinoxylans and endoxylanases in wheat flour bread-making. Journal of Cereal Science 35, 225–243.

Courtin, C.M., Gys, W., Delcour, J.A., 2006. Arabinoxylans. Journal of the Science of Food and Agriculture 86, 1587–1595.

Hille, J.D.R., Schooneveld-Bergmans, M.E.F., 2004. Hemicellulases and their synergism in breadmaking. Cereal Foods World 49, 283–286.

Izydorczyk, M.S., 2009. Arabinoxylans. In: Phillips, G.O., Williams, P.A. (Eds.), Handbook of Hydrocolloids. CRC Press, Boca Raton, pp. 653–692.

Kale, M.S., Pai, D.A., Hamaker, B.R., Campanella, O.H., 2010. Structu8re-function relationships for corn bran arabinoxylans. Journal of Cereal Science 52, 368–372.

Rose, D.J., Patterson, J.A., Hamaker, B.R., 2010. Structural differences among alkali-soluble arabinoxylans from maize (*Zea mays*), rice (*Oryza sativa*), and wheat (*Triticum aestivum*) brans influence human fecal fermentation profiles. Journal of Agricultural and Food Chemistry 58, 493–499.

Beta-glucans

Duss, R., Nyberg, L., 2004. Oat soluble fibers (β-glucans) as a source for healthy snack and breakfast foods. Cereal Foods World 49, 320–325.

Izydorczyk, M.S., Dexter, J.E., 2008. Barley β-glucans and arabinoxylans: molecular structure, physicochemical properties, and uses in food products – a review. Food Research International 41, 850–868.

Stevenson, D.G., 2009. Cereal β-glucans. In: Phillips, G.O., Williams, P.A. (Eds.), Handbook of Hydrocolloids. CRC Press, Boca Raton, pp. 616–652.

Polydextrose

Auerbach, M.H., Mitchell, H., Moppett, F.K., 2011. Polydextrose. In: O'Brien Nabors, L. (Ed.), Alternative Sweeteners, fourth ed. CRC Press, Boca Raton, pp. 489–505.

Craig, S.A.S., 2001. Polydextrose: analysis and physiological benefits. In: McCleary, B.V., Prosky, L. (Eds.), Advanced Dietary Fibre Technology. Blackwell Science, London, pp. 503–508.

Starch and Health

Ashwar, B.A., Gani, A., Shah, A., Wani, I.A., Masoodi, F.A., 2016. Preparation, health benefits and applications of resistant starch – a review. Starch Staerke 68, 287–301.

Birt, D.F., Boylston, T., Hendrich, S., Jane, J.-l., Hollis, J., Li, L., McClelland, J., Moore, S., Phillips, G.J., Rowling, M., Schalinske, K., Scott, M.P., Whitney, E.M., 2013. Resistant starch: promise for improving human health. Advances in Nutrition 4, 587–601.

Emmambux, M.N., Taylor, J.R.N., 2016. Starch: nutritional and health aspects. Chap. 11. In: Eliasson, A.-C. (Ed.), Carbohydrates in Food. CRC Press, Boca Raton.

Fuentes-Zaragoza, E., Riqueline-Navarrete, M.J., Sánchez-Zapata, E., Pérez-Alvarez, J.A., 2010. Resistant starch as functional ingredient: a review. Food Research International 43, 931–942.

Fuentes-Zaragoza, E., Sánchez-Zapata, E., Sandra, E., Sayas, E., Navarro, C., Fernández-López, J., Pérez-Alvarez, J.A., 2011. Resistant starch as prebiotic: a review. Starch Staerke 63, 406–415.

Homayuni, A., Amini, A., Keshtiban, A.K., Mortazavian, A.M., Esazadeh, A.M., Pourmoradian, S., 2014. Resistant starch in food industry. Starch Staerke 66, 102–114.

Ma, Z., Boyce, J.I., 2018. Research advances on structural characterization of resistant starch and its structure-physiological function relationship: a review. Critical Reviews in Food Science and Nutrition 58, 1059–1083.

McOrist, A.L., Miller, R.B., Bird, A.R., Keogh, J.B., Noakes, M., Topping, D.L., 2011. Fecal butyrate levels vary widely among individuals but are usually increased by a diet high in resistant starch. Journal of Nutrition 141, 883–889.

Nasrin, T.A.A., Anal, A.K., 2014. Resistant starch: properties, preparations and applications in functional foods. In: Noomhorm, A., Ahmad, I., Anal, A.K. (Eds.), Functional Foods and Dietary Supplements: Processing Effects and Health Benefits. John Wiley & Sons, Hoboken, pp. 227–253.

Priebe, M.G., Eelderink, C., Wachters-Hagedorn, R.E., Vonk, R.J., 2018. Starch digestion and applications of slowly available starch. In: Sjöö, M. (Ed.), Starch in Food, second ed. Woodhead Publishing, Duxford, pp. 805–826.

Raigond, P., Ezekiel, R., Raigond, B., 2015. Resistant starch in food: a review. Journal of the Science of Food and Agriculture 95, 1968–1978.

Svihus, B., Hervik, A.K., 2016. Digestion and metabolic fates of starch, and its relation to major nutrition-related health problems: a review. Starch Staerke 68, 302–313.

Shi, Y.-C., Maningat, C.C. (Eds.), 2013. Resistant Starch: Sources, Applications and Health Benefits. Wiley-Blackwell, Hoboken.

Prebiotics

Bornet, F.R.J., 2001. Fructo-oligosaccharides and other fructans: chemistry, structure and nutritional effects. In: McCleary, B.V., Prosky, L. (Eds.), Advanced Dietary Fibre Technology. Blackwell Science, London, pp. 480–493.

Broekaert, W.F., Courtin, C.M., Verbeke, K., van de Wiele, T., Verstraete, W., Delcour, J.A., 2011. Prebiotic and other health-related effects of cereal-derived arabinoxylans, arabinoxylan-ligosaccharides, and xylooligosaccharides. Critical Reviews in Food Science and Nutrition 51, 178–194.

Charalampopoulos, D., Rastall, R.A. (Eds.), 2009. Prebiotics and Probiotics Science and Technology. Springer-Verlag, New York.

Cho, S.S., Finocchiaro, E.T., 2009. *Handbook of Prebiotics and Probiotics Ingredients: Health Benefits and Food Applications*. CRC Press, Boca Raton.

Flamm, G., Glinsman, W., Kritchevsky, D., Prosky, L., Roberfroid, M., 2001. Inulin and oligofructose as dietary fiber: a review of the evidence. Critical Reviews of Food Science and Nutrition 41, 353–362.

Fuentes-Zaragoza, E., Sánchez-Zapata, E., Sendra, E., Sayas, E., Navarro, C., Fernández-López, J., Pérez-Alvarez, J.A., 2011. Resistant starch as prebiotic: a review. Starch Staerke 63, 406–415.

Gibson, G.R., Roberfroid, M. (Eds.), 2008. Handbook of Prebiotics. CRC Press, Boca Raton.
Hidaka, H., Adachi, T., Hirayama, M., 2001. Development and beneficial effects of fructo-oligosaccharides (Neosugar). In: McCleary, B.V., Prosky, L. (Eds.), Advanced Dietary Fibre Technology. Blackwell Science, London, pp. 471–479.
Hursh, H., Martin, J., 2005. Low-carb and beyond: the health benefits of inulin. Cereal Foods World 50, 57–60.
Raben, A., Macdonald, I.A., Fogelholm, M., 2016. Health aspects of mono- and disaccharides. Chap. 3. In: Eliasson, A.-C. (Ed.), Carbohydrates in Food, third ed. CRC Press, Boca Raton.
Rastall, B., Gibson, G. (Eds.), 2006. Prebiotics. Wiley-VCH, Hoboken.
Swennen, K., Courtin, C.M., Delcour, J., 2006. Non-digestible oligosaccharides with prebiotic properties. Critical Reviews in Food Science and Nutrition 46, 459–471.
Tomasik, P.J., Tomasik, P., 2003. Probiotics and prebiotics. Cereal Chemistry 80, 113–117.
Yoo, J.Y., Kim, S.S., 2016. Probiotics and prebiotics: present status and future perspectives on metabolic disorders. Nutrients 8, 173/1–173/20.

Some Other Individual Poly- and Oligosaccharide Components That Are Beneficial to Human Health

Finley, J.W., Soto-Vaca, A., Helmbach, J., Rao, T.P., Juneja, L.R., Slavin, J., Fahey, G.C., 2013. Safety assessment and caloric value of partially hydrolyzed guar gum. Journal of Agricultural and Food Chemistry 61, 1756–1771.
López-Franco, Y., Higuera-Clapara, I., Goycoolea, F.M., Wang, W., 2009. Other exudates: tragacanth, karaya, mesquite gum and larch arabinogalactan. In: Phillips, G.O., Williams, P.A. (Eds.), Handbook of Hydrocolloids. CRC Press, Boca Raton, pp. 495–534.
Ohkuma, K., Wakabayashi, S., 2001. Fibersol-2: a soluble, non-digestible, starch-derived dietary fibre. In: McCleary, B.V., Prosky, L. (Eds.), Advanced Dietary Fibre Technology. Blackwell Science, London, pp. 509–523.
Schoterman, H.C.M., 2001. Galacto-oligosaccharides: properties and health aspects. In: McCleary, B.V., Prosky, L. (Eds.), Advanced Dietary Fibre Technology. Blackwell Science, London, pp. 494–502.

Bulking Agents (See also Polydextrose)

Auerbach, M., Dedman, A.-K., 2012. Bulking agents – multi-functional ingredients. In: O'Donnell, K., Kearsley, M. (Eds.), Sweeteners and Sugar Alternatives in Food Technology, second ed. Wiley-Blackwell, Ames, pp. 435–470.
Vasiljevic, T., Varzakas, T., 2012. Bulking and fat-replacing agents. In: Varzakas, T., Labropoulos, A., Anestis, S. (Eds.), Sweeteners. CRC Press, Boca Raton, pp. 395–418.

Fat Mimetics

Deis, R.C., 2014. Other low-calorie ingredients: fat and oil replacers. In: O'Brien-Nabors, L. (Ed.), Alternative Sweeteners, fourth ed. CRC Press, Boca Raton, pp. 507–523.
Vasiljevic, T., Varzakas, T., 2012. Bulking and fat-replacing agents. In: Varzakas, T., Labropoulos, A., Anestis, S. (Eds.), Sweeteners. CRC Press, Boca Raton, pp. 395–418.

Nonenzymic Browning and Formation of Acrylamide and Caramel

18

Chapter Outline

Key information and skills that should be obtained from the study of this chapter will enable you to

1. Define and/or identify:
   - Maillard reaction
   - Nonenzymic (nonenzymatic) browning
   - Maillard browning
   - Amadori rearrangement
   - Amadori product/compound
   - Hydroxymethylfurfural (HMF; 5-hydroxymethyl-2-furaldehyde)
   - Melanoidins
   - Caramel
   - Caramelization
   - Levoglucosan
2. List the three basic stages in the nonenzymic (Maillard) browning reaction sequence.
3. List the five factors affecting nonenzymic (Maillard) browning and describe how each affects the reaction.
4. List four ways to control nonenzymic (Maillard) browning.
5. Describe how bisulfite ion works to retard nonenzymic (Maillard) browning.
6. Describe the roles of sulfite and ammonium ions in caramelization.
7. Explain the origin of acrylamide in foods.
8. List five factors governing acrylamide formation in foods.
9. Explain why acrylamide levels in deep-fried potato products are especially high.

Carbohydrate Chemistry for Food Scientists. https://doi.org/10.1016/B978-0-12-812069-9.00018-2

Introduction

Under some conditions, reducing sugars in foods that have been heated to high temperatures produce desirable brown pigments, flavors, and/or aromas, but other brown colors, flavors, aromas, and/or other compounds obtained upon heating or during long-term storage of foods containing reducing sugars are undesirable. Both types of these products and the processes that produce them are the subject of this chapter.

Maillard browning

Common browning of foods on heating or storage is usually due to a chemical reaction between reducing sugars (most often D-glucose) and a primary amino group (most often on a protein molecule). The reaction is called the **Maillard reaction** (named after the chemist who first studied and described it). The overall process is called **nonenzymic** (or **nonenzymatic**) **browning** (sometimes **Maillard browning**) to differentiate it from the more rapid, enzyme-catalyzed browning commonly observed in freshly cut fruits and vegetables (such as apples and potatoes).

Browning occurs via several reactions and can take several different routes through the reaction pathway, but the reaction sequence can be divided into three basic stages:

1. Condensation of a reducing sugar with an amine followed by a molecular rearrangement.
2. Molecular dehydrations and cyclizations.
3. Condensations of the products from stage 2 with themselves and with amino compounds.

Condensation of a reducing sugar with a compound containing an amino group and molecular rearrangement of the product

When aldoses or ketoses are heated in solution with amines, several reactions occur. These reactions produce numerous compounds (some of which are flavors, aromas, and dark-colored polymeric materials). They form rapidly at high temperatures and low water activities (such as are obtained during frying, roasting, and baking). (The same reactions that form the flavors, aromas, and colors may also occur during storage but much more slowly.) In foods, the amino groups are most often provided by protein molecules. Peptides and amino acids may also provide amino groups. With proteins, the primary amino group of L-lysine is the primary reactant. Reactions with the secondary amino group of the guanidino group of L-arginine also occur.

The reducing sugar reacts reversibly with the amino group to form a Schiff base (an imine, RN = CHR'), which may cyclize in the same way that an aldose cyclizes (Chapter 1) to form a compound called a glycosylamine[1] (as illustrated with D-glucose in Fig. 18.1). The Schiff base form may also undergo a reaction called an **Amadori**

[1] Sometimes erroneously called an N-glycoside. Nucleic acids, nicotinamide dinucleotide (NAD), and coenzyme A contain glycosylamine structures.

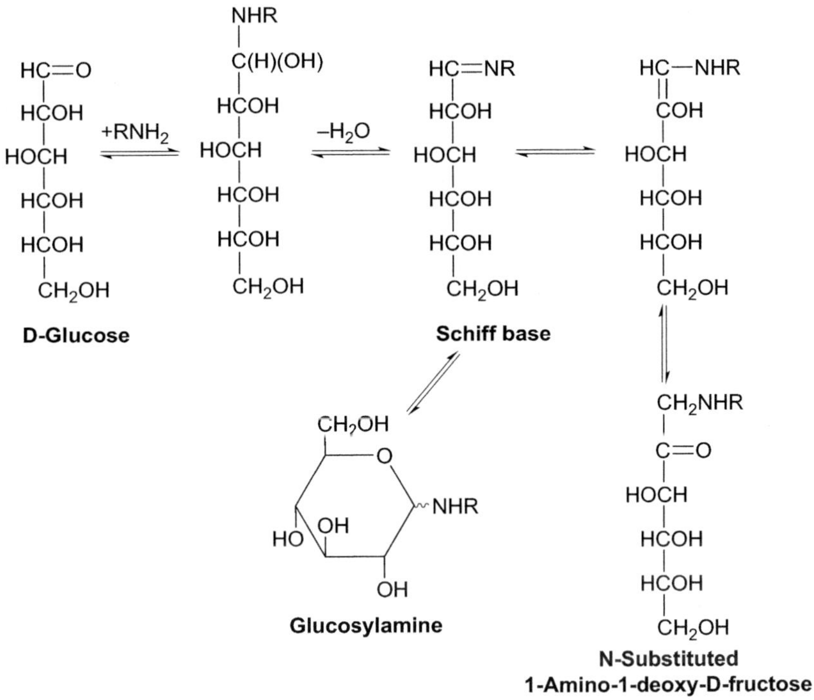

Figure 18.1 Condensation of D-glucose with a compound containing an amino group to form a Schiff base (imine) followed by cyclization of the imine into a glycosylamine or its isomerization to form an *N*-substituted 1-amino-1-deoxy-D-fructose compound (an Amadori compound). All reactions in this sequence are reversible.

rearrangement to give, in the case of D-glucose, a derivative of 1-amino-1-deoxy-D-fructose, which is known as an **Amadori product** (or **compound**). The mechanism for the Amadori rearrangement is presented in Fig. 18.2.

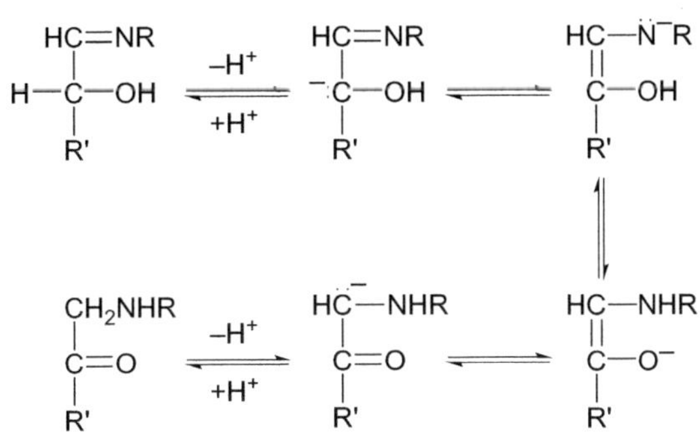

Figure 18.2 Basic mechanism of the Amadori rearrangement. Compared with the mechanism for isomerization of aldoses and ketoses (Fig. 1.4).

Amadori compounds produced by reaction of a reducing sugar (usually D-glucose) with various amino acids have been detected in a variety of foods, usually after heating or upon long-term storage. Among these are dried fruits, dried vegetables, dried milk products, condensed milk, evaporated milk, baked goods, soy sauce, and roasted nuts, cocoa beans, and coffee beans. Amadori reaction products (especially from reaction of D-glucose with hemoglobin) are found in the blood (especially in the blood of patients with diabetes mellitus). In fact, the resulting glycosylated protein is a better indicator of the overall blood sugar status of the patient over time than is determination of blood glucose.[2] And the Amadori reaction is involved in reactions that result in the natural cross-linking of the protein (notably collagen) that occurs with aging.

Amadori compounds are early intermediates in the browning reaction sequence. The process continues, especially at pH 5 or lower, through a complex series of reactions starting with dehydrations as shown below.

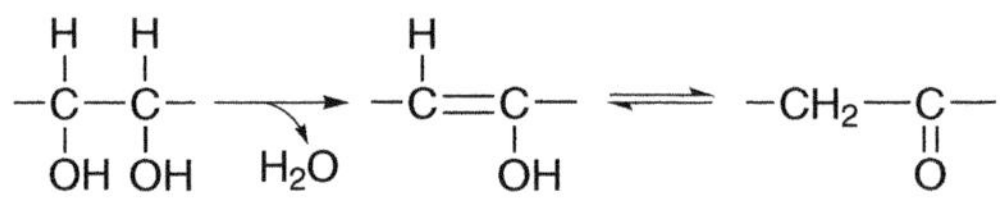

Dehydrations and cyclizations

Amadori compounds undergo rearrangements and eliminations to produce 1-, 3-, and 4-deoxydicarbonyl compounds, usually known by their common names (namely 1-, 3-, and 4-deoxyosones) (Fig. 18.3). Osone-forming reactions occur most readily at pH 4–7. The products (α-dicarbonyl compounds) are quite reactive. They can cyclize in the same way that aldoses and ketoses do. They will also lose HOH from adjacent carbon atoms (especially at high temperatures) and form conjugated unsaturated systems.

The primary products formed next from 3-deoxyosones by cyclization and elimination of two molecules of water are **5-hydroxymethyl-2-furaldehyde** (commonly known as **HMF**) from aldohexoses (Fig. 18.4). The empirical formula for HMF is that of an aldohexose minus 3 mol of water or $C_6H_6O_3$. Another furan (furfural [furaldehyde]) is formed from aldopentoses. (Furfural has the same structure as HMF, but without the hydroxymethyl [-CH_2OH] group.) Under less acidic conditions ($pH > 5$), the reactive cyclic compounds (HMF, furfural, and others) polymerize with compounds containing amino groups to form a dark-colored insoluble material containing nitrogen called **melanoidin** (Fig. 18.5). Amino acids and furans are incorporated into the polymeric end products constituting melanoidin. The individual components of the brown-to-black polymers constituting melanoidin vary in color, composition, molecular weight, solubility, and nitrogen content.

[2] Determination of the product (hemoglobin A1c [HbA1c]) in the blood gives an indication of the average blood glucose concentrations over the previous 2–3 months.

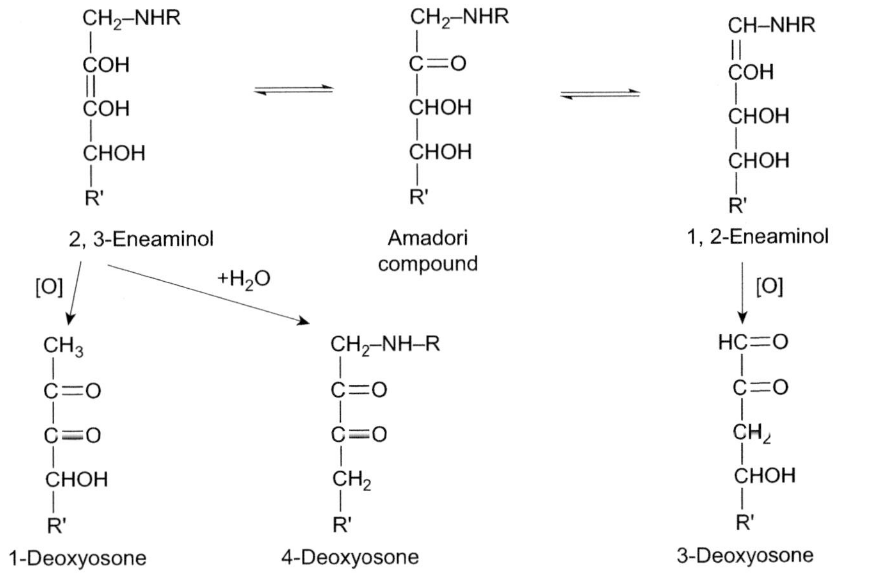

Figure 18.3 Generation of deoxyosones from Amadori products via elimination of H_2O and/or H_2NR. ([O] = a symbol indicating oxidation).

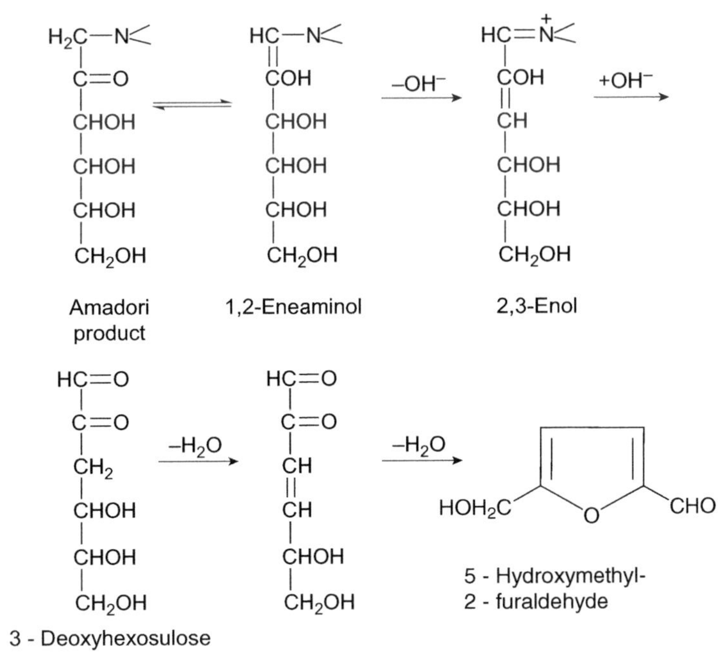

Figure 18.4 Isomerization of an *N*-substituted 1-amino-1-deoxyhexulose (an Amadori product from a hexose) followed by dehydration to form a 3-deoxyhexosulose, which undergoes another dehydration to form hydroxymethylfurfural (HMF; 5-hydroxymethyl-2-furaldehyde).

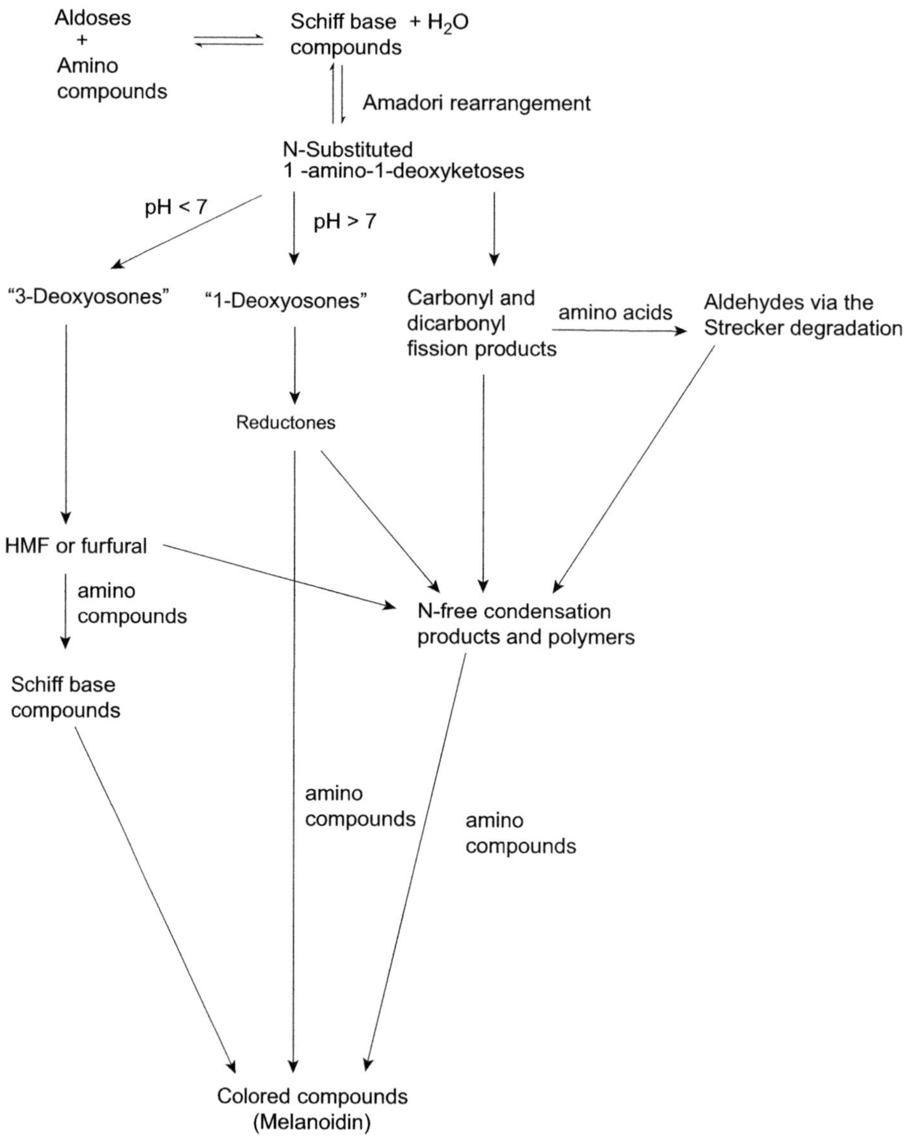

Figure 18.5 A condensed and simplified summary of reactions and products that may occur during nonenzymic browning.

A few examples of compounds formed from 3-deoxyosones are shown in Fig. 18.6. When higher concentrations of ammonia or primary amines (including the L-lysine side chain of proteins) are present, the primary products are pyrroles, where R′ can be H or CH_2OH as in furfural and HMF, respectively, and R is related to the amino compound (often lysine) that reacted to form the pyrrole. Pyridine compounds of the type shown below can also be formed as secondary products by a slightly different mechanism from common intermediates (a key one of which is the acylic structure shown).

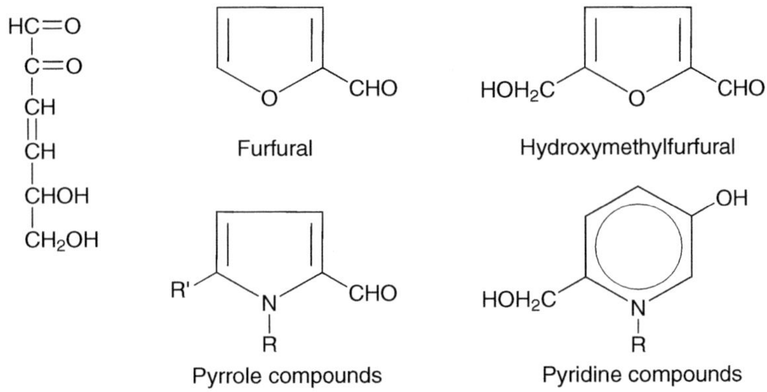

Figure 18.6 Some products formed from 1-deoxyosones.

Products formed from 1-deoxyosones include maltol (3-hydroxy-2-methylpyran-4-one) and isomaltol (3-hydroxy-2-acetylfuran), both of which contribute to the aroma and flavor of bread. Maltol and isomaltol are formed from disaccharides such as lactose (from the milk used to make the dough) and maltose (produced by amylase-catalyzed hydrolysis of starch [Chapter 7] during the fermentation step effected by yeast). In the structure below, isomaltol is shown as a substituted compound in which R is a sugar unit attached via a glycosidic bond.

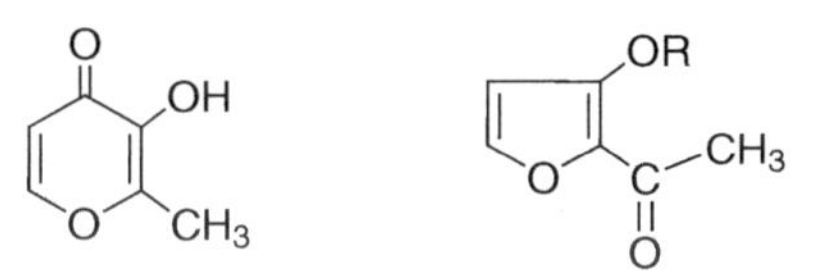

Maltol　　Substituted isomaltol

The browning product intermediates termed reductones in Fig. 18.5 can be either acyclic or cyclic compounds. In either case, they have the general structures given below. Reductones are weak acids and reducing agents. As a result of being reducing agents, they are antioxidants. (Vitamin C [ascorbic acid] has a reductone structure.) A reductone in which there is one or more carbon atoms without an oxygen atom bonded to it (as the result of dehydration [that is, loss of HOH]) is called a deoxyreductone. Because reductones and deoxyreductones can be involved in redox reactions, a variety of other intermediates can also be formed.

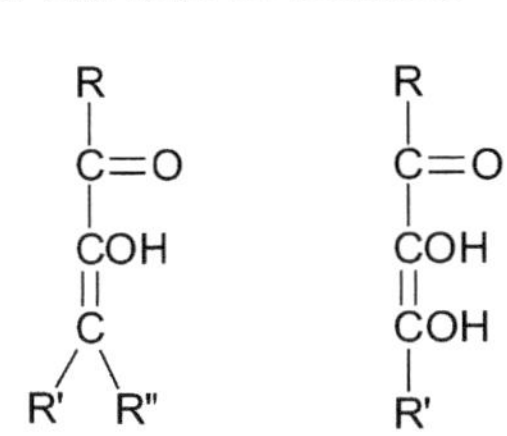

Reductones

Osones and deoxyosones will also undergo cleavage, either between the two carbonyl groups or at the site of an enediol (–COH=COH–, formed from $-\overset{O}{\overset{\|}{C}}-CHOH-$ by proton migration). Such a cleavage forms shorter chain products (that is, fission products), primarily aldehydes that can undergo reactions similar to what have already been discussed.

Another important reaction of dicarbonyl compounds (osones and deoxyosones) is the Strecker degradation (Fig. 18.7). Reaction of one of these compounds with an α-amino acid (R″–$CHNH_2$–CO_2H results sequentially in a Schiff base being formed, then decarboxylation (releasing CO_2), dehydration, and elimination to produce an aldehyde that is one-carbon atom shorter than the original amino acid.

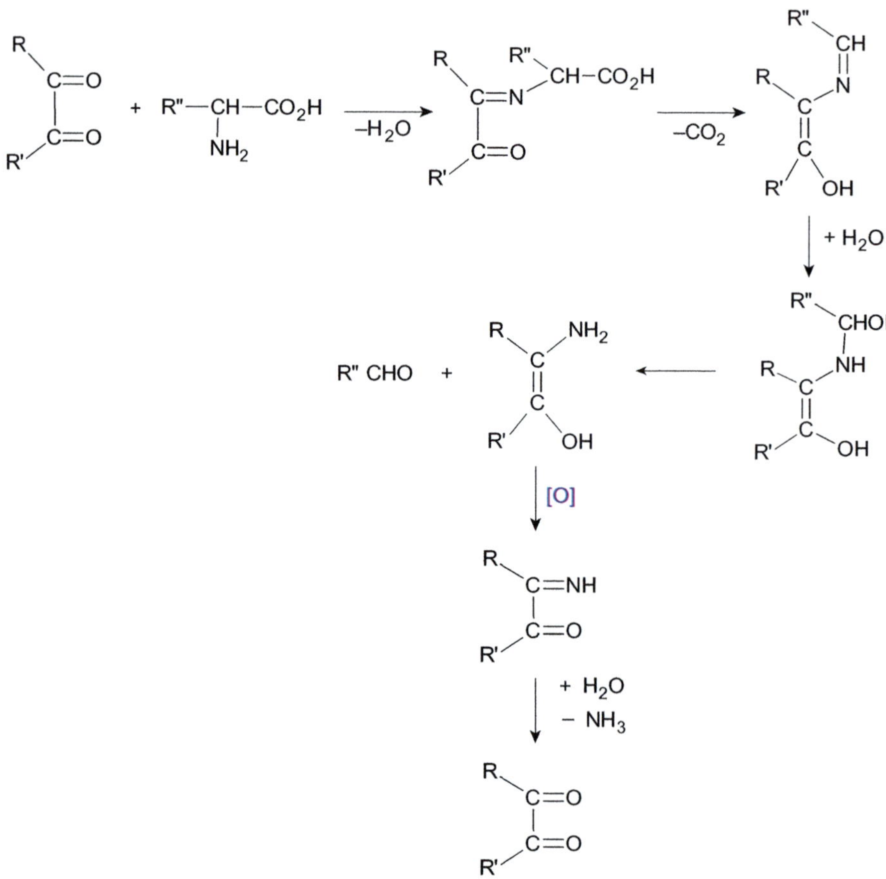

Figure 18.7 The mechanistic pathway of the Strecker reaction. In this reaction, an α-amino acid is degraded to an aldehyde that has one less carbon atom than did the original amino acid. Carbon dioxide and ammonia are released in the process. The ammonia can participate in pigment-forming reactions.

(Oxidation, addition of water, and loss of ammonia regenerate the original dicarbonyl compound from the remaining intermediate.) The aldehydes produced from amino acids often are major contributors to the aroma produced during baking. Aroma compounds produced in this way include 3-methylthiopropanal (methional, $CH_3-S-CH_2-CH_2-CHO$) from L-methionine, phenylacetaldehyde ($Ph-CH_2-CHO$) from L-phenylalanine, methylpropanal [$(CH_3)_2-CH-CHO$] from L-valine, 3-methylbutanal [$(CH_3)_2-CH-CH_2-CHO$] from L-leucine, and 2-methylbutanol [(CH_3-CH_2) $(CH_3)-CH-CHO$] from L-isoleucine. And because they are highly reactive compounds, osones and deoxyosones can react with other food components, especially in nucleophilic reactions, to produce a variety of other addition products that together with the combination products described here are sometimes called advanced glycation end products.

Condensations of the products of dehydrations and cyclizations

A variety of colored compounds (collectively called **melanoidins**) are formed in the nonenzymic browning process. Melanoidins can have molecular weights as great as 100,000. A variety of structures and molecular weights are found because of the variety of intermediates and the variety of possible condensation reactions. Some compounds that constitute the melanoidins contain nitrogen; some contain only carbon, hydrogen, and oxygen atoms. Due to their complexity, polydispersity,[3] and polymolecularity,[3] their characterization has been only cursory at best.

The structure of one low molecular weight condensation product that has been identified is given below as being typical of the low molecular weight, nitrogen-free products formed. This product absorbs light (that is, is colored) because of its eight conjugated double bonds. Such products can and do undergo further condensations, resulting in formation of brown-to-black pigments.

Modified proteins as Maillard reaction products

Products of the Maillard browning reaction can also be modified proteins. Modification of proteins is primarily the result of their reaction (especially reaction of the

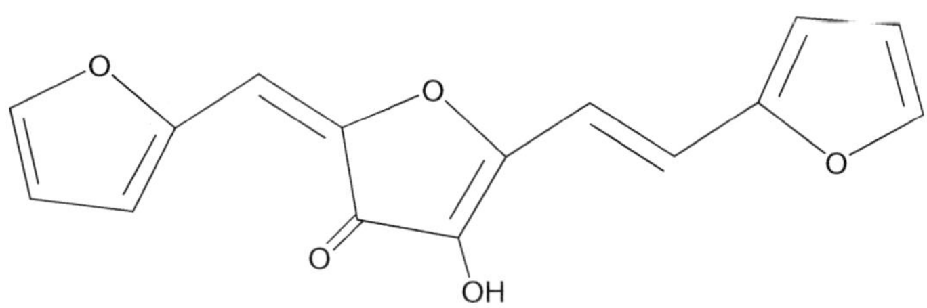

[3] For definitions, see Chapter 4 (Chemical Structures).

side chains of their L-lysine and L-arginine units) with carbonyl group–containing compounds such as reducing sugars, osones, furfural, HMF, and pyrrole derivatives. For example, reaction of the ε-amino group of a unit of L-lysine in a protein molecule followed by the Amadori rearrangement converts the L-lysine unit into a unit of N-fructofuranosyl-lysine. Further reaction via a 4-deoxyosone intermediate yields a unit of furosine. When the reaction proceeds via a 3,4-dideoxyosone intermediate, a pyrraline unit is formed.

Information about the previous heat treatment of milk can be obtained by enzyme-catalyzed hydrolysis of the proteins followed by quantitative analysis of the amino acids released. Approximate ranges of furosine detected in milk products are as follows (in ppm [mg/kg protein]): unpasteurized milk (35–55), pasteurized milk (50–75), ultrahigh temperature-processed (UHT) milk (500–1800), and powdered milk (1800–12,000). During thermal processing, some of the furosine undergoes cleavage, forming carboxymethyl-L-lysine (LysNH–CH_2CO_2H) units, which can also be used as an indicator of the extent of previous heat treatment of a food or food ingredient containing protein and reducing sugar.

Pyrraline units are produced in foods that have been heated to higher temperatures for longer times. Examples of approximate ranges of pyrraline detected in foods are as follows (in ppm [mg/kg protein]): UHT milk (2–5), condensed milk (30–135), pretzels (220–230), bread crust (540–3700), cookies (1000–1300). The range of furosine concentrations in bakery products is approximately 200–6000 mg/kg.

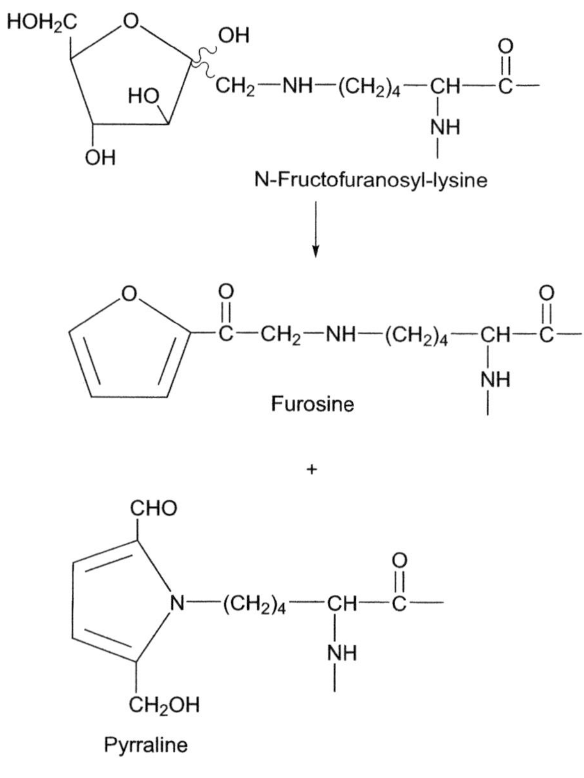

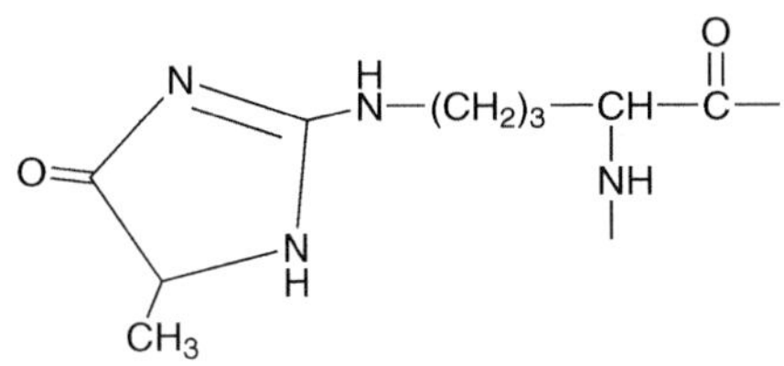

L-Arginine is modified in a similar way, producing protein units like that given below.

Reactions such as those described above lead to destruction of the amino acid unit reacted. Both L-lysine and L-arginine are classified as essential amino acids, and their destruction in this way (particularly in products baked at an alkaline pH) can be significant. For example, as much as two-thirds of the total L-lysine plus L-arginine can be destroyed in pretzel crusts. However, in less-alkaline products that are roasted or baked, losses of 15%–40% are more common.

Another result of Maillard reactions on proteins can be protein-protein cross-linking. Such a reaction is fairly easy to understand. It has already been pointed out that the reactions that result in the formation of furosine and pyrraline units from L-lysine go through 4-deoxyosone and 3,4-dideoxyosone intermediates, respectively. These dicarbonyl compounds can react with the amino groups on L-lysine and L-arginine units in other protein molecules or between two portions of the same molecule. Cross-linking results in some decrease in digestibility of the protein.

Factors affecting the extents of Maillard browning reactions

Different sugars undergo nonenzymic browning at different rates. In terms of classes of sugars, aldoses are more reactive than are ketoses and reactivity generally follows the order pentoses > hexoses > disaccharides. Because the reaction has a relatively high energy of activation, application of heat is generally required. Browning occurs at ever-increasing rates during baking, frying, roasting, etc., as water at the surface of the food is driven off, allowing surface temperatures to rise. The reaction is fastest when the system is not very acidic, primarily because, as pH values drop below the isoelectric pH of amino groups (about 8.5), the amino groups become increasingly protonated ($RNH_2 \rightarrow RNH_3^+$), and in the protonated form, amino groups are not able to participate in nucleophilic reactions. The rate of the Maillard reaction is also a function of the water activity (a_w) of a food product, reaching a maximum at a_w values in the range 0.6–0.7. Thus, for some foods, Maillard browning can be controlled by controlling water activity as well as by controlling reactant concentrations, time, temperature, and pH. Sulfur dioxide and bisulfite ions react with aldehydic groups (forming addition compounds) and thus will retard Maillard browning by reducing the concentration of at least some of a reactant (reducing sugar, HMF, furfural, etc.).

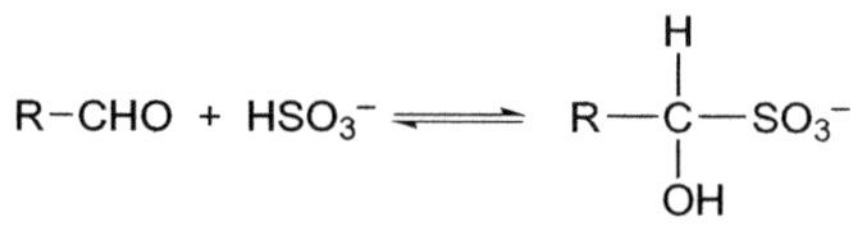

In summary, the product mixture formed is a function of temperature, pH, time, the nature of the reducing sugar, and the nature of the amino compound. Color, taste, and aroma are, in turn, determined by the product mixture. Reaction variables that can be controlled to increase or decrease the Maillard browning reaction are as follows:

1. Temperature (Decreasing the temperature reduces the reaction rate.)
2. pH (Decreasing the pH reduces the reaction rate.)
3. Adjustment of water content (Maximum reaction rate occurs at water activity values of 0.6–0.7 [about 30% moisture].)
4. The specific sugar (related to the percentage in the acyclic form) (More reactive sugars, with or without specific amino acids, may be employed to create browning reactions. Nonreducing carbohydrates such as sorbitol may be used to avoid browning. Bisulfite ions somewhat lower the concentration of reducing sugar.)
5. Presence of transition metal ions, such as Fe(II) and Cu(I) ions, undergo a one-electron oxidation under energetically favorable conditions. (A free-radical reaction may be involved near the end of the pigment-forming process.) The presence of metal ions is the most difficult to control, but fortunately is the least important factor.

Maillard browning and foods

Maillard browning products (including soluble and insoluble polymers/pigments) are found in products in which, during their preparation, mixtures of reducing sugars and amino acids, proteins, and/or other nitrogen-containing compounds are heated (for example, in soy sauce[4] and bread crusts). Browning is desired in most bakery products. The volatile compounds produced by nonenzymic browning (the Maillard reaction) provide the desirable aromas produced by baking, frying, and roasting. The Maillard reaction also produces flavor compounds. Aromas and flavors produced by this reaction are prominent in such products as beer, bread (crusts), chocolate, cocoa, cookies, roasted peanuts, popcorn, and cooked potatoes, and include pyrroles and pyrazines. The Maillard reaction is used to create chicken, bacon, beef, and pork flavors. Protein sources used are hydrolyzed vegetable protein (primarily soy protein), autolyzed yeast extracts, and meat and poultry extracts. Amino acids (primarily cysteine, alanine, and methionine) may also be used. Sugars used are D-glucose/dextrose, D-fructose, D-xylose, and others. Time, temperature, and pH are

4 Soy sauce is made from ground soy beans and wheat by a fermentation process. The microbial enzymes hydrolyze the proteins to amino acids and peptides and the starch to mono-saccharides and oligosaccharides. There is also a pasteurization step. So the conditions are favorable for nonenzymic browning reactions.

variables. The products may be in the liquid form or in the form of powders produced by pan drying (roasted flavor) or spray drying (nonroasted flavor). Maillard reaction products formed via reactions of reducing sugars with milk proteins are important contributors to the flavors of caramels, milk chocolate, fudges, and toffees.

As described above, the volatile compounds produced by Maillard reactions often are the source of desirable flavors and aromas. However, the reaction can also be the source of undesired off-flavors and -aromas. Off-flavors and -aromas are most likely to be produced during pasteurization of milk, production of concentrated milk products, and dry milk powders and during storage of dehydrated foods. And while the bitter substances formed in the process are desired in coffee, they are unwanted in most grilled, fried, baked, or roasted foods. Also reaction of reducing sugars with amino acids destroys the amino acid (because the amino group is lost from it). As previously mentioned, this is especially important with L-lysine, an essential amino acid whose ε-amino group can react while the amino acid is a unit of a protein molecule. Other essential amino acids (namely, L-arginine, L-cysteine, and L-methionine) may also be lost. In addition, potentially mutagenic heterocyclic amines formed by Maillard reaction sequences have been isolated from broiled and fried meat and fish and beef extracts, as well as from model systems. Finally, the off-color (brown) can be undesirable in products such as dry milk powders.

Acrylamide

A Maillard-type reaction is responsible for the formation of acrylamide (3-aminopropanamide) in certain foods that have been heated to a high temperature (greater than 120°C [250°F]) during processing or subsequent meal preparation. The list of products containing acrylamide is extensive and includes potato products (such as french fries and potato chips [crisps]), cereal-based products (such as breads, toast, crackers, cookies [biscuits], and breakfast cereals), coffee, and roasted nuts. Common to these products is the fact that they all have been subjected to heating at high temperatures (as occur during frying, baking, toasting, grilling, and roasting) during processing or preparation. Levels found in such foods are typically less than 1.5 ppm. Acrylamide is not formed in boiled foods (such as boiled potatoes) because the temperature during boiling does not go much above 100°C (212°F)). Nor is acrylamide detected (or it is detected at only very low levels) in canned or frozen fruits, vegetables, and vegetable protein products (such as vegetable burgers and related products). An exception is certain pitted, sliced, and canned ripe olives in which measured amounts have been as high as 1925 (μg/kg). Most green and ripe olives have levels of acrylamide near 0 ppb. There is no direct evidence that acrylamide is a health risk to humans in amounts typically consumed. Nevertheless, efforts continue to reduce levels in foods (discussed later). Such efforts begin with an understanding of how it is formed.

It is known from model systems that there are five factors associated with acrylamide formation by the primary pathway. (1) Free L-asparagine must be present in significant amounts. (2) A reducing sugar must be present. (3) The food must be exposed to high temperatures. (4) Only small amounts of moisture can be present (in order to achieve the high temperature required). (5) The optimal pH range is 5.5−7.0). The acrylamide-formation reaction is second order.

The natures of all intermediates in the reaction pathway have not been established, but the pathway given below is a reasonable one. The first step in the reaction sequence (Fig. 18.8) is probably formation of a Schiff base intermediate. The Schiff base product

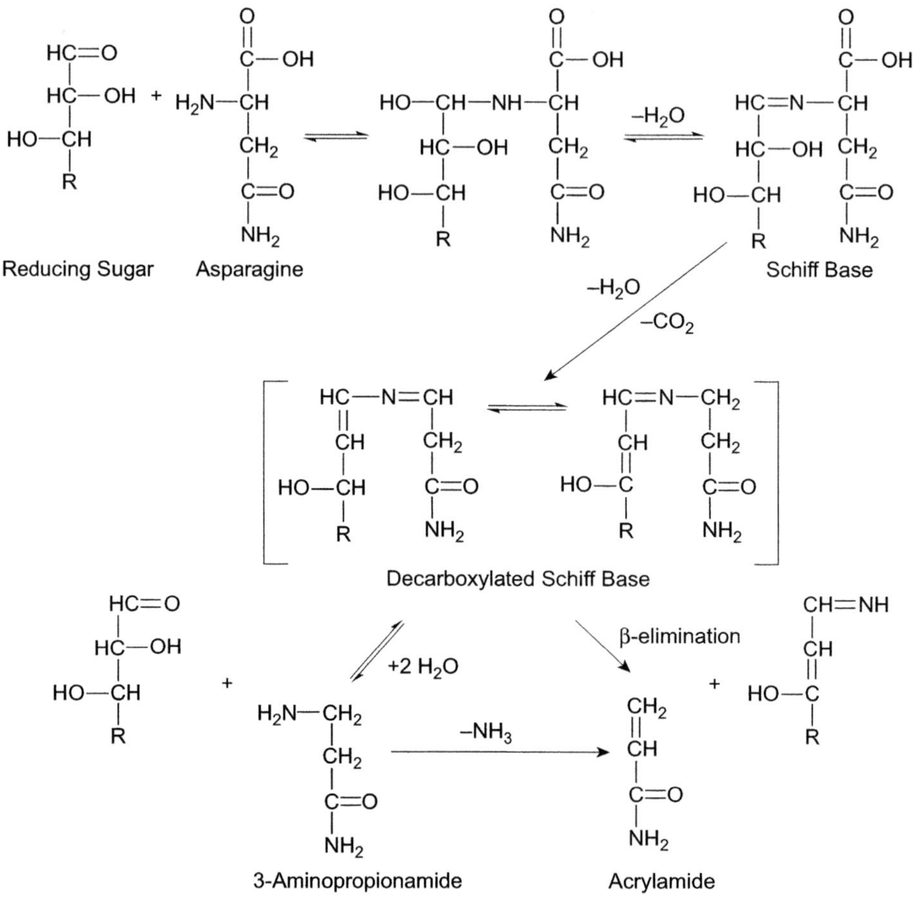

Figure 18.8 A mechanistic scheme for the formation of acrylamide from D-glucose and L-aspargine (as proposed by Huber, K.C., BeMiller, J.N., 2017. Carbohydrates. Chap. 3. In: Damaodoran, Parkin (Eds.), Fennema's Food Chemistry, fifth ed. CRC Press; based on mechanisms previously proposed by others). It is known with certainty that the three carbon atoms of acrylamide arise from carbon atoms 2, 3, and 4 of asparagine, that the carboxyl carbon atom (C1) is lost as carbon dioxide, and that assistance of a reducing sugar is required. This mechanism is consistent with those facts. However, parts of the mechanism are speculative, but the speculation is based on known chemistry. Undoubtedly, the reaction is a concerted reaction in which the proposed carbanion intermediate never actually forms. Two final steps to give the final product are shown. Either one or both may occur.

probably undergoes decarboxylation, followed by carbon-carbon bond cleavage to form acrylamide (since it is known that the atoms of acrylamide all come from the L-asparagine). Acrylamide is not the favored product from the reaction of a reducing sugar with asparagine (the reaction efficiency for its formation being only about 0.1%), but it does accumulate to detectable levels in food products subjected to heating at high temperatures.

The reaction pathways in complex food systems may be more complicated, but it is evident that deep-fried potato products (such as potato chips [crisps] and french fries) are particularly likely to contain acrylamide because potatoes contain both free L-asparagine and D-glucose (a reducing sugar). Particularly in potatoes subjected to cold storage (3–4°C [37–39°F]), starch (Chapter 6) is converted to sucrose (Chapter 3), which in turn is split to release D-glucose and D-fructose. Commercially, a solution of glucose/dextrose is applied to blanched potato strips prior to par-frying (before freezing) to promote formation of a brown color. (Even with the addition of glucose to the strip surface, only about 0.6% of the total asparagine lost during the frying reaction was converted into acrylamide.) It has been suggested that (in deep-fried potato products) Maillard browning intermediates (such as the dicarbonyl deoxyhexuloses and deoxyosones and, perhaps HMF) react with L-asparagine and contribute to acrylamide formation. In the case of bakery products and ready-to-eat breakfast cereals, syrups derived from starch that contain reducing sugars (Chapter 7), honey (Chapter 19), and sucrose (Chapter 3) (which is hydrolyzed during dough fermentation, releasing D-glucose and D-fructose) are added during preparation and are the source of reducing sugars.

Acrylamide levels increase rapidly in the latter stages of heating processes as the water at the surfaces of foods is removed by evaporation, allowing surface temperatures to increase above 100°C (212°F). Acrylamide formation requires a minimum temperature of 120°C (250°F), which means that it cannot occur in high-moisture foods. Formation is favored at temperatures approaching 200°C (390°F). Products with a large amount of surface area (such as potato chips, french fries, and breakfast cereal flakes) are among those foods whose surfaces are subjected to high temperatures during processing and are likely to contain relatively high contents of acrylamide. With extended heating at temperatures above 200°C (390°F), acrylamide levels may actually decrease via thermal degradation.

Food levels of acrylamide are also impacted by pH. Acrylamide formation is favored as the pH is increased over the range of 4–8. Reduced acrylamide formation in the acid range is thought to be due, at least in part, to protonation of the α-amino group of L-asparagine, reducing its nucleophilic potential. Furthermore, acrylamide itself appears to undergo increased rates of thermal degradation as the pH decreases.

As already stated, the risk of acrylamide exposure by consumption of foods is very small, no adverse effects have been reported from consumption of foods with "high" levels of acrylamide, and there seems to be no cause for alarm related to ingestion of acrylamide in foods, but even so efforts continue to reduce amounts of acrylamide in processed foods. The factors associated with the primary pathway for acrylamide formation form the basis of approaches to reduce its amounts in foods: (1) a reduction in or elimination of either one or both of the substrates, (2) a change in processing conditions, and (3) removal of acrylamide from the food after processing. Up to a 60% reduction in amounts of acrylamide in processed potato products (such as french fries and potato chips (crisps)) have been achieved by blanching[5] or soaking the potato pieces in water to remove reducing sugars and free asparagine prior to frying. Lowering the pH (for example, by adding citric acid to the blanching or washing water) results in greater protonation of the amino group of asparagine, making it less nucleophilic (that is, effectively reducing the amount of reactive asparagine). Another approach is conversion of L-asparagine to L-aspartic acid using the enzyme asparaginase. Yet another approach is addition of competing substrates that do not form acrylamide (such as amino acids other than L-asparagine and proteins). Where possible, control of temperature and treatment time during thermal processing may be effective in reducing acrylamide formation. A long-term approach to reducing the amounts of the two reactants is development and use of potato and cereal grain cultivars that contain less of them.

Caramel formation

Caramel is produced commercially both as a coloring and as a flavoring ingredient. Most commercial caramel is used to impart brown color to food products (that is, as a coloring ingredient). To make caramel, a carbohydrate is heated alone or in the presence of an acid, a base, or a salt (but in the absence of compounds containing amino groups) to the point that much of the carbohydrate is destroyed by pyrolysis/thermolysis) in a process called **caramelization**. (As in nonenzymic browning, caramelization involves a complex series of reactions.) The carbohydrate most often used is sucrose (Chapter 3), but D-fructose, D-glucose (dextrose), invert sugar (Chapter 3), glucose syrups, high-fructose syrups, malt syrups (Chapter 19), and molasses (Chapter 3) may also be used. Food-grade sulfuric, sulfurous, phosphoric, acetic, and citric acids may be used. Bases that may be used are ammonium, sodium, potassium, and calcium hydroxides. Salts that may be used are ammonium, sodium, and potassium carbonates, bicarbonates, phosphates (both monobasic and dibasic), sulfates, and bisulfites. As a result, there are a very large number of variables in caramel manufacture, including the source of the carbohydrate, the temperature employed, the time of heating, and the acid, base, and/or salt used. And as with nonenzymic browning, the final product (caramel) contains a complex mixture of polymeric compounds, most of which are formed from unsaturated, cyclic (five- and six-membered ring) compounds, and low molecular weight flavor and aroma compounds.

There are four classes of caramel colors. Class I caramel (also called plain caramel or caustic caramel) is prepared by heating a carbohydrate in the absence of

[5] Blanching involves plunging the food substance into boiling water, removing it after a brief, timed period, and finally treating it with cold water to stop the cooking process.

ammonium[6] or sulfite ions (an acid or a base may be employed). This caramel is used to color high-proof alcoholic beverages. Class II caramel (also called caustic sulfite caramel) is prepared by heating a carbohydrate in the presence of sulfite ions, but in the absence of ammonium ions (an acid or a base may be employed). This caramel is reddish brown, contains colloidal particles with slightly negative charges, and has a solution pH of 3–4. It has only limited food and beverage applications. Class III caramel (also called ammonium caramel) is prepared by heating a carbohydrate in the presence of ammonium ions, but in the absence of sulfite ions (an acid or a base may be employed). This caramel is used to color bakery products, beer, puddings, sauces, and soups. It is reddish brown, contains colloidal particles with positive charges, and gives a solution pH of 4.2–4.8. Class IV caramel (also called sulfite ammonium caramel) is prepared by heating a carbohydrate in the presence of both sulfite and ammonium ions (an acid or a base may be employed). This caramel is used as a coloring agent in baked goods, cake mixes, candies, cola soft drinks and other acidic beverages, pancake syrups, pet foods, and dry seasonings. It is brown, contains colloidal particles with negative charges, and gives a solution pH of 2–4.5.

In the preparation of caramel from sucrose, an acid or acidic (bisulfite) salt catalyzes cleavage of the glycosidic bond of sucrose, and ammonia[6] participates in the Amadori rearrangement and subsequent reactions. Caramelization is not confined to a food ingredient manufacturing facility; it may also occur during food preparation, especially when sugar is present and high temperatures (above 100°C [212°F]) are employed (that is, when moisture levels are low). It occurs, along with nonenzymic browning, during the preparation of chocolate and fudge. Both nonenzymic browning and caramelization reaction sequences produce brown-colored compounds, flavors, and aromas, but the actual colored polymers and the flavor and aroma compounds are different.

Heating effects dehydration (loss of HOH) of sugar molecules and either the introduction of double bonds or the formation of anhydro rings (as in **levoglucosan**, a dicyclic acetal[7]). As in the Maillard browning reaction, introduction of double bonds leads to unsaturated rings, such as those of furans. Reaction of intermediate 3-deoxyosones with ammonia,[6] which is often used in caramel color production, gives (from hexoses) pyrazine and imidazole derivatives of the type shown in Fig. 18.9. Conjugated double bonds absorb light (produce color). Often unsaturated rings will condense to polymers yielding desirable colors or flavors. Catalysts increase the reaction rate and are used to direct the reaction to specific types of caramel colors, solubilities, and acidities.

As already stated, like the pigments produced in the Maillard reaction, caramel pigments are polymeric molecules with complex, variable, and unknown structures. These polymers form colloidal particles. The polymers contain (in addition to ordinary hydroxyl groups) carboxyl, carbonyl, and enolic- and phenolic-like hydroxyl groups, giving the polymer molecules anionic (negative) charges.[8] Their rate of formation

[6] Ammonia is released upon heating ammonium salts.

[7] An acetal is a carbon atom with two OR groups attached.

[8] There are also special caramel color products that have a positive charge that may be required when the color is used in products containing protein (for example, in beer) in order to avoid flocculation.

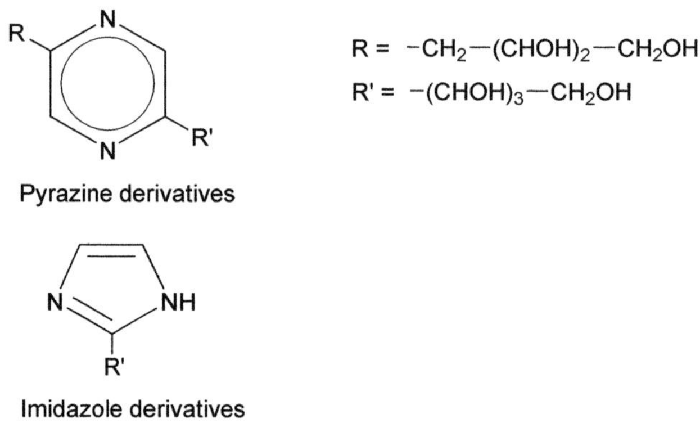

Figure 18.9 Examples of compounds formed by reaction of 3-deoxyosones with ammonia during caramel color production.

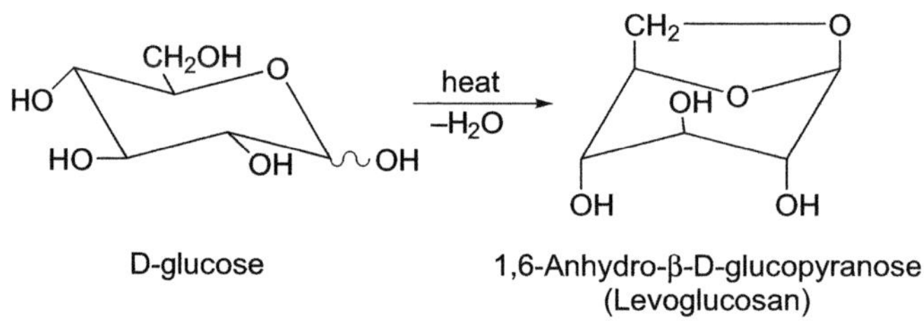

increases with increasing temperature and pH. In the absence of buffering salts, larger amounts of a bitter polymeric substance called humin are formed. Formation of humin is generally minimized. In fact, good caramel color will have little flavor. However, in the preparation of products such as crème brûlée and crème caramel, a slight bitter taste is desired.

The pH of caramel color solutions is usually acidic (pH 2—5), although solutions of products that have been neutralized prior to spray drying may have pH values as high as 8. Those products with pH values below 5 have good microbiological stability. Caramel color is used in a wide range of food products (including alcoholic beverages, bakery products, barbeque and other sauces, breakfast cereals, cola soft drinks, confections, and pancake syrups).

Caramel flavor is also important in food products. Certain pyrolytic reactions produce unsaturated ring systems that have unique flavors and aromas, in addition to colored substances. For example, caramelization is the source of the color and flavor of maple syrup (produced during boiling down of maple sap to produce maple syrup). Maltol (3-hydroxy-2-methylpyran-4-one) and isomaltol (3-hydroxy-2-acetylfuran), both products of caramelization (also of nonenzymic browning; see above), contribute to the flavor of bread. Maltol has a sweet, malty, caramel-like flavor,

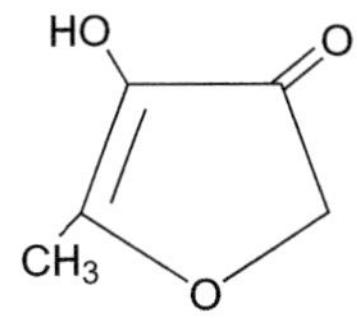

2H-4-Hydroxy-5-methylfuran-3-one

but more importantly, enhances or modifies other flavors and aromas. 2H-4-Hydroxy-5-methylfuran-3-one has a burnt flavor (as in cooked meat), but enhances various flavors and sweeteners. These and nonenzymic browning and pyrolysis products are present in such foods as toasted bread.

Additional reading

Maillard Browning

Corzo-Martinez, M., Corzo, N., Villamiel, M., del Castillo, M.D., 2012. Browning reactions. In: Simpson, B.K. (Ed.), Food Biochemistry and Food Processing, second ed. Wiley-Blackwell, Oxford, pp. 56–83.

Croguennec, T., 2016. Non-enzymatic browning. In: Jeantet, R., Croguennec, T. (Eds.), Handbook of Food Science and Technology 1: Food Alteration and Food Quality. John Wiley & Sons, Hoboken, pp. 133–157.

Fayle, S.E., Gerard, J.A. (Eds.), 2002. The Maillard Reaction. The Royal Society of Chemistry, Cambridge.

Glomb, M.A., Henning, C., 2016. Formation of reactive fragmentation products during the Maillard degradation of reducing sugars – a review. In: Granvogl, M., Peterson, D., Schieberle, P. (Eds.), Browned Flavors: Analysis, Formation, and Physiology, ACS Symposium Series, vol. 1237. American Chemical Society, Washington, pp. 117–131.

Granvogl, M., Peterson, D., Schieberle, P. (Eds.), 2016. Browned Flavors: Analysis, Formation, and Physiology. ACS Symposium Series, vol. 1237. American Chemical Society, Washington.

Lund, M.N., Ray, C.A., 2017. Control of Maillard reactions in foods: strategies and chemical mechanisms. Journal of Agricultural and Food Chemistry 65, 4537–4552.

Malgorzata, M., 2016. Saccharide-derived flavor compounds. Chap. 5. In: Jeleń, H. (Ed.), Food Flavors. CRC Press, Boca Raton.

Mottram, D.S., 2007. The Maillard reaction: source of flavor in thermally processed food. In: Flavours and Fragrances. Springer-Verlag, Berlin, pp. 269–283.

Nursten, H.E. (Ed.), 2005. The Maillard Reaction: Chemistry, Biochemistry and Implications. The Royal Society of Chemistry, Cambridge.

O'Brien, J. (Ed.), 2005. The Maillard Reaction in Foods and Medicine. Woodhead Publishing, Cambridge.

Rannou, C., Laroque, D., Renault, E., Prost, C., Serot, T., 2016. Mitigation strategies of acrylamide, furans, heterocyclic amines and browning during the Maillard reaction in foods. Food Research International 90, 154–176.

Soria, A.C., Villamiel, M., 2012. Non-enzymatic browning in cookies, crackers and breakfast cereals. In: Simpson, B.K. (Ed.), Food Biochemistry and Food Processing, second ed. Wiley-Blackwell, Oxford, pp. 584–593.

Thomas, M.C., Forbes, J., 2010. *The Maillard Reaction: Interface Between Aging and Nutrition.* The Royal Society of Chemistry, Cambridge.

Acrylamide

Anese, M., Manzocco, L., Calligaris, S., Nicoli, M.C., 2013. Industrially applicable strategies for mitigating acrylamide, furan, and 5-hydroxymethylfurfural in food. Journal of Agricultural and Food Chemistry 61, 10209–10214.

Arvanitoyannis, I.S., Dionisopoulou, N., 2014. Acrylamide: formation, occurrence in food products, detection methods, and legislation. Critical Reviews in Food Science and Nutrition 54, 708–733.

Benford, D., Ceccatelli, S., Cottrill, B., Di Novi, M., Dogliotti, E., Edler, L., Farmer, P., Fuerst, P., Hoogenboom, L., Katrine Knutsen, H., et al., 2015. Scientific opinion on acrylamide in food. EFSA Journal 13, 1–321.

Bethke, P.C., Bussan, A.J., 2013. Acrylamide in processed potato products. American Journal of Potato Research 90, 403–424.

Erkekoglu, P., Baydar, T., 2014. Acrylamide neurotoxicity. Nutritional Neuroscience 17, 49–57.

Friedman, M., 2015. Acrylamide: inhibition of formation in processed food and mitigation of toxicity in cells, animals, and humans. Food & Function 6, 1752–1772.

Halford, N.G., Curtis, T.Y., 2016. Reducing the acrylamide-forming potential of wheat, rye and potato: a review. In: Granvogl, M., Peterson, D., Schieberle, P. (Eds.), Browned Flavors: Analysis, Formation, and Physiology, ACS Symposium Series, vol. 1237. American Chemical Society, Washington, pp. 35–53.

Rannou, C., Laroque, D., Renault, E., Prost, C., Serot, T., 2016. Mitigation strategies of acrylamide, furans, heterocyclic amines and browning during the Maillard reaction in foods. Food Research International 90, 154–176.

Riboldi, B.P., Vinhas, A.M., Moreira, J.D., 2014. Risks of dietary acrylamide exposure: a systematic review. Food Chemistry 157, 310–322.

Vijay, P., Ezekiel, R., Pandy, R., 2016. Acrylamide in processed potato products. Acta Physiologiae Plantarum 38, 1–23.

Virk-Baker, M.K., Nagy, T.R., Barnes, S., Groopman, J., 2014. Dietary acrylamide and human cancer: a systematic review of literature. Nutrition and Cancer 66, 774–790.

Carbohydrate and Noncarbohydrate Sweeteners

19

Chapter Outline

Carbohydrate Chemistry for Food Scientists. https://doi.org/10.1016/B978-0-12-812069-9.00019-4

Key information and skills that can be obtained from the study of this chapter will enable you to

1. Define
   nutritive sweetener
   low-calorie/reduced-calorie sweetener
   high-potency/high-intensity sweetener
2. List the factors that influence the intensity and quality of the sweet taste.
3. Discuss the importance of sucrose/sugar
   delaying the gelatinization and pasting of starch
   in maintaining the moistness of cakes
   as a preservative
   in the creaming of shortening
   as a whipping aid for egg-based foams
   in inducing surface cracking on cookies
   in providing good volume and crumb texture in bakery products
   in maintaining the shape and texture of canned fruits
   in contributing to the color, flavor, and aroma of bakery products
4. Describe why replacing sugar with another nutritive or a nonnutritive sweetener is not simply a matter of providing equivalent sweetness (except, perhaps, in some beverages).
5. Describe the major difference in food products (other than beverages) formulated with a high-fructose syrup (HFS) in place of sucrose.
6. Describe the relation of dextrose equivalence (DE) to sweetness and other properties of glucose syrups.
7. Describe in general terms the caloric content of polyols.
8. Explain why some place polyols in a separate category of low-calorie or reduced-calorie sweeteners.
9. Describe the importance of heat of solution to sensory characteristics.
10. Give (1) the method of preparation and (2) the chemical structures of sorbitol, maltitol, hydrogenated starch hydrolyzates (HSHs), xylitol, lactitol, erythritol, isomalt, and/or tagatose.

Introduction

Much work has been done to determine the relationship of structure to sweet taste, to develop a theory of sweetness, to map the various sweet taste receptors, to determine biochemical mechanisms of sweetness, to develop methods to compare relative sweetness values, to determine genetic variability in the perception of sweetness, and to determine the relationship of sweetness to other flavors, especially the close relationship between sweetness and bitterness.[1] These aspects of sweetness are not covered in this book, but references to discussions of them are found at the end of the chapter.

Nutritive/carbohydrate sweeteners

Nutritive sweeteners are substances that both impart a sweet taste and provide significant calories (energy). Sweeteners that are classified as nutritive sweeteners are sucrose, D-fructose, D-glucose (dextrose), syrups (glucose, high-fructose, maple, and other syrups), maltodextrins, honey, and fruit juice concentrates. As discussed later, the sugar alcohols (D-glucitol, mannitol, hydrogenated starch hydrolyzates, xylitol) and neosugar (Chapter 3) also contribute to the caloric content of a food, but in reduced amounts for an equal degree of sweetness compared to the truly nutritive sweeteners. (Sucrose and other nutritive carbohydrate sweeteners that are contributors to the caloric content of the human diet were presented in Chapter 3 (sucrose, oligosaccharides related to sucrose, lactose) and Chapter 6 (glucose/dextrose, fructose, glucose syrups, high-fructose syrups (HFS), maltodextrins). Discussions of the chemistry of these nutritive sweeteners given in Chapters 2, 3, and 6 should be consulted in connection with this chapter.

There are differences among the above-named substances in both the intensity of sweetness and the quality of the sweet taste. Sweetness is application dependent and relative sweetness values depend on a number of factors, primarily among which are temperature and concentration, with the effect of each varying between sweeteners. Other factors influencing both the intensity and the quality of the sweet taste include the system pH, the viscosity of the system, and the concentrations of other sweeteners and flavors. Even so, relative sweetness values of nutritive sweeteners compared to sucrose as the standard can be determined and are basically as given in Table 19.1.

Sweetness responses also vary with time (Fig. 19.1). Therefore, in food and confectionery products, both the intensity and duration of sweetness can be controlled by using mixtures of sweeteners. Different sweeteners also have different effects on other ingredients. For example, the gelatinization temperatures of starches and the viscosities and gel strengths of products containing starch may vary with the type of sweetener when different sweeteners are used at the same concentration (by weight). Production and properties of individual nutritive sweeteners have been discussed in earlier chapters.

Sucrose is the standard to which all other sweeteners are compared. Although the nature of its sweetness is what is preferred by humans, functionalities imparted by sucrose (Table 19.2) are often more important than the sweet taste it imparts. The effect of sucrose (sugar) on the gelatinization of starch is a good example of the fact that its function is much more than to provide a sweet taste. Cake making can be used as an

[1] Most people are surprised that not all carbohydrates are sweet. Some have no taste. Some, such as D-mannose, have a somewhat bitter taste.

Table 19.1 Approximate relative sweetness values of nutritive sweeteners

Carbohydrate	Relative sweetness[a]
Crystalline D-fructose	120–170
HFS[b] 90	120–160
HFS[b] 55	>100
Invert sugar	>100
Sucrose	100
HFS[b] 42	100
Xylitol	100
Tagatose	92
Maltitol	70–90
D-Glucose	70–80
70-DE glucose syrup	70–75
Erythritol	70
Mannitol	70
D-Glucitol (Sorbitol)	50–70
42-DE glucose syrup	~50
Isomalt	45–65
Isomaltulose	50
Lactitol	40
Maltose	30–50
Lactose	20

HFS, high-fructose syrups.
[a]Relative values on an equal weight basis.
[b]High-fructose syrup. The number indicates the percent of carbohydrate that is D-fructose.

example. The gelatinization/pasting temperature of wheat starch is 50–55°C (120–130°F). As the starch gelatinizes and pastes, intense thickening occurs and the cake structure is established. However, cake making involves chemical leavening, and decomposition of the leavening agent to produce gas does not occur until the temperature reaches 85–90°C (185–195°F). It is essential for maximum gas production and starch gelatinization and pasting to occur concurrently to produce the desired texture. Sucrose delays the onset of starch swelling, gelatinization, and pasting (that is, it delays the temperature at which the cake batter changes from a liquid to a semisolid until the time of maximum gas production).The result is the desired crumb structure. This example illustrates why replacing sucrose with a nonnutritive or a less-caloric nutritive sweetener is not simply a matter of providing a similar sweet taste.

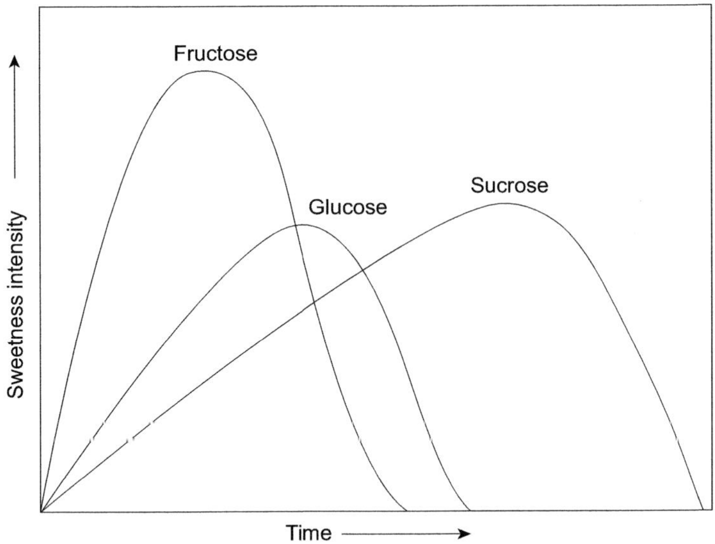

Figure 19.1 The relative time-dependent sweetness responses of sucrose, D-fructose, and D-glucose.

Noncaloric or less-caloric ingredients added in attempts to provide the physical attributes provided by sucrose or sweet syrups are called bulking agents (Chapter 17).

Another functionality imparted by sucrose and/or syrups that may have to be imparted by a bulking agent in combination with a nonnutritive sweetener, if a replacement for sucrose is desired, is the ability to provide sufficient shelf-life. The hydrophilicity of sucrose and syrups provides humectancy. Their incorporation, therefore, may prevent or slow undesirable migration of water molecules, slow increases or decreases in the moisture content of a product, and/or reduce water activity. In the latter case, they may, therefore, function as preservatives. By reducing water activity, nutritive sweeteners may also affect the hydration of flour proteins and development of gluten, achieving a delay in gluten development that is often desirable for producing a tender crumb texture.

Sucrose also allows appropriate spread when making cookies. There are many different kinds of cookies, but a common feature is that they are made with more sugar, more shortening, and less water than are cakes. Just as in preparing shortened cakes, sugar is used in the creaming process[2] to introduce air into the batter. About half the sugar remains undissolved at the end of the mixing process because of the shortage of water. As the temperature of the cookie dough increases during baking, the shortening melts. Also with increasing temperature, the undissolved sugar dissolves, and the sugar solution increases in volume. Both processes result in a more fluid dough, allowing it

[2] The term creaming, as applied to baking, refers to the mixing of shortening and crystalline sugar so that air is incorporated. The air cells, which are stabilized by small sugar crystals at the air–fat interface, expand during baking when filled with gas from a leavening agent.

Table 19.2 Some functions of sugar (other than providing sweetness) in some processed food products

Product	Function	Result
Baked goods		
Breads, yeast	Ties up water, prevents full hydration of gluten, and retards gluten development	Gluten maintains optimum elasticity. Final baked product has good volume and crumb texture
	Yeast substrate	Hastens leavening process, allowing dough to rise at a faster and more consistent rate
	Upon hydrolysis, provides reducing sugar for Maillard browning reactions	Contributes to brown color of the crust and the aroma
	Caramelization	Contributes to the brown color of the crust and its taste and aroma
Cakes, shortened	Creaming aid	Helps to provide "lightness" by incorporating air into the chemically leavened batter.[a]
	Ties up water, prevents full hydration of gluten, and retards gluten development	Produces a fine crumb texture and good volume.[a]
	Ties up water and delays gelatinization of the starch and "setting" of the batter	Tenderizes (see text)[a]
	Caramelization	Contributes to the brown color and flavor of the surface
	Upon hydrolysis, provides reducing sugar for Maillard browning reactions	Contributes to the brown color of the surface and aroma
	Humectancy	Maintains a moist product
Cakes, unshortened	Whipping aid that stabilizes beaten egg foams	Makes egg foam more elastic so that air cells can expand[a]
	Raises temperature at which egg proteins "set"	Provides tender texture and good volume[a]
	Ties up water, prevents full hydration of gluten, and retards gluten development	Produces a fine crumb texture and good volume[a]

Table 19.2 Some functions of sugar (other than providing sweetness) in some processed food products—cont'd

Product	Function	Result
	Ties up water and delays gelatinization of starch and "setting" of the batter	Tenderizes (see text)[a]
	Caramelization	Contributes to the brown color of the surface, flavor, and aroma
	Upon hydrolysis, provides reducing sugar for Maillard browning reactions	Contributes to the brown color of the surface and aroma
	Humectancy	Maintains a moist product
Cookies	Creaming aid	Incorporates air into the chemically leavened dough
	Absorbs water	Induces spread (some cookies)
	Crystallizes	The relatively high concentration of sugar and the relatively low content of water results in sugar crystallization, particularly on the surface of the cookie. As the sugar crystallizes, water and heat is released. The water evaporates, resulting in surface cracking (some cookies)
	Caramelization	Contributes to the brown color, flavor, and aroma (of some cookies)
Barbecue sauce		
	Enhances tomato, vinegar, and lime flavors.	
	Contributes to the browning and caramelization processes	Provides color and caramel and sweet tastes
Confections/candies (made primarily with sucrose)		
	Formation of highly supersaturated solutions	Key to most candy making
	Crystallization in which crystal size can be controlled	A few candies contain perceptible crystals. Others (cream candies) contain crystals that are too small to be detected by the tongue (the more common type). Still others are based on a glass made from sucrose; these are noncrystalline or amorphous candies.

Continued

Table 19.2 Some functions of sugar (other than providing sweetness) in some processed food products—cont'd

Product	Function	Result
Custards		
	Delays coagulation of egg proteins	Raises the temperature at which the custard "sets," resulting in a smooth consistency and reduced syneresis
Fruits		
Canned	Increases osmotic strength	By more or less equalizing the solute concentration inside and outside the fruit, helps to maintain the fruit's shape and texture (that is, prevents it from becoming too mushy).
Frozen	Delays surface discoloration	Reduces the size of ice crystals, providing a smooth, creamy texture
Ice creams and other frozen desserts		
	Lowers the freezing point Increases viscosity Enhances flavors Balances the acidity of added fruits	Imparts a thick, creamy mouthfeel
Icings		
	Provides sweetness, flavor, bulk, and structure (Functions are similar to those in confections.)	
Jams, jellies, preserves, and marmalades		
	Preservative (in full-sugar products)	The high-soluble solid content (at least 65%) required for gelation of pectin reduces the water activity to the point that spoilage is prevented.
	Helps retain the color of the fruit	
	Regulates gelling of HM pectin (Chapter 15)	
Meringues		
	Foam stabilizer	Produces a stiffer and more stable egg white foam

Table 19.2 Some functions of sugar (other than providing sweetness) in some processed food products—cont'd

Product	Function	Result
Puddings, pie fillings, and sauces		
	Dispersant	When mixed with starch, flour, and/or a food gum to be added to a hot liquid (in a ratio of at least two parts of sugar to one part of thickener), allows thickener particles to disperse evenly and prevents lumping.
Salad dressings		
	Balances bitter, sour, and spicy flavors and makes them taste better	

[a]In general, in cakes, the flour (starch and protein/gluten) and egg protein provide structure and the shortening and sugars have a tenderizing effect. A cake batter is an emulsion, a foam, and a suspension with rheological attributes of a gel.

to spread. In addition, sugar generates appropriate color and flavor via browning and caramelization reactions (Chapter 18), and sugar ensures appropriate surface cracking. Sucrose crystallizes at the surface as moisture is lost in the oven. Crystallization releases more water, which also evaporates. The result is a shrinking of the surface as it dries and surface cracking.

The preparation of many batters and doughs involves aeration in which sucrose plays a critical role. It functions as a whipping aid when egg whites or whole eggs (with added sugar) are whipped into a light foam, and as already mentioned, it helps incorporate air into fat in the creaming process.

Some properties of crystalline sucrose and imparted functionalities in food products are listed in Table 19.2. Certain of these functions can be provided by glucose syrups, HFS, liquid sugar, and other carbohydrates. However, the drawback to use of liquid sugar and syrups (in most cases) is the addition of extra water, and any other carbohydrate is usually deficient in one or more of the desired properties. As is evident from Table 19.2, sugar is also used in a wide range of nonsweet foods (from canned vegetables to brined and salt-cured meat products) to complement salty flavors, to blend with acids added to produce a tangy taste, to mellow the acidity of tomato and other products, to minimize starchy flavors, to contribute to browning, and to add flavor as a result of caramelization (Chapter 18). Nutritive sweeteners also produce viscosity and, therefore, provide body and the accompanying mouthfeel to liquids. They provide texture to semisolids. They lower the freezing points of ice creams and frozen desserts, a property important to making quality products with a thick, creamy mouthfeel. They are substrates for the fermentation required in the preparation of bread, other yeast-leavened baked goods, pickles, alcoholic beverages, and other products utilizing fermentation as part of the process.

These are but a few examples of why the physical and chemical properties of sucrose in particular, but also of other nutritive sweeteners (such as D-glucose, glucose syrups, and HFSs), are at least as important, if not more important, than the intensity and quality of sweetness they impart. The examples presented illustrate why it is such a challenge to replace nutritive sweeteners (macroingredients) with high-potency, nonnutritive sweeteners (microingredients) in many products (with the primary exception of beverages, in which the lost body can be provided by a very small amount of a hydrocolloid and in products such as yogurt, where the only functionality desired is sweetness).

In summary, if the goal is replacement of sucrose in a food product for diets requiring calorie restriction or for prevention of dental caries, the substance or substances used as a substitute, whether a nutritive compound, a nonnutritive compound, or a combination of the two, should be economical and have the following attributes:

- the same quality of sweetness as sucrose,
- no (or at least significantly reduced) caloric value,
- the same (or very similar) functional properties, such as providing bulk, crystallization, hygroscopicity (humectancy), solubility, melting temperature, production of viscosity, reduction of water activity, and impartation of other colligative properties,
- colorless,
- odorless,
- stable under processing, storage, and food preparation conditions, and
- noncariogenic.

It is no surprise that no alternative substance or combination of substances meets all these criteria. A brief overview of properties is given following a brief review of sucrose.

Sucrose (Chapter 3)

Sucrose occurs in only small amounts in natural foods consumed by humans. Even honey, which contains about 40% D-fructose, 35% D-glucose, and 18% water (so it is essentially an invert sugar [Chapter 3]), contains only a small amount of sucrose. Fruits also contain small amounts. Commercial sucrose is obtained from sugar cane and sugar beets. It is available to food processors as a concentrated solution (called "liquid sugar") and in a variety of crystalline forms. Sucrose is the standard for intensity and quality of sweetness to which all other sweet substances are compared.

Sucrose is a nonreducing disaccharide. As a result, it cannot participate in nonenzymic browning reactions (Chapter 18) unless it is first hydrolyzed to D-glucose and D-fructose, which occurs when products that are even slightly acidic are heated. It will undergo caramelization (Chapter 18).

As described in the introductory part of this section, sucrose is an important functional ingredient in many food products. In fact, in many food products, the functional properties provided by sucrose are at least as important as, if not more important than, the sweet taste it provides. It was already pointed out that this fact makes it difficult to replace sucrose with a high-intensity sweetener. There are also food products in which it would be desirable to have more sucrose because of its functional properties, but in which more sucrose would make the product too sweet. To address this need, a product has been developed that contains sucrose, a maltodextrin product (Chapter 7), and a compound that blocks the sweet taste receptors on the tongue.

Invert sugar (Chapter 3)

Invert sugar is the equimolar mixture of D-glucose and D-fructose formed by hydrolysis of sucrose. Some unhydrolyzed sucrose may also be present in invert sugar preparations. HFS products made from starch (below) have similar compositions.

High-fructose syrups (Chapter 7)

This category includes both the high-fructose syrups (HFS)/high-fructose corn syrups (HFCS) made from starch and invert sugar syrups (Chapter 3). Both D-fructose and D-glucose can bind/hold more water than does sucrose at the same concentration (by weight); neither will mixtures of the two sugars crystallize, so foods such as cookies and granola bars formulated with sweeteners high in D-fructose and/or D-glucose generally will remain soft and moist longer than will the same product made with sucrose. Because D-fructose is sweeter than is sucrose, HFS containing 55% or more of D-fructose and invert sugar require use of less carbohydrate to achieve the same degree of sweetness. As a result, pancake syrups made with HFS containing 55% or more of D-fructose have lower caloric values than ones made with sucrose and can be labeled as being "lite" syrups. The total carbohydrate in some such products can contain as little as one-half the normal amount of carbohydrate.[3] The saccharide compositions of two common HFS products are as follows:

Fructose (%)	**Glucose (%)**	**Oligosaccharides (%)**
42	51	7
55	41	4

About 90% of HFS/HFCS is used in beverages, especially carbonated beverages (soft drinks). It is also used as an ingredient in some bakery products, fruit fillings, pancake syrups, and other processed foods. It is generally not used in applications in which browning is a problem.

Some have claimed that HFS/HFCS use is uniquely responsible for the increased prevalence of obesity—even though HFS/HFCS and hydrolyzed/digested sucrose are essentially the same. However, all scientific evidence points to the fact that

[3] Because such syrups contain less dissolved solids, they are of low viscosity. Therefore, hydrocolloid is added to increase the viscosity to what the consumer expects of a pancake syrup.The hydrocolloid used is often CMC because of, among other attributes, the clarity of its solutions.

HFS/HFCS and sucrose have essentially equivalent effects on body weight, blood triglyceride levels, and metabolism.

Glucose syrups (Chapter 7)

Glucose syrups provide adhesiveness, body, cohesiveness, and gloss to a variety of products. Glucose syrups are solutions of glucose plus other saccharides derived from starch via acid- and/or enzyme-catalyzed hydrolysis. The dextrose equivalence (DE) values (Chapter 7) of different syrups range from DE 20 to DE 95. Those with the lower DE values provide the best foam stabilization, prevention of sugar crystallization, and the highest viscosity. Those with the higher DE values provide the most sweetness and the greatest freezing point depression, boiling point elevation (both colligative properties), browning reaction, hygroscopicity, and flavor enhancement. All glucose syrups are purified products containing up to 85% dissolved solids. Corn syrup solids (Chapter 7) are basically dried glucose syrups of lower DE values, as only syrups with very low DE values can be converted to a powdered form.

Dextrose (D-glucose) (Chapter 1)

Dextrose, which is the commercial name for D-glucose, is available as both monohydrate and anhydrous crystals, a microcrystalline product, and in an agglomerated form. (Of course, a 90–95 DE glucose syrup contains mostly D-glucose). The sweetness of D-glucose is 70%–80% that of an equal concentration by weight of sucrose. Applications of dextrose include baking, brewing, and canning. It is also used as an ingredient in confections, dry mixes, and prepared meats. It is the starting material for the preparation of sorbitol, glucono-δ-lactone, and ascorbic acid (vitamin C) (Chapter 2).

Fructose (Chapter 1)

D-Fructose is available both as a 90% syrup and in a crystalline form. The perceived sweetness of β-D-fructopyranose (the predominate form in an aqueous solution) is 1.6–1.8 times that of sucrose; that of a 10% solution of fructose is about 1.2–1.7 times as sweet as a 10% solution of sucrose. (Sweetness varies with temperature and pH.) D-Fructose is more soluble than is sucrose and dissolves faster. D-Fructose provides an early burst in sweet taste and then a rapid drop (Fig. 19.1). It enhances flavors such as cinnamon, chocolate, fruit flavors, and vanilla. It also enhances acid perception. Fructose finds use as an ingredient in bakery products, confections, dry mixes (including beverage mixes), fruit fillings, and yogurts.

In the introduction to this section, the effect of sucrose on the pasting of starch and the value of delayed pasting was presented. All nutritive sweeteners are carbohydrates that compete with starch for water and have this property to different degrees because their water binding/holding capacities differ. Fructose, as compared to sucrose, enhances the hydration of starch granules and increases peak and final viscosities (Chapter 6). HFS/HFCS 42 has the same effect.

High-maltose syrups (Chapter 7)

Compared to a glucose syrups of the same DE value, high-maltose syrups contain 40%–65% maltose and provide better heat and color stability, provide better flavor transfer, are less hygroscopic, and result in less browning.

Maltodextrins (Chapter 7)

Maltodextrins are mixtures of oligosaccharides derived from starch (that is, they are maltooligosaccharides). Maltodextrins have average DE values in the approximate range 5–18. (By definition and regulation, the DE values of maltodextrin products is less than 20 [average degree of polymerization (DP) more than 5].) Maltodextrins are available as dried products and as concentrated solutions. Liquid products have properties similar to those of glucose syrups, and dried products have properties similar to those of glucose (corn) syrup solids. The properties of all three products are related to DE values in the same way as are glucose syrups (Section on Glucose Syrups in this chapter). In general, maltodextrins are only moderately sweet. The degree of sweetness (taste intensity) decreases as the average chain length of the maltodextrin increases (that is, as the DE value decreases).

Reduced-calorie carbohydrate sweeteners

Some carbohydrates are either not absorbed or only partially absorbed through the walls of the small intestine or are not completely digested to monosaccharides that can be absorbed, so they pass to the large intestine (colon) where they participate in fermentation. Therefore, they are components of dietary fiber that are categorized both as soluble dietary fiber and as prebiotics (Chapter 17). As prebiotics, they contribute some calories via production and absorption of short-chain fatty acids. While the U.S. Food and Drug Administration (FDA) categorizes them as nutritive sweeteners (because they do provide some calories), others place them in a category between nutritive sweeteners and high-potency sweeteners and call them **reduced-** or **low-calorie sweeteners**.

Polyols (Chapters 2 and 3)

Polyols (also known as alditols, sugar alcohols, and polyhydric alcohols) Chapters 2 and 3 have slightly sweet tastes (to different degrees) and provide fewer calories than do nutritive sweeteners. Both their sweetness and prebiotic attributes are desirable. However, while, for example, it is desirable for most humans to ingest larger amounts of dietary fiber, ingestion of macronutrient quantities of monomeric polyols results in the flatulence and laxation associated with nonabsorbed, fermentable carbohydrates (to different degrees with different polyols and to different extents in different individuals). Polyols made from oligosaccharides are either not digested or slowly and incompletely digested. Even though polyols present challenges with regard to caloric-value classification, they are valuable food ingredients. They are often used in combination with high-intensity sweeteners to provide bulk, texture, and mouthfeel.

As already pointed out, the caloric values of polyols are not values that are agreed upon by all regulatory agencies. That is because energy is obtained from polyols in two ways: (1) partial absorption from the small intestine and subsequent catabolism and (2) fermentation in the large intestine (colon) (Chapter 17), and the distribution between the two pathways is dependent on the quantity of the polyol consumed and the individual. Generally accepted values for their caloric content (as compared to D-glucose, which has a caloric content of 4.0 Kcal/g) are given in Table 19.3. (Note that, as with sweetness values, these values are also given on a "per gram" rather than a "per mole" basis.) (The caloric content of sucrose is also 4.0 Kcal/g.)

Crystalline polyols have negative heats of solution that are much greater than that of sucrose (Table 19.4). As a result, they produce a cooling sensation in the mouth.

Table 19.3 Approximate caloric content of polyols[a]

Polyol	Kcal/g
Hydrogenated starch hydrolyzates	3.0
Sorbitol	2.6
Xylitol	2.4
Maltitol	2.1
Lactitol	2.0
Isomalt	2.0
Mannitol	1.6
Erythritol	0.2

[a]The accepted caloric values that are used to calculate the caloric value of a serving of a food product vary from regulatory agency to regulatory agency.

Table 19.4 Heats of solution of polyols and sucrose

Polyol	cal/g
(Sucrose)	−4.3
Isomalt	−9.4
Lactitol monohydrate	−12.4
Lactitol dihydrate	−13.8
Maltitol	−18.9
Sorbitol	−26.6
Mannitol	−28.9
Xylitol	−36.6

Polyols, therefore, can be described as having sweet-and-cool sensory attributes. The approximate solubilities (in g/100 g of water) of some polyols are as follows: mannitol (20), erythritol (37), xylitol (64) (about the same as sucrose), and sorbitol (70). Because polyols do not have a reducing (carbonyl) group, they do not participate in nonenzymic (Maillard) browning reactions (Chapter 18).

The U.S. FDA allows a health claim to be made that polyols do not promote dental caries.[4] Because it has been determined that polyols pose no risk to human health, no acceptable daily intake value has been assigned to them.

Sorbitol (Chapter 2)

Of the polyols, sorbitol (D-Glucitol) is used in the greatest amounts. It is the most soluble of the common polyols and is widely used for its humectant properties in baked goods and chocolate products to keep them from becoming dry and hard. It is also widely used in sugar-free chewing gum and candies. Sorbitol is about 60% as sweet as sucrose (Table 19.1) and is absorbed more slowly from the digestive tract than is glucose. Sorbitol is available in both crystalline and solution forms. Sorbitol is categorized as a Generally Recognized as Safe (GRAS) ingredient, but the FDA requires the statement "Excess consumption may have a laxative effect" on the labels of food products whose consumption may cause the daily consumption of sorbitol to exceed 50 g."

Mannitol (Chapter 2)

Mannitol is the least soluble of the common polyols. Because it is not hygroscopic, mannitol is used as a dusting powder for chewing gum (to prevent it from sticking to the manufacturing equipment and wrappers). Because of its high melting point, mannitol is used in chocolate-flavored coatings for ice creams and confections. FDA requires the same statement required for sorbitol on the labels of food products whose consumption might lead to a daily intake of mannitol exceeding 50 g.

Xylitol (Chapter 2)

Xylitol (made by reduction of D-xylose) is a crystalline substance with essentially the same solubility and sweetness as sucrose. Xylitol is not only noncariogenic, it is anticariogenic (that is, it will prevent the formation of dental caries[4]). Xylitol contributes to improved dental health both by interfering with the formation of new cavities and by inhibiting the progression of existing cavities. In fact, xylitol has been claimed to be more effective than fluoride in preventing tooth decay in young children because it prevents oral bacteria from producing acid and makes it harder for them to adhere to teeth and gums.

In addition to its anticariogenic activity, another notable attribute is its high negative heat of solution (Table 19.4). This means that it produces a strong cooling

[4] Tooth decay or cavities.

sensation in the mouth. For this reason, the largest use of xylitol is in breath mints, other compressed mint tablets (sometimes in combination with sorbitol), hard candies (where citric acid enhances its cooling effect), hard coatings, and sugarless chewing gum.

Erythritol (Chapter 2)

To date, erythritol is the only polyol not made commercially by reduction; it is made by a fermentation process.[5] Erythritol has a clean sweet taste and is about 70% as sweet as sucrose. It is nonhygroscopic. Consumption of erythritol does not produce laxation. To date, erythritol has largely been used in combination with other substances (for economic reasons). A 2:3 blend of erythritol and sorbitol is reported to improve chewing gum coatings. A blend of erythritol, inulin Chapter 17), and isomalt (Section on Isomalt below) is reported to be effective in production of sugar-free chocolate products with improved digestive tolerance (as compared to maltitol) in addition to producing a very low glycemic response (Chapter 17). Erythritol is used together with the high-potency sweetener stevia (Section on High-potency Sweeteners below) in soft drinks both to add bulk and to provide a sweet taste similar to that of sucrose. A blend of erythritol, tagatose (below), and a maltodextrin mimics the taste, texture, and mouthfeel of sucrose. Erythritol crystallizes easily and is not as soluble as sucrose.

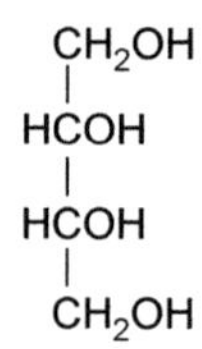

Erythritol

Maltitol (Chapter 3)

Maltitol (made by reduction of maltose) is about 80% as sweet as sucrose (Table 19.1) and results in a slower rise in blood sugar and insulin levels as compared to D-glucose or sucrose. It is incompletely digested. After administration of 19 g to humans, 4%−22% of intact maltitol and 9%−16% of sorbitol were recovered at the terminal ileum. The intestinal discomfort produced by maltitol is relatively mild.

Crystalline maltitol is the most widely used ingredient as a sucrose (sugar) replacer in the manufacture of sugar-free chocolate products. Its low hygroscopicity, relatively high degree of sweetness, and modest heat of solution give the final chocolate product properties and sensory qualities similar to those made with sucrose. In addition, because its melting point is close to that of sucrose and it does not increase the

[5] A factory to produce erythritol from glucose via an electrochemical process is under construction.

viscosity during the conching process,[6] process changes are not required when making chocolate with maltitol. Maltitol also provides a creamy texture to brownies, cakes, and cookies by producing crystals that are not perceived on the tongue. Because it is a disaccharide, its colligative properties, for example, its effect on water activity, are the same as those of sucrose. Maltitol is also available as a reduced high-maltose syrup.

Lactitol (Chapter 3)

Lactitol (made by reduction of lactose) is about 40% as sweet as sucrose (Table 19.1). Its time-intensity profile (Fig. 19.1) is essentially the same as that of sucrose. It is said to have a "cleaner" sweetness than does sorbitol. Both crystalline forms (the monohydrate and the dihydrate) are essentially nonhygroscopic and highly soluble in water. Both forms have low negative heats of solution (Table 19.4). Lactitol passes through the small intestine essentially intact, so it is a prebiotic (Chapter 17) and produces the flatulence and laxation associated with nondigestible, nonabsorbed, low -molecular-weight carbohydrates if consumed in sufficient amounts to cause these effects. The symptoms are much the same as those occurring in individuals with lactase deficiency (Chapter 3). Neither is it utilized significantly by the microorganisms in the mouth. Because very little acid is produced from it, it does not contribute to dental caries. Because it is a disaccharide, its colligative properties are the same as those of sucrose. It is a suitable polyol for a range of food products.

Isomalt (Chapter 3)

To make isomalt, sucrose is converted into isomaltulose (6-*O*-α-D-glucopyranosyl-D-fructose) in an enzyme-catalyzed reaction. Isomaltulose is then catalytically hydrogenated to isomalt, a mixture of 6-*O*-α-D-glucopyranosyl-D-glucitol and 6-*O*-α-D-glucopyranosyl-D-mannitol dihydrate.

Isomalt has a sweetness about half that of sucrose (Table 19.1). Among the polyols, it has the lowest negative heat of solution (Table 19.4) and, hence, very little cooling effect. Isomalt has a low degree of hygroscopicity. Because it is a mixture of two disaccharides, the colligative properties of isomalt are the same as those of sucrose. It does not promote dental caries. About half of the ingested isomalt is digested in (and the hydrolysis products absorbed from) the small intestine. It has been used in baked goods, chewing gum, chocolates, cough drops, hard candies, ice cream, pan-coated products, toffees, and throat lozenges, often in combination with a high-potency sweetener and more extensively in Europe than in North America. Isomaltulose itself has about the same sweetness as isomalt, but isomalt is used more extensively.

[6] The conching process consists of kneading cacoa at a relatively high temperature (49–82°C [120–180°F]) for an extended time (up to 24 h) in the chocolate making process. The high temperature produces a caramel (Chapter 18) flavor and promotes the Maillard reaction (Chapter 18) in milk chocolate.

Hydrogenated starch hydrolyzates (Chapter 7)

An hydrogenated starch hydrolyzate (HSH) is made by hydrogenating a low-DE glucose or high-maltose syrup in the presence of a catalyst in the same way that a solution of D-glucose (dextrose) is hydrogenated to make sorbitol. That means that HSH products are mixtures of sorbitol, maltitol, and reduced maltooligosaccharides. Because a variety of syrups can be made from starch by varying both the means of hydrolysis (acid and/or different amylases and mixtures of amylases) and the extent of hydrolysis, a variety of HSH products with different compositions can be made. Those that contain more than 50% sorbitol are usually labeled sorbitol syrups; those that contain more than 50% maltitol are usually labeled maltitol syrups, and the others are called hydrogenated starch hydrolyzates. The sweetness of HSH is determined by its composition and can range from low to approaching that of maltitol. HSH products can serve as bodying agents, crystallization inhibitors, cryoprotectants, humectants, rehydration aids, sweeteners, viscosifiers, and carriers for flavors, colors, and enzymes. Because they are excellent humectants, they are used in bakery products. Because, in general, they are excellent humectants that do not crystallize, they can be used to make sugar-free confections using the same processing equipment and methods used to make confections containing sugar. The glycemic responses (Chapter 17) of HSH are less than those of the parent syrups from which they are made. A prominent use of HSH is in making prepared foods for diabetics because of their property of slowly raising blood glucose levels (that is, raising them over an extended time period, rather than producing a hyperglycemic spike).

Other carbohydrate reduced–calorie sweeteners

D-Tagatose (Chapter 1)

D-Tagatose is a ketohexose. It is related to D-galactose in the same way that D-fructose is related to D-glucose and is made by isomerization of D-galactose. Tagatose is about 92% as sweet as sucrose. Tagatose has about the same degree of hygroscopicity as does sucrose. It can participate in Maillard browning reactions (Chapter 18).

Because only about 20% of D-tagatose is absorbed from the small intestine, it is a reduced-calorie sweetener and, does not produce a glycemic response. It produces less acid in the oral cavity than does sucrose and is a prebiotic. Tagatose acts synergistically with other sweeteners and, if used, it is used in combination with other sweeteners where it contributes to the overall sensory attributes of the food product and acts as a prebiotic (Chapter 17). It is a GRAS substance.

Allulose (Chapter 1)

What the company that produces the compound calls allulose (not an incorrect name) is more commonly known as psicose. The actual scientific name is D-*ribo*-hexulose. Allulose is about 70% as sweet as sucrose, but provides only about 10% as many calories as sucrose. (Some is absorbed from the small intestine but not metabolized in the body.)

It provides browning during baking. It is claimed that it functions well in combination with sucralose and stevia (Section on High-potency Sweeteners in this chapter). The compound has GRAS status.

"Natural sweeteners"

So-called natural sweeteners are also available. Sucrose, D-glucose, and D-fructose are, of course, natural sweeteners. Tagatose, allulose, sorbitol, xylitol, and other sweeteners are also natural substances, but none of these are included in this designation. What is included in this category are such things as honey, malt syrups, maple syrup, molasses (both from sugar cane and from sorghum), and various fruit juice concentrates, especially grape juice concentrates, all of which contain sucrose, glucose, and fructose. Also included are rice syrups, brown rice syrups, and oat syrups, which are made basically in the same way as are the syrups made from corn, wheat, or any other starch, but the latter are not included. The only difference between other syrups made from starch and brown rice syrup is that brown rice syrup is made from rice that has had none or only a portion of the bran removed.

Honey was the only available sweetener until humans learned to boil down sweet wines or grape juices and, more importantly, until humans learned to grow and harvest sugar cane and to extract the sugar from it – a process that required the development of sophisticated technology.

To make malt, barley seeds are subjected to steeping (soaking in water), then germinated under controlled conditions. The germinated seeds are then dried. During germination, many enzymes are produced or activated, including proteolytic enzymes, enzymes involved with respiration, and enzymes involved with the breakdown of starch, primary among which is β-amylase (Chapter 7), which is known in the malting and brewing industries as diastase.[7] If dried properly (in a process known as kilning), the enzymes of malt will not be denatured (inactivated). The heat applied during drying also results in flavor and color development via Maillard browning reactions (Chapter 18).

Next the malted barley seeds are ground and mixed with water in a special vat called a mash tun, in which the temperature is controlled. During its time in the mash tun, some of the starch is converted into low molecular weight saccharides (primarily into maltose) by β-amylase and the other amylases produced by the germinating barley seeds. So-called adjuncts can be added at this stage to produce more mellow and sweeter flavors. The adjunct most often used is corn grits.[8] (Addition of corn grits results in a liquid portion that contains 5%–8% more reducing sugar and 2%–4% less protein than an all-barley-malt extract.) When the desired degree of starch and protein breakdown has been achieved in the mash tun, the mixture is transferred to another special vessel called a lauter tun. During the time in the lauter tun, proteins and starch

[7] The entire mixture of enzymes that act to break down starch is known as a diastatic system.

[8] Corn grits are coarse particles (more coarse than those of corn meal) produced by dry milling of corn kernels.

are broken down further by additional enzymic activity. Finally, the grain residue is removed by passing the mixture through the slotted false bottom of the lauter tun. The liquid phase may be transferred directly to a brew kettle for production of an alcoholic beverage (beers, whiskeys, etc.), in which case it is known as wort. Alternatively, for food industry use, in which case it is known as malt extract, the liquid phase may be filtered and transferred to a vacuum pan for evaporation to a syrup of about 80% solids. Depending on the temperatures used, malt extracts and syrups with from high to no enzyme activity can be produced. Other variables are color, flavor, and protein content. A concentrated extract may also be spray dried. Typical carbohydrate compositions of extracts/syrups would be 39%−42% maltose, 25%−30% maltooligosaccharides of DP 4 and higher, 10%−15% maltotriose, 7%−10% D-glucose, 1%−3% sucrose, and 1%−2% D-fructose.

When used as an ingredient in doughs, diastatic malt[9] adds amylase activity to the wheat flour and thereby creates additional fermentable carbohydrates. It also provides sweetness (maltose), soluble proteins, dough-conditioning enzymes, salts, and nutritive substances that result in optimum yeast activity. Malt syrups also add flavor, aroma, and color to finished baked products. Dark breads use malts of darker color. In cracker production, addition of a diastatic malt syrup improves fermentation, aids in sheeting and laminating, improves color, and adds flavor.

Nondiastatic malt extracts (that is, those with no enzymic activity) are used in the production of ready-to-eat breakfast cereals, malt pancakes, and malt waffles.

Other food products made with malt as an ingredient include bagels, chocolate products, confections, cookies, granola and granola bars, infant foods, pet foods, pretzels, puddings, soy milk, and vinegar.

An interesting sweetener, Gaz of Khunsar, is a sweetening agent that, like honey, is produced by an insect. Gaz of Khunsar, which is named after a town in the producing area in Iran, is exuded by the last instar nymph of a small insect and is collected from a shrub that grows wild in west central Iran. Gaz of Khunsar contains about 41% reducing sugars (mainly fructose), about 31% polysaccharides, about 2% sucrose, about 16% moisture, and about 10% other substances, including ash. It is used to make a traditional confection, Gaz of Isfahan (named after a large city in central Iran), which is made with Gaz of Khunsar, egg white, starch, sugar, and pistachios or almonds.

High-potency sweeteners

Although this is a book about carbohydrates, a very brief section on noncarbohydrate sweeteners is presented with references to additional resources at the end of the chapter because of their obvious relationship to carbohydrate use.

The sweetness of **high-potency sweeteners** (also known as **high-intensity sweeteners**) ranges from about 30 to about 10,000 times the sweetness of sucrose

[9] Diastatic malt contains enzymic (amylase) activity.

(Table below). Even those that can be digested are, for practical purposes, nonnutritive sweeteners because they are substances that impart sweet tastes without adding appreciable calories to the products in which they are used as an ingredient. For example, peptide high-potency sweeteners, such as aspartame, are digested by and catabolized in the human body, but because they are high-potency sweeteners and, therefore, used in only very small amounts, they do not provide appreciable calories to diets and could be considered to be nonnutritive. Small amounts of high-potency sweeteners replace much greater amounts of sugar/sucrose. They are used in products consumed by persons concerned about the level of carbohydrates (particularly sugar) and/or calories in their diets. High-potency sweeteners do not cause dental caries.

Just as nutritive sweeteners differ in sweet taste intensity and quality, so do the high-potency sweeteners. And like the nutritive sweeteners, taste intensities and qualities are a function of temperature, pH, and the presence of other compounds, especially other flavors. In short, the perception of sweetness is always application dependent. There are, however, two major differences between nutritive and high-potency sweeteners; both safety and the qualities of the sweet taste are much greater concerns in the case of high-potency sweeteners than they are with nutritive sweeteners. Desired is a clean sweet taste without an aftertaste. In some cases, blends of high-potency sweeteners or blends of a high-potency sweetener and a nutritive sweetener are used to more closely mimic the taste profile of sucrose (the standard). Some combinations also allow use of less total high-intensity sweetener, resulting in a cost savings. Because safety has not been established in all applications, some high-potency sweeteners are approved for use in only a limited number of food products. Major high-potency sweeteners and their approximate relative sweetness values are as follows:

High-potency sweetener	Approximate relative sweetness
Advantame	20,000
Neotame	10,000
Alitame	2000
Sucralose	700
Stevia	30–300
Saccharin	300
Aspartame	200
Acesulfame-K	200
Cyclamate	30
Sucrose	1

Aspartame

Aspartame is the high-potency sweetener used in greatest amounts. It is approved for general use. Aspartame is a dipeptide methyl ester, specifically L-aspartyl-L-phenylalanine methyl ester. It is categorized as a nutritive sweetener by the U.S. FDA because, on an equal weight basis (like other peptides and proteins) it has approximately the same caloric content as a carbohydrate. However, because 1 g of aspartame will replace about 200 g of sucrose (that is, since it is used in amounts of only about 0.5% that of sucrose to achieve the same degree of sweetness), aspartame is used as a microingredient rather than a macroingredient and adds negligible calories to products in which it is used.

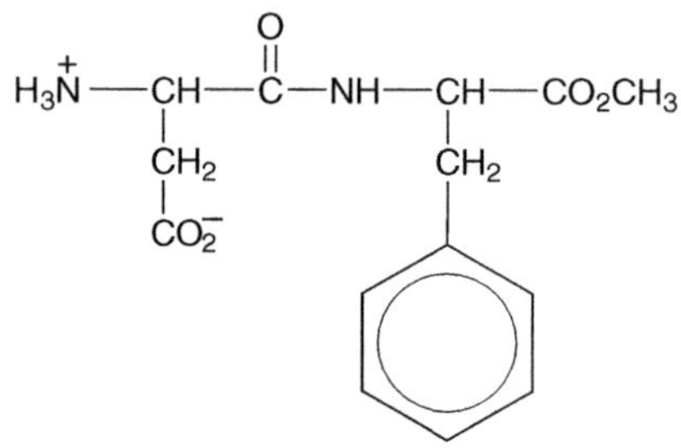

Aspartame

Aspartame is claimed to have a sucrose-like taste. It has a relatively slow sweetness onset and a sweet aftertaste. It acts synergistically with acesulfame-K, saccharin, cyclamate (all presented below), and malic acid and enhances and extends fruit flavors. Its maximum stability is in the pH range 3–5, which presents no problem for most food and beverage products. However, in products such as soft drinks, acid-catalyzed hydrolysis over time removes the methyl ester group, leading to loss of sweetness. Heating at pH values above 5 results in the conversion of some aspartame molecules to a diketopiperazine, also resulting in loss of sweetness. However, aspartame can withstand the UHT heat-processing regimes used for dairy products and fruit juices and in aseptic processing. A granulated, encapsulated form, in which the protective coating acts as a time/temperature release system, is available for use in bakery products. Aspartame is sometimes used in combination with other sweeteners.

Many controlled scientific studies have found aspartame to be safe, as could be predicted because it is the methyl ester of a dipeptide of two natural amino acids. However, because it contains L-phenylalanine, the U.S. FDA requires that products containing it contain the warning statement "Phenylketonurics: contains phenylalanine."

Advantame

Advantame is structurally related to aspartame. The only difference in their structures is that in advantame the -NH_3^+ group in the aspartame structure is replaced by a -NH-vanillin group in advantame. Advantame is about 20,000 times sweeter than sugar. It is claimed to have greater stability at higher pH and temperature conditions than is aspartame. Advantame has been approved as a GRAS substance.

Neotame

Neotame is also closely related to aspartame. The only structural difference is that the -NH_3^+ group in the aspartame structure is replaced by a -NH-alkyl group in neotame. Neotame is the most potent sweetener known, being 7000–13,000 times sweeter than sucrose. Neotame, being a peptide-like aspartame, has essentially the same stability as aspartame, perhaps being a little more stable to heating. It is stable as a dry substance but slowly loses its potency in soft drinks due to their acidic nature. It is completely eliminated from the body. Neotame is said to provide a clean sweet taste devoid of bitter or metallic taste overtones and to enhance other flavors. It is reported to extend the sweet taste of chewing gum and extend the time during which other flavors are perceived to a greater extent than do other high-potency sweeteners. It provides synergistic sweetness when used with nutritive and other nonnutritive sweeteners (for example, blends of neotame and saccharin are 14%–24% sweeter than predicted by simply adding the sweetness intensities of the two).

Alitame

Alitame is yet another dipeptide analog. It is composed of L-aspartic acid and D-alanine and contains a thioalkyl amide group at the carboxyl end. It has a clean flavor profile (that is, its sweet taste is described as sucrose-like without bitter or metallic taste overtones). It is reported to be more stable to heat and pH (in the range 5–8) than is aspartame. It acts synergistically with acesulfame-K and cyclamate. Because it does not contain phenylalanine (as does aspartame), alitame is safe for use by those with phenylketonuria. Alitame is not now approved for use in the United States, but is approved for use elsewhere.

Acesulfame-K

Acesulfame, whose potassium salt is known as acesulfame-K or, less commonly, as acesulfame potassium, also contains an *N*-sulfonyl amide ($-N^{-}-SO_2-O-R-$) unit. Acesulfame-K is most often used in combination with other high-intensity sweeteners, particularly aspartame, with which it acts synergistically. It is heat stable and pH tolerant. It provides a quick onset of sweetness. Some perceive it as having a metallic and bitter aftertaste, but people differ with respect to sensitivity to this characteristic.

Cyclamate

Cyclamate was the most widely used artificial sweetener in the 1960s. Its use in the United States was prohibited in 1970, but it is approved for use in more than 40 other countries. Cyclamate contains the $-NH - SO_3^-$ (sulfamate) structural unit common to several compounds with a sweet taste. It is sold as a sodium or calcium salt. Cyclamate is soluble and stable, has a sugar-like taste, and acts synergistically with saccharin and aspartame. In countries where it is allowed, it is almost always used in combination with other sweeteners. It is good at masking bitter tastes. It enhances fruit flavors and can mask the tartness of some citrus fruits. Where it can be used, it is used in a broad range of food products. Cyclamate is partially absorbed but not catabolized.

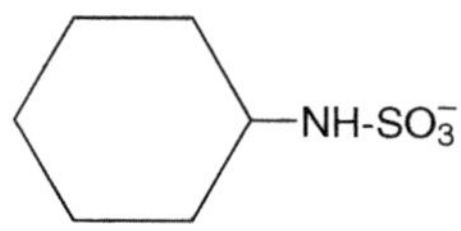

Cyclamate

Saccharin

Saccharin contains the -$CONHSO_2$- (*N*-sulfonyl amide) structural unit common to several compounds with a sweet taste. Saccharin is available in three forms: acid saccharin, sodium saccharin, calcium saccharin (the latter two being the sodium and calcium salt forms, respectively). For about one-third of the population, the sweet taste of saccharin is accompanied by significant metallic and bitter aftertastes; the remaining two-thirds of the population perceives these off tastes to degrees that range from moderate to zero. Saccharin is stable to heat and acids. It acts synergistically with aspartame and cyclamate. It is not metabolized by humans. Saccharin is the most economical of the high-intensity sweeteners.

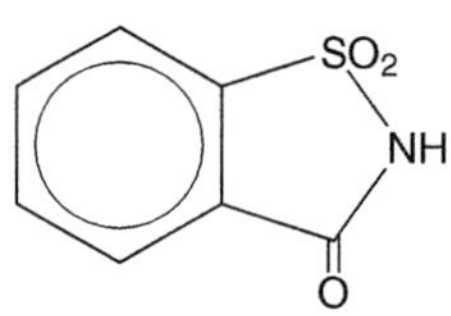

Saccharin

Stevia

The term stevia refers to one or more of the sweet-tasting steviol glycosides. These glycosides are obtained from the leaves of *Stevia rebaudiana,* a South American plant that grows in the border region of Paraguay and Brazil and which is cultivated elsewhere, primarily in China. Leaves of this plant are extracted with hot water. The extract is clarified and then passed through a column of an adsorptive resin. The steviol glycosides are eluted from the column with alcohol, and the eluate is dried. The dried eluate is then redissolved in methanol or aqueous ethanol for crystallization (that is, additional purification).

Ten different steviol glycosides have been identified in extracts. All 10 glycosides have a common aglycon (Chapter 1) called steviol. They differ from one another in the type and number of sugars attached (either as mono- or oligosaccharides) at two different locations on the steviol molecule. Most of the carbohydrate portions are chains of from 1 to 3 β-D-glucopyranosyl units, but a few of the oligosaccharides contain xylosyl or rhamnosyl units (Chapter 1). The water solubilities of the 10 steviol glycosides range from 0.03% to 1.7%. Their sweetness values (relative to sucrose) range from 30 to 300. As glycosides, they are subject to acid-catalyzed hydrolysis but are quite stable in the pH range of 4–8, even to UHT processing.

One crystalline glycoside (rebaudioside A, also called rebiana, that is 200 times sweeter than sucrose) and high-purity mixtures containing it are commercially available. These products have a clean sweet taste when used at low concentrations, but are perceived as having some bitter and licorice tastes at higher use levels. As a result, the products are most often used in combination with other sweeteners, especially erythritol and natural flavor systems, and in beverages to mask the bitterness and improve mouthfeel. Their current primary use is in zero-calorie soft drinks in a blend with a caloric or noncaloric sweetener such as acesulfame-K, aspartame, cyclamate, erythritol, fructose, glucose, HFS, sucralose, saccharin, sorbitol, sucrose, and xylitol. Stevia is a GRAS substance.

Sucralose (Chapter 3)

Sucralose is perceived by most persons as providing a flavor profile closest to that of sucrose. It has a clean and quickly perceptible sweet taste. It also has good solubility and heat and acid stabilities. It is used in a broad range of products from soft drinks to bakery products. Sucralose passes rapidly through the body without change.

Thaumatin

Thaumatin is a protein extracted from the fruit of a West African plant. The fruit is known both as Katemfe fruit and the miraculous fruit of the Sudan. Thaumatin has a molecular weight of about 22,000. It is readily soluble in cold water, and solutions of more than 60% concentration can be made. It is stable at temperatures used in retorting and UHT processes, even in systems below pH 2. Thaumatin enhances the response to certain flavors and, in addition, reduces the negative notes related to them. It acts synergistically with savory flavor enhancers (monosodium glutamate and 5′-nucleotides). It also enhances responses to other sweeteners. Thaumatin masks the bitter, unpleasant tastes of sodium, potassium, and iron ions. Because thaumatin also has a licorice-like taste, its use as a sweetener is normally restricted to very small amounts acting synergistically with other sweeteners, allowing significant reductions in the amount of sweetener used.

Blends

Each nonnutritive, low-calorie sweetener performs better in some products than in others. In some cases, blends of high-intensity sweeteners or blends of a high-intensity sweetener and a nutritive sweetener more closely mimic the taste profile of sucrose. Another advantage of blends is that the two sweeteners often act synergistically, resulting in a more intense sweet taste and, therefore, use of less total sweetener. From 1910 (when it was first used commercially) until the 1960s, saccharin was the only nonnutritive sweetener available. Cyclamate was introduced in the 1950s. In the 1960s, these two sweeteners were often used together in diet soft drinks, tabletop sweeteners, and other low-calorie food and beverage products. Saccharin boosts the sweetening power of the less-potent cyclamate; cyclamate eliminates most of the aftertaste of saccharin. Currently, a 10:1 mixture of cyclamate to saccharin is used extensively in Europe. Other useful combinations are aspartame + acesulfame-K, sucralose + acesulfame-K, aspartame + saccharin, neotame + saccharin, and stevia + erythritol. Nonnutritive and nutritive sweeteners are also used together. It is important to customize blends for the specific application; for example, perceptions of flavor attributes of different sweetener blends differ when they are used in cola drinks versus fruit-flavored drinks.

Future

Compounds that are not themselves sweet but which enhance the sweet taste of sucrose—thus allowing less of it to be used—are being sought.

Additional resources

General

Kroger, M., Meister, K., Kava, R., 2006. Low-calorie sweeteners and other sugar substitutes: a review of safety issues. Comprehensive Reviews in Food Science and Food Safety 5, 35–47.

O'Brien-Nabors, L. (Ed.), 2016. Alternative Sweeteners, fourth ed. CRC Press, Boca Raton.

O'Donnell, K., Kearsley, M. (Eds.), 2012. Sweeteners and Sugar Alternatives in Food Technology, second ed. Wiley-Blackwell, Ames (Includes a chapter on bulking agents).

Pietrzycki, W., 2004. Saccharide sweeteners and the theory of sweetness. In: Tomasik, P. (Ed.), Chemical and Functional Properties of Food Saccharides. CRC Press, Boca Raton, pp. 57–71.

Weerasinghe, D.K., DuBois, G.E. (Eds.), 2009. Sweetness and Sweeteners: Biology, Chemistry and Psychophysics (ACS Symposium Series 979). American Chemical Society, Washington.

Nutrative sweeteners

Sucrose

Black, R.M., January 1993. Sucrose in health and nutrition - facts and myths. Food Technology 130–133.

Chinachoti, P., January 1993. Water mobility and its relation to functionality of sucrose-containing food systems. Food Technology 134–140.

Jeffery, M.S., January 1993. Key functional properties of sucrose in chocolate and sugar confectionery. Food Technology 141–144.

Mathouthi, M., Reiser, P. (Eds.), 1995. Sucrose. Properties and Applications. Blackie Academic & Professional, London.

Pennington, N.L., Baker, C.W. (Eds.), 1990. Sugar. A User's Guide to Sucrose. AVI, New York.

Fructose, High-fructose, and Glucose Syrups

Buck, A.W., 2016. High fructose corn syrup. In: O'Brien-Nabors, L. (Ed.), Alternative Sweeteners, fourth ed. CRC Press, Boca Raton, pp. 403–422.

Hull, P., 2010. *Glucose Syrups: Technology and Applications*. Wiley-Blackwell, Oxford.

White, J.S., 2016. Crystalline fructose. In: O'Brien-Nabors, L. (Ed.), Alternative Sweeteners, fourth ed. CRC Press, Boca Raton, pp. 379–402.

Reduced-calorie sweeteners

Erythritol

Boesten, D.M.P.H.J., den Hartog, G.J.M., de Cock, P., Bosscher, D., Bonnema, A., Bast, A., 2015. Health effects of erythritol. Nutrafoods 14, 3–9.

de Cock, P., Embuscado, M.E., Patil, S.K., 2011. Erythritol. In: O'Brien-Nabors, L. (Ed.), Alternative Sweeteners, fourth ed. CRC Press, Boca Raton, pp. 249–264.

Isomalt and isomaltulose

Sentko, A., Bernard, J., 2011. Isomalt. In: O'Brien-Nabors, L. (Ed.), Alternative Sweeteners, fourth ed. CRC Press, Boca Raton, pp. 275–298.

Sentko, A., Bernard, J., 2011. Isomaltulose. In: O'Brien-Nabors, L. (Ed.), Alternative Sweeteners, fourth ed. CRC Press, Boca Raton, pp. 423–438.

Lactitol

Zacharis, C., Stowell, J., Mesters, P.H.J., van Velthuijsen, J.A., Brokx, S., 2011. Lactitol. In: O'Brien-Nabors, L. (Ed.), Alternative Sweeteners, fourth ed. CRC Press, Boca Raton, pp. 315–332.

Maltitol and Hydrogenated Starch Hydrolyzates

Deis, R., 2011. Maltitol syrups and polyglycitols. In: O'Brien-Nabors, L. (Ed.), Alternative Sweeteners, fourth ed. CRC Press, Boca Raton, pp. 265–274.

Kearsley, M.W., Boghani, N., 2011. Maltitol. In: O'Brien-Nabors, L. (Ed.), Alternative Sweeteners, fourth ed. CRC Press, Boca Raton, pp. 333–348.

Sorbitol

Jamieson, P.R., Lee, A.S., Mulderrig, K.B., 2011. Sorbitol and mannitol. In: O'Brien-Nabors, L. (Ed.), Alternative Sweeteners, fourth ed. CRC Press, Boca Raton, pp. 333–348.

Tagatose

Vastenavond, C., Bertelsen, H., Hansen, S.J., Laursen, R.S., Saunders, J., Eriknauer, K., 2011. Tagatose. In: O'Brien-Nabors, L. (Ed.), Alternative Sweeteners, fourth ed. CRC Press, Boca Raton, pp. 423–438.

Xylitol

de Silva, S.S., Chandel, A.K. (Eds.), 2012. Xylitol: Fermentative Production, Application and Commercialization. Springer, New York.

Zacharis, C., Stowell, J., Olinger, P.M., Pepper, T., 2011. Xylitol. In: O'Brien-Nabors, L. (Ed.), Alternative Sweeteners, fourth ed. CRC Press, Boca Raton, pp. 349–378.

"Natural sweeteners"

Hickenbottom, J.W., 1996. Processing, types, and uses of barley malt extracts and syrups. Cereal Foods World 41, 788–790.

Hoseney, R.C., 1994. An overview of malting and brewing. Cereal Foods World 39, 675–679.

Rybak-Chmielewska, H., 2004. Honey. In: Tomasik, P. (Ed.), Chemical and Functional Properties of Food Saccharides. CRC Press, Boca Raton, pp. 73–79.

Gaz

Grami, B., 1998. Gaz of Khunsar: the manna of Persia. Economic Botany 52, 183–191.

High-potency sweeteners

Acesulfame-K

Klug, C., von Rymon Lipinski, G.-W., 2011. Acesulfame-K. In: O'Brien-Nabors, L. (Ed.), Alternative Sweeteners, fourth ed. CRC Press, Boca Raton, pp. 13–30.

Advantame

Bishay, I.E., Bursey, R.G., 2011. Advantame. In: O'Brien-Nabors, L. (Ed.), Alternative Sweeteners, fourth ed. CRC Press, Boca Raton, pp. 31–46.

Alitame

Auerbach, M.H., Locke, G., Hendrick, M.E., 2011. Alitame. In: O'Brien-Nabors, L. (Ed.), Alternative Sweeteners, fourth ed. CRC Press, Boca Raton, pp. 47–56.

Aspartame

Abegaz, E.G., Mayhew, D.A., Butchko, H.H., Stargel, W.W., Comer, C.P., Andress, S.E., 2011. Aspartame. In: O'Brien-Nabors, L. (Ed.), Alternative Sweeteners, fourth ed. CRC Press, Boca Raton, pp. 57–76.

Cyclamate

Hunt, F., Bopp, B.A., Price, P., 2011. Cyclamate. In: O'Brien-Nabors, L. (Ed.), Alternative Sweeteners, fourth ed. CRC Press, Boca Raton, pp. 93–116.

Monellin

Kinghorn, A.D., Compadre, C.M., 2011. Less common high-potency sweeteners. In: O'Brien-Nabors, L. (Ed.), Alternative Sweeteners, fourth ed. CRC Press, Boca Raton, pp. 223–246.

Neotame

Mayhew, D.A., Meyers, B.I., Stargel, W.G., Comer, C.P., Andress, S.E., Butchko, H.H., 2011. Neotame. In: O'Brien-Nabors, L. (Ed.), Alternative Sweeteners, fourth ed. CRC Press, Boca Raton, pp. 423–438.

Saccharin

Bakal, A.I., O'Brien-Nabors, L., 2011. In: O'Brien-Nabors, L. (Ed.), Alternative Sweeteners, fourth ed. CRC Press, Boca Raton, pp. 151–158.

Sucralose

Grotz, V.L., Molinary, S., Peterson, R.C., Quinlan, M.E., Reo, R., 2011. Sucralose. In: O'Brien-Nabors, L. (Ed.), Alternative Sweeteners, fourth ed. CRC Press, Boca Raton, pp. 181–196.

Thaumatin

Kinghorn, A.D., Compadre, C.M., 2011. Less common high-potency sweeteners. In: O'Brien-Nabors, L. (Ed.), Alternative Sweeteners, fourth ed. CRC Press, Boca Raton, pp. 223–246.

Blends

Bakal, A.I., 2011. Mixed sweetener functionality. In: O'Brien-Nabors, L. (Ed.), Alternative Sweeteners, fourth ed. CRC Press, Boca Raton, pp. 473–505.

Fry, J.C., Meyers, B.I., Mayhew, D.A., 2011. Aspartame-acesulfame. In: O'Brien-Nabors, L. (Ed.), Alternative Sweeteners, fourth ed. CRC Press, Boca Raton, pp. 77–92.

Summary of Carbohydrate Functionalities

20

Carbohydrates used as food ingredients can be grouped into the following types.

Carbohydrate types	Examples
Low molecular weight carbohydrates	
Reducing sugars (can participate in browning reactions)	D-Glucose, D-fructose, lactose, maltose, high-DE maltodextrins
Nonreducing sugars (cannot participate in browning reactions)	Sucrose, all polyols (sugar alcohols), low-DE maltodextrins
Lactone	GDL
Polysaccharides	
Uncharged/Neutral	
Unbranched	Cellulose, inulin, konjac glucomannan, curdlan, MC, HPMC, HPC
Branched	Guar gum, LBG, starches, and modified starches
Charged/Anionic	
Unbranched	CMC, gellans, carrageenans, algins, pectins, furcellaran
Branched	Modified starches, xanthans, exudate gums

It has been obvious throughout this book that carbohydrates are used as food ingredients primarily because of the functionalities they impart and that, therefore, much of the chemistry of carbohydrates in foods is physical chemistry. It has been obvious that each carbohydrate component or ingredient in a food product (monosaccharides, sucrose, oligosaccharides, starch-based and invert syrups, polyols, starches and modified starches (Table 5.6), hydrocolloids (Table 5.6), other polysaccharides) has unique properties, both as a category and individually, giving a wide range of properties and functionalities. And it has been obvious that it is difficult to classify food carbohydrates other than by using the very broad structural classes given above and those in

Carbohydrate Chemistry for Food Scientists. https://doi.org/10.1016/B978-0-12-812069-9.00020-0

Table 4.1 because of their variety of structures, including modified structures. This diversity of structures is the reason why food carbohydrates have such a wide range of chemical and physical properties. Table 20.1 is an attempt to group food carbohydrates by the functionalities they provide. No claim is made for completeness of the tables, either in functionalities, applications, or carbohydrates to consider.

Table 20.1 Some[a] carbohydrates to consider when needing specific functionalities

Functionality or application	Carbohydrateb[b,c,d,e]
Acidification	GDL
Acid stability	Cross-linked starch products,[f] HM pectin, PGA, xanthan
Anticaking	MCC
Body	Maltodextrins, starches[f]
Bulk	Maltodextrins, starches[f]
Carrier for flavors and spices	Cyclodextrins, maltodextrins
Compatibility with acids (low pH)	PGA, cross-linked starch products,[f] HM pectins, xanthan
Compatibility with high-temperature processing	Cross-linked starch products,[f] xanthan
Compatibility with milk products (Ca^{2+})	Carrageenans[g] (for gelling, stabilizing, thickening, suspension stabilizing)
Compatibility with salts	CMC, guar gum, xanthan
Creaminess	PGA, starches[f]
Crystallization inhibition	CMC
Dietary fiber[h]	Cellulose, FOS, β-glucan, hydrolyzed guar gum, high-amylose starch, inulin, specially prepared maltodextrins, MC, MCC, polydextrose
Emulsification	Gum arabic, MC, PGA, OS starch
Emulsion stabilization (flavor oils)	Gum arabic, OS starch
Emulsion stabilization (vegetable oils)	MCC, OS starch, PGA, xanthan
Film formation	Low-viscosity type of CMC, carrageenans, MC

Table 20.1 Continued

Flavor oil encapsulation	Cyclodextrins, gum arabic, maltodextrins, OS starch
Foam formation	HPC, MC
Foam stabilization	HPC, MCC, PGA, xanthan
Freeze–thaw stability	CMC, gellan, stabilized and cross-linked starch products[f]
Frozen food stabilization	CMC, gellan, ι-type carrageenans, stabilized and cross-linked starch products[f]
Frost (on breakfast cereals)	Sucrose
Functionality without heating	CMC, guar gum, pregelatinized and cold-water-swelling starch products
Gelation with acids	Sodium alginates, HM pectins + sugar
Gelation with calcium ions	Sodium alginates, ι-type carrageenans, LM pectins
Gelation with potassium ions	κ-Type darrageenans
Gelation with any cation	Gellan
Gelation upon cooling of hot solutions	Agar, calcium alginates, calcium ι-carrageenates, potassium κ-carrageenates, κ-carrageenans + LBG, gellan, HM pectins + acid + sugar, LM pectins + Ca^{2+}, xanthan + agar, xanthan + κ-carrageenans, xanthan + LBG
Gelation upon heating of solutions	Curdlan, MC (reversible)
Gelation with milk proteins	κ-Carrageenans, furcellaran
Gels, firm, sliceable, chewy	Alginates, gum arabic, high-amylose starch dextrins, LBG + xanthan, pectins
Gels, heat-stable	MCC
Glass formation	Sorbitol, sucrose, xylitol
Glaze	Glucose syrups, sucrose
Icings, heat-stable	Agar
Low-glycemic-index products	Polyols, including hydrogenated glucose syrups and maltodextrins (HSH)

Continued

Table 20.1 Continued

Moisture retention/hygroscopicity	Glucose syrups, HFS, sorbitol, most hydrocolloids
pH/acid stability	Cross-linked starch products[f], HM pectins, PGA, xanthan
Prebiotic	FOS, other nonstarch oligosaccharides
Protein stabilization, in milk products	Carrageenans[g]
Protein stabilization, in meat products	Sodium alginates, carrageenans[g]
Protein stabilization, soy protein	CMC
Spray drying	Gum arabic, OS starch
Stabilization, of coffee whiteners	Carrageenans[g]
Stabilization, of dairy products	Carrageenans[g]
Stabilization, of flavor oil emulsions	Gum arabic, OS starch
Stabilization, of flavors	Cyclodextrin, gum arabic, maltodextrins
Stabilization, of foams	MCC, HPC, xanthan
Stabilization, of frozen foods	CMC, ι-carrageenans, gellan, stabilized and cross-linked starch products[f]
Stabilization, of glassy confections	CMC, glucose syrups, HFS, invert sugar
Stabilization, of ice creams and frozen desserts	Alginates, CMC, carrageenans,[g] guar gum, LBG, MC, starches,[f] xanthan
Stabilization, of low-fat margarines	Carrageenans[g]
Stabilization, of low-pH systems	Cross-linked starch products, HM pectins, PGA, xanthan
Stabilization, of meat proteins	Sodium alginates, carrageenans[g]
Stabilization, of milk proteins	Carrageenans,[g] LM pectins
Stabilization, of soy protein	CMC
Stabilization, of suspensions	Carrageenans,[g] gellan, xanthan
Stabilization, of vegetable oil emulsions	MCC, PGA, xanthan
Structured foods	Alginates

Table 20.1 Continued

Suspension stabilization	Carrageenan, gellan, pectins, xanthan
Sweetening, low-caloric	Sorbitol, xylitol, other polyols
Sweetening, nutritive, crystallizing	Maltitol, sucrose
Sweetening, nutritive, non-crystallizing	Glucose syrups, HFS
Syneresis inhibition	CMC, guar gum
Thickening, delayed (during thermal Processing)	Cross-linked starches
Thickening, delayed, thermoreversible	MC
Thickening, of milk systems	Carrageenans,[g] furcellaran, modified starches
Thickening, of water systems	All hydrocolloids and starches,[f] except low-viscosity types of hydrocolloids
Water binding, in soft cheese products (to improve yield and consistency)	CMC, carrageenans,[g] guar gum, LBG
Water binding, in meat products	Alginates, carrageeans,[g] guar gum

[a]Not all carbohydrates and functionalities are included. Given in broad terms are major examples of each as currently practiced. There are always exceptions and the table is not all-inclusive.
[b]Only carbohydrates added as ingredients are listed. The table does not include flours or inherent ingredients such as starch, pectin, fiber, hemicelluloses, or sucrose.
[c]Only those displaying positive attributes are listed. Carbohydrates to avoid in certain applications are not given.
[d]Examples are listed in alphabetical order rather than by extent of use. Cost per functionality is not considered.
[e]Abbreviations used are *CMC*, carboxymethylcellulose; *FOS*, fructooligosaccharides; *GDL*, glucono-δ-lactone; *HFS*, high-fructose syrup; *HM pectin*, high-methoxyl pectin; *HPC*, hydroxypropylcellulose; *HSH*, hydrogenated starch hydrolyzate; *LBG*, locust bean gum; *LM pectin*, low-methoxyl pectin; *MC*, methylcellulose and hydroxypropylmethylcellulose; *MCC*, microcrystalline cellulose; *OS* starch, octenylsuccinylated starch; *PGA*, propylene glycol alginate.
[f]There is such a plethora of available modified starch products that each cannot be listed. Therefore, the term "starches" indicates both native and modified starches.
[g]The term "carrageenans" includes the κ-, ι-, and λ-type carrageenan families.
[h]All carbohydrates other than sucrose, lactose, starches, and starch-derived products are indigestible and contribute to dietary fiber. Listed are some ingredients, other than whole grains, commonly added to foods and beverages to increase the dietary fiber content.

Additional reading

Lai, V.M.F., Lii, C.-Y, 2004. Role of saccharides in texturization and functional properties of foodstuffs. In: Tomasik, P. (Ed.), Chemical and Functional Properties of Food Saccharides. CRC Press, Boca Raton, pp. 159–179.

Index

'*Note:* Page numbers followed by "f" indicate figures, "t" indicate tables and "b" indicate boxes.'

E

I

9780128120699